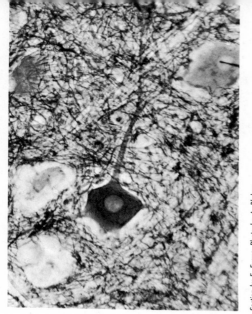

Section by E. Evans; Photo by A. Blaker

ABOUT THE COVER PHOTOGRAPH Section through spinal cord (yellow-red-brown). This section shows a large ventral horn cell of the spinal cord. The cell body is clearly seen with an axon extending from it. This axon leaves the spinal cord and carries impulses to muscles. The field is interlaced with great numbers of branches and twigs from other nerve axons.

PRENTICE-HALL BIOLOGICAL SCIENCES SERIES
William D. McElroy
and
Carl P. Swanson
editors

modern cell biology

cell biology modern

cell biology

cell biology

modern

biology modern cell

modern cell biology

modern

modern
cell biology
SECOND EDITION

WILLIAM D. McELROY

Chancellor
University of California, San Diego

CARL P. SWANSON

Department of Botany
University of Massachusetts

PRENTICE-HALL, INC., Englewood Cliffs, N.J.

Library of Congress Cataloging in Publication Data

McElroy, William David (date)
   Modern cell biology.

   (Prentice-Hall biological sciences series)
   Issued also as pt. 1 of Biology and man in 1975.
   Includes bibliographies and index.
   1.   Cytology.   I.   Swanson, Carl P.,   joint author.
II. Title.
QH581.2.M27   1976   574.8'7   75-22218
ISBN 0-13-589614-2

modern cell biology

second edition
William D. McElroy and Carl P. Swanson

© 1976, 1969 by PRENTICE-HALL, INC.,
Englewood Cliffs, New Jersey
© 1975 by PRENTICE-HALL, INC.
as part 1 of BIOLOGY AND MAN—
by McElroy, Swanson, and Macey.

10 9 8 7 6 5 4 3 2 1

Printed in the United States of America

contents

x

preface

Our readers may think that the title of this volume, *Modern Cell Biology,* does not strictly describe the contents. That is, they may feel that the title suggests a narrower coverage than the contents, and particularly the last four chapters, would imply. This, we readily admit, is a matter of judgment, but we feel comfortable with our choice of title.

As cell biologists, we are dealing with the structure, function, and biochemistry of cells, topics which have been enormously enriched in recent decades as new instruments, new techniques, and new methods of approach have made cells more readily accessible to study at more and more refined levels of investigation and understanding. But we are concerned not only with cells as such, but also with their relation to the larger problems of inheritance, development, evolution, and the origin of life. In addition, to help the reader visualize cells in the total scheme of time and place and change, we range in subject from atoms to galaxies, and from viruses and bacteria to humans and redwoods.

The cell doctrine, one of the central concepts in biology, was made possible by the development of the microscope, which enabled us to see smaller objects than the naked eye would permit. The cell, therefore, becomes the middle ground for our study of life. In the domain of smaller structures, cells are the objects wherein the chemistry of life is being enacted, where atoms aggregate into molecules, and molecules into organelles; in multicellular form, the cell is the object that enables us to see and describe the visible world of life that is so much more familiar to us.

Our emphasis, as a consequence, remains cellular even though our immediate topic may be molecules or humans. Where there are no cells, life is absent; the cell is the smallest unit of organized matter exhibiting all of those activities that, collectively, distinguish the living from the nonliving. The cell, through its informational content and its behavior in division, fertilization, and reproduction, provides the physical basis for the transmittal of characteristics from one individual to another, and from one generation to the next. These characteristics express themselves through the medium of cells; it is the multiplication, movement, and differentiation of cells that lead to the development and maturation of an individual, whether that individual be a single cell or a multicellular organism. If we think of cells functioning through time, displaying the heritable changes that occur continuously but randomly, seeing these changes gain expression in individuals, and visualizing their spread through populations, we are then dealing with the evolution of living systems. It is through these processes that we gain an understanding of the diversity as well as the continuity and unity of life.

Cells can also be viewed as the entities through which energy is continuously being manipulated; that is, the cells are so constructed that energy is trapped, converted to more usable forms, stored, and eventually utilized, after which it is returned to the external world as chemical or heat energy. This is all accomplished in an organized and orderly fashion, and it is the organization and structure of cells that permit this to take place.

As humans we are the most important living species because we live in an anthropomorphic world of our own creation. Like all other organisms, we are a collection of cells, organized as to type, number, and location in such a way as to bring about the appearances, and display the behaviors, of a human being. To emphasize the human race as the principal species for focus in the cellular world that we are describing is merely to provide additional relevance to our examination of cells and their structure and function. No other justification is needed.

This volume, like the first edition, is a portion of a larger volume. The first edition was derived from the *Foundations of Biology,* authored by McElroy, Swanson, Galston, Buffaloe, and Macey (Prentice-Hall, 1969). When the *Foundations of Biology* was revised, its title and authorship was altered (*Biology and Man,* McElroy, Swanson, and Macey) as was its contents. This edition of *Modern Cell Biology* has already appeared as part I of BIOLOGY AND MAN. It reflects many of the changes that have occurred since the first edition, and we like to think that it is also an improved version.

THE AUTHORS

modern cell biology

introduction
chapter 1

Each of us, whatever our training, whatever our interests, is an observer of the many worlds about us. Through sight, sound, taste, odor, and touch we are in some way, consciously or unconsciously, constantly responding to our environment and what goes on within that environment: a Cape Cod saltwater marsh, or the grandeur of snow-capped mountain peaks; the night sky sequined with thousands of stars, or the play of light filtering through a forest canopy; a tumbling mountain, the awesomeness of a raging sea, or a puddle left by a recent rain; the tranquility of a rural scene, or the bustle, noise, and excitement of a modern city; an athletic contest, or the tragedy and horror of contending armies; the individuals, groups, and communities that exist around us. As we look and see, listen and hear, with varying degrees of responsiveness, we perceive a universe abounding in diversity of form and action. And as there is diversity in what we perceive, so, too, is there diversity in our individual abilities to perceive. Each of us, because of inheritance and experience, time and circumstance, has a different image of himself and the world about him.

The universe is an enormously varied complex of matter and energy. There is nothing else. But the diversity around us stems from the manner in which matter and energy are organized, managed, manipulated, and expressed. At one extreme are the largest units of matter of which we are aware and can comprehend, the galaxies and supergalaxies. The bowl of the Big Dipper in the northern skies appears featureless and empty to the unaided eye, but this is due to our limits of resolution. Through a large telescope, nearly half a million galaxies, each containing billions of stars, can be seen to lie framed within the bowl. Like the individual stars that form them—some red, some blue, some giants, and some dwarfs—the galaxies vary in size, structure, shape, age, and in the amount of energy they emit, some of which reaches our earth as electromagnetic waves: visible light, x-rays, or cosmic rays (Fig. 1.1). The great number of galaxies visible in only one small patch of sky gives us an idea of the immensity of the universe. By comparison, our own galaxy, which has a diameter of about 100,000 light-years, is dwarfed.

At the other extreme is one of the smallest units of matter, the hydrogen atom, with a diameter of about one angstrom unit. Halfway along this cosmic yardstick, measured on a log scale, stands man the observer—curious, wondering, experimenting, and asking questions. His ability to ask questions is one aspect of his behavior that sets him apart from the other animals. He can also turn his sights inward and ask questions about himself. On the basis of his answers he can then alter his behavior and consider the consequences of doing so. He can also alter his environment to suit his present or future needs and wishes. Man can transform a barren desert into a lush tropical garden. Any other species, plant or animal, would have to adapt or become extinct.

Man is also a user and maker of tools. His mastery in technology has enabled him to construct artificial microenvironments such as bathyspheres and space probes, in which he can journey to the deepest trenches of the sea floor or into interplanetary space (Fig. 1.2). In the process of molding his environment and making replicas of it to meet his own ends, as no other animal has ever done, man at the present time is the most successful animal on the face of the earth. Yet an animal he is, as much a product of evolution as the other species he observes. It is as biological for man to dream, build, paint, and sing as it is for a muskrat to build a shelter of mud, sticks, and grass.

Despite his capabilities and his uniqueness, man, too, is an organized expression of matter and energy. His environment consists of three zones: the *lithosphere,* the *hydrosphere,* and the *atmosphere.* Man partakes of each, for his body, like that of all living things, consists of minerals, water, and gases, through which energy in various forms is accepted, transformed, stored, or used. It is the manner of organization of matter and of energy manipulation that led to the emergence of new qualities, that of life and of the biosphere, with all their varied manifestations.

Figure 1.1 We live in many environments. Our stellar environment, a galaxy similar to Andromeda (see artist's rendering above), consists of many billions of stars. One of those stars, the Sun, supports the only life we are aware of in the Universe, the life that we know on our own planet.

Ⓒ *California Institute of Technology*

4

Figure 1.2 A view of the Earth taken during the Apollo 17 lunar voyage. Africa, somewhat obscured by cloud cover, is to the left; Asia to the middle and upper right; the Mediterranean Sea is at the top; and the Antarctic polar ice cap is at the bottom.

And from the living world, and with the coming of man, has developed what the great French theologian and scientist Teilhard de Chardin has called the *noosphere*, the realm of the mind, inseparable from, created by, and peculiar to the body of mankind. Each of these spheres evolved out of the past, and each continues to evolve as time moves on into the future.

CHANGE
AND EVOLUTION

Evolution is the most subtle yet the most significant process determining man's relation to his environment. We commonly think of evolution as a series of changes taking place in plants and animals over many millions of years, but evolution operates in nonliving systems of matter as well.

Out near the edge of our home galaxy, part of which we see as the Milky Way, is our local star, the Sun. Compared with the 100 billion other stars in the galaxy, the Sun is not especially luminous or large; it is best described as a "typical" star in middle

5

age. To us, however, it is the most important member of the galactic community, for it is the source of all energy that sustains life on our planet. Some 5 billion years ago the Sun and its planets were formed out of a great cloud of interstellar gas and dust. Ever since it began shining, the Sun has been radiating light, heat, and other forms of energy at a steady rate, and, to the best of our knowledge, it will continue to do so for many additional billions of years. The conditions for life on our planet have been right for a long time in the past and, so far as we can tell, they will continue to be right for a long time to come.

But the Sun cannot go on shining forever. It heats us and otherwise sustains life on Earth by converting its vast store of hydrogen into helium deep within the solar core. Eventually, however, as the Sun continues to evolve, its hydrogen supply will be used up. When that day arrives the Sun will increase in size and pour forth much more energy than it does now, so much that the oceans will boil away and life on our planet will no longer be possible. Long after that event takes place, many billions of years from now, the Sun will end its life as a feebly radiating dwarf star, a spent member of a galaxy gradually growing dark because its stellar embers are going out one by one.

The Earth has evolved along with the Sun. As the nebular material that was to become the Earth packed itself into a sphere, it condensed. The hydrogen atoms and dust material then began to undergo chemical evolution which, eventually, produced the large variety of chemical elements known to us today. As the atoms of different elements combined in seemingly endless variety, they formed molecules, some of which dazzle us with their crystalline beauty, and others of which gradually became organized into living systems. Chemical evolution, then, gave rise to biological evolution, and the thread of life began.

The first living things appeared on our planet some 3 billion years ago. The predecessors of man possessing an upright stature made their appearance 2.5 million or more years ago, but *Homo sapiens*, modern man, has been in existence for less than 100,000 years. Over the thousands of millions of years since the first living things arose, the chain of life has not once been broken, to our knowledge, and it is likely to continue unbroken until the Sun enters the next stage of its evolution.

We cannot see evolution taking place, either among the stars or amid the diverse living and nonliving things on the Earth. The rate of change is much too slow, far slower even than the gradual rotting into dust of a giant redwood tree. It takes about 1 million years for a new species of plant or animal to arise, or so our best estimates tell us. Although this is but a tick of the cosmic clock in relation to the age of the Sun, it is a long time in relation to the recorded history of man, only about 5,000 years. Although we cannot see evolution taking place, we can see an expression of it in diversity (Fig. 1.3).

Figure 1.3 The Grand Canyon of the Colorado River, with the layered cliffs of sedimentary rock rising from the river's channel. The layers of limestone, sandstone and shale, laid down in the sea over the course of many years, and eroded by the cutting action of water over additional millions of years, differ in hardness and hence in resistance to erosion. This diversity in the inorganic world is paralleled by a comparable degree of diversity in the organic realm.

From Omikron

DIVERSITY AND EVOLUTION No matter where we look we find diversity—from atoms and antelopes to zircons and Zulus. An awareness of diversity in its varied aspects—animal behavior, cell function and structure, and so on—sharpens our powers of observation and heightens our sense of wonder and appreciation of the worlds around us. And our ability to understand diversity and the evolutionary processes that spawn it help shape our attitudes toward ourselves, toward other people, and toward life itself (Fig. 1.4).

Sheer size is perhaps the most striking form of diversity. The dinosaurs are gone, becoming extinct long before man appeared on the scene. We know them only as fossils, but these highly successful animals dominated the landscape with their tremendous bulk and splendid variety for about 100 million years. Even though these mammoth creatures are no longer around, we still have the blue whale, which reaches 100 tons in weight and 100 feet in length. We also have the elephant, largest of land mammals. Among plants, the giant redwoods tower 300 feet into the air and live 3,000 to 4,000 years. Many individual redwoods still standing today have lived throughout the greater part of man's recorded history.

There are only two species of elephants, three of redwoods, and only a few more species of whales (Fig. 1.5). Something about bigness is not favorable to continued evolutionary success. The blue whale, elephants, and redwoods are all on the danger list

7

Jerry Focht from
Leonard Lee Rue Enterprises

Leonard Lee Rue III

Figure 1.4 Two examples of diversity in the insect world: the prometheus moth on the left and the 17-year cicada on the right.

Figure 1.5 An old stand of Redwoods (*Sequoia sempervirens*) in Del Morte County, California.

U.S. Forest Service

of extinction, although for these species the blame rests with man. So, too, are large members of the cat family (tiger, lion, cougar, leopard), and those of the hooved group (bison, moose, and many varieties of deer). It is among smaller organisms that richness in numbers and variety abounds. Many are too small or too secretive in their habits to be observed easily. Of those that are easily observed, insects are the most striking in coloration, camouflage, and diversity of behavior. They live in the air, under water, beneath and on the ground, and they feed on everything organic. As a group, they are highly successful, living everywhere over the globe, from ice-covered Antarctica to the tropics and deserts.

If we think of recent examples of evolutionary success in the plant world, we turn to the herbs; in terms of diversification of form, habitat, and reproductive capability, they are the counterpart of the insects in the animal world. Although we may think of trees as the more conspicuous part of the landscape, trees are, in many parts of the world, being replaced gradually by herbs, with common weeds being among the more successful: dandelions, quack grass, nut grass, hawkweed, chickweed, and chicory (Fig. 1.6).

So far, we have been discussing diversity on the macrobiological level, that is, among the plants and animals that we can see and describe without the aid of instruments. In the microbiological world, revealed through the use of sophisticated instruments, diversity also abounds. The smallest unit of living matter is the cell. A redwood tree or a blue whale is an orderly aggregation of millions upon millions of cells, the great majority of them highly specialized in form and function. These massive organisms, like all others originating through sexual reproduction, are derived from only two cells, the male and female reproductive cells,

Figure 1.6 An abandoned farmhouse surrounded by plants that have invaded the once-cultivated areas.

each specialized to engage in the act of fertilization. Many of the intimate details that lead to cell specialization remain obscure, but there is no doubt that answers, when found, will involve the orderly and sequential chemical activities that take place within each cell. In addition, there is a vast array of microscopic organisms, plants and animals alike, that consist of but a single cell, yet these single cells carry out the wide variety of activities that qualify them as "living" entities.

The last 20 years have witnessed a revolution in biology, with the greater portion of information being related to what goes on in the living cell, or to how parts of cells perform their functions in cell-free systems within a test tube. We have come to recognize the internal diversity of the molecule that is responsible for the biological diversity that exists at other levels. This is the molecule deoxyribonucleic acid, DNA. It is found in all cells and determines whether an organism is to be a bacterium, a cow, or a man; and among men, whether a person is to be an idiot, a genius, or just an average member of the human race.

matter, energy, and information

All scientific descriptions, collections of data, hypotheses, and laws are concerned with matter and energy in some form. It makes little difference whether we are dealing with events or things in galaxies, solar systems, the earth, an organism, a cell, or an atom, or whether we are dealing with man as an animal or as a human being surrounded with all aspects of his culture—we are dealing with a dynamic system that can be made intelligible in terms of amounts and transformations of matter and energy. This being so, the rules of physics and chemistry apply equally well to animate as well as to inanimate systems, although animate systems are likely to be far more complicated and hence more difficult to analyze and to understand. The question of whether there is a vital force, independent of matter and energy and hence of supernatural origin and character, governing the activity of organisms has long been dismissed as unanswerable, and therefore of no value to the biologist when he functions as a scientist. This mechanistic point of view does not, of course, preclude the possibility that laws peculiar to living things, and unknown in, or inapplicable to, the inanimate world, may be discovered, but science clearly cannot function or move ahead on the basis of undiscovered facts or on intuitive, but unsupported, ideas.

Living organisms, on the other hand, differ profoundly from nonliving systems in that the management of matter and energy is governed by predetermined sources of information possessed by each organism. This coded information resides in the nucleic acids of the cell and sets limits to the kind, amount, and use of matter and energy that can be manipulated for purposes of growth, maintenance, activity, and reproduction. With the exceptions of radiant energy from the sun captured by photosynthetic organisms, and the conversion of chemical to electrical energy in such forms as the electric eel, all activities of cells, organs, and organisms are based on chemical reactions. Since organisms, particularly warm-blooded ones, can function only within a limited range of temperatures, energy in the form of heat can be used only indirectly.

The continued success of organisms during their lifetimes, and species over longer periods of time, depends, therefore, on their success in drawing matter and energy from the environment, and the subsequent manipulation of matter and energy under the coded instructions of the internal organization of the organism. We should, as a consequence, expect to find—and, indeed, we do find—that each species, as a unique system, possesses mechanisms for the acquisition, transformation, storage, and utilization of matter and energy (Fig. 1.7). There is, of course, a limit to the energy upon which an organism can draw, and a limit to its ability to make use of even that energy which is available. A modern automobile engine, for example, cannot use the energy available in gasoline unless its parts are organized and work in a particular way. Individuals must be similarly organized if they are to make use of their energy sources. Also, just as engines differ in the way that they extract energy from gasoline, so do organisms differ in their management of energy. This is, in part, a reflection of the individuality and diversity of organisms, and inevitably it leads to competition for energy sources.

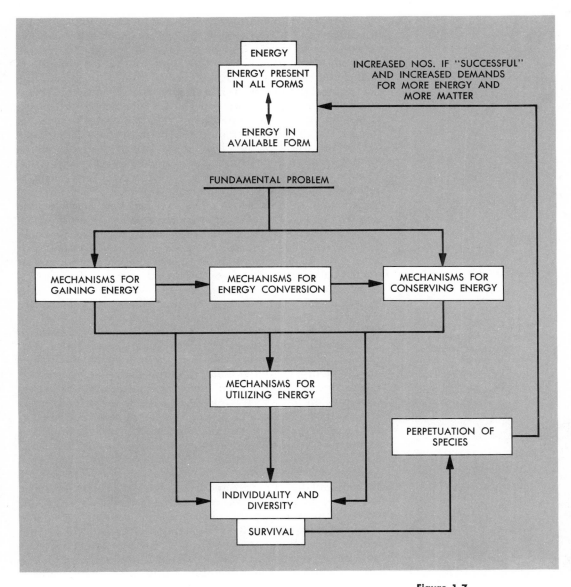

Figure 1.7

The successful acquisition and management of energy determines survival, and survival, coupled with success in reproduction, leads to the perpetuation of species. Continued success results in an increase in the number of individuals, and this factor, in turn, leads to greater demands on the environment for more matter and more energy. Since neither matter nor energy is available in unlimited amounts, competition for them limits both the numbers and kinds of organisms. Evolution is the inevitable result.

There is another way in which we might examine the use of matter and energy by organisms. In a general way, we have no difficulty in identifying one species from another: roses from marigolds, cows from horses, men from apes. And we recognize further that like begets like, that roses produce more roses and not marigolds, and men produce more men through reproduction. There is, therefore, order in the manner by which each species handles matter and energy. This implies a measure of control, and the situation is quite different from that taking place outside the world of life. The second law of thermodynamics informs us that the entropy of the universe is increasing; that is, matter is becoming more and more dispersed and energy is becoming more and more uniformly distributed. The ultimate result is total randomness, or chaos, or a lack of orderliness. Life reverses this process, creating order out of disorder, and raising energy to higher levels of concentration. In doing so, however, living systems do not constitute an exception to the second law; order, whether of matter or of energy concentration, can be created only at the expense of additional energy. Life, in all its manifestations, from daily existence to evolution through vast geological ages, is, therefore, a continual competition for the free energy available in the environment—radiant energy from the sun or chemical energy bound up in, but extractable from, usable molecules. Life involves, on the other hand, a good deal of constant wear and tear, a breakdown in the orderliness of structure and function; a continual input of energy and matter is required to counteract this breakdown, and a good deal more when stress is imposed on the system. Only death reverses the process, interfering with the input of energy and turning order into disorder. But death is necessary for life; the death of one organism is necessary for the life of another. Life has a way of recycling itself, although its form may change with each complete cycle.

The success of man as a species is a function of his success as a manipulator of matter and energy. In this he does not differ from any other species. But man, unique in the living world, has succeeded in tapping sources of energy and matter unavailable to other species (Fig. 1.8). His methods of agriculture for the production of high-energy foodstuffs, his increasingly sophisticated tools, and his use of wind, water, fossil fuels, and nuclear energy for power, all have combined to give him extraordinary control over, and management of, his environment. However, the fact that unlimited matter and energy are not within his grasp poses problems for him as a species, and for his way of life.

Figure 1.8 This vast wheat field is just one example of man's use of agricultural techniques to provide foodstuffs for his own survival.

Jack Dermid, Wilmington, N.C.

THE GROWTH OF SCIENCE

Before getting to the heart of biology, let us first take a brief look at science in general. We live in an *Age of Science,* an age during which we are gaining knowledge of ourselves and the world around us more rapidly than at any other time in our history. Think of the commonplace things that are a part of your everyday world and about which your parents, at your age, had no knowledge whatsoever: the jet airplane, transistor radio, synthetic fibers of many kinds, packaged frozen foods, "wonder" drugs, electronic computers, hydrogen bombs, artificial satellites, and spaceships. Each year it becomes possible to add many items to this list. Often the new replaces the old even before we become fully acquainted with the old. Just as you live differently from the way your parents did at your age, so your children will grow up in a world that will be vastly different from yours.

The things we have listed are products of the **applied sciences**—engineering, medicine, and agriculture (Fig. 1.9). But behind each lie ideas developed by the **basic sciences.** What is an applied science as contrasted with a basic science? How does one depend upon the other? The discovery and use of the antibiotic penicillin may help to illustrate this point. In 1928, Sir Alexander Fleming, the English bacteriologist, discovered that a mold (fungus), *Penicillium notatum,* had contaminated some of his bacterial cultures. Not only did the mold kill the bacteria it was touching, but it was killing bacteria at some distance away also. From this observation, it was possible to conclude that a product formed by the mold diffused, or spread outward, killing the bacteria with which it came into contact. After more than a decade of further research and testing, medical men were able to isolate, purify, and use penicillin to cure many bacterial infections in man. In this sense, they *applied* a *basic* idea to solve a particular problem confronting them.

Although this illustration is clear cut, it is not always so easy to distinguish between these two aspects of science and their relation to each other. In the treatment of cancer, as in so many instances, the applied precedes the basic information. We know how to control some forms of cancer, if caught in time, but we still lack a basic knowledge of *why* cancer develops and behaves as it does. Ultimately, this knowledge may come from basic research, or from applied research, or from a combination of the two.

One thing we can be certain of is that knowledge of ourselves and the world around us advances as basic scientific discoveries are made. This has been proved time and again. And use of this knowledge is a form of power: to improve and increase our food supply, to control and eliminate disease, to educate the peoples of the world, to communicate faster and more precisely. Fleming's work with antibiotics changed the whole complexion of medical science just as the molecular aspects of structure, growth, and behavior have changed the complexion of biology over the past few years. The unique power of science is that man can now transform the world if he chooses to do so. What is needed is planning, cooperation, and a vision of what he wishes the world to be for himself and his descendants.

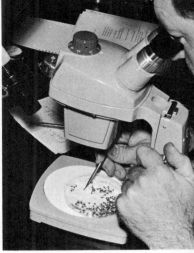

Paul Knipping

Figure 1.9A An entomologist separating species of mosquitoes in an effort to determine which species transmits disease-producing viruses. He will crush the mosquitoes and then add the "brei" to growing chick embryos to search for the presence of viruses.

Figure 1.9B A field biologist searching a stream for organisms that transmit disease-producing viruses, in this case a virus that causes vesicular stomatitis. The disease is characterized by blisters on the tongue, lips, and membranes of the mouth in horses, cattle, and swine, and can be transmitted to man.

Paul Knipping

13

One reason for the rapidity of change today may be gained by considering the recent increase in the number of scientists. We think of modern science as having its origins sometime during the 16th and 17th centuries when men, in their effort to understand the world around them and to come to terms with it, turned from a life of comtemplation to one in which greater and greater reliance was being given to direct observation and experimentation. This time was the beginning of the scientific revolution, which is still going on. Yet of all the scientists who have ever lived, nearly 90 per cent are alive and active today. In recent decades the scientific population has also been given greater financial support by individual donors, governments, industry, and foundations. No wonder, then, that the volume of information and ideas produced by scientists today is proportionately greater. It has been estimated that our store of scientific knowledge doubles every 10 years or less. Scientific discoveries made during the last 20 years alone far exceed all that man had discovered up to that time. Most important, the information obtained has not been haphazard, but directed to the solution of specific problems chosen because of their importance. No person can possibly read all this printed information. Even the devoted scholar finds it difficult to keep himself fully informed about all the new findings in his own field. Indeed, we are in danger of being drowned in an ocean of information unless sophisticated systems of information storage and retrieval are established.

The situation reminds one of Alice's conversation with the Queen in *Through the Looking Glass*. "Well, in our country," said Alice, "you'd generally get to somewhere else, if you ran very fast for a long time as we've been doing."

"A slow sort of country," said the Queen. "Now here, you see, it takes all the running you can do to keep in the same place. If you want to get somewhere else, you must run at least twice as fast as that."

THE WAYS OF SCIENCE It has often been stated that science is a search for truth. This is an overstatement since truth is an absolute and we have no certain way of recognizing truth when it appears. A scientific statement about some aspect of reality is provisional, the best that can be stated at a given moment in time. Being provisional, it is also likely to be ephemeral, to be replaced by something more definitive or more encompassing as additional information is forthcoming. **Scientific truth,** therefore, can best be defined as a bit of information or an idea accepted as reasonable by reasonable investigators. Such a consensus is necessary, for one needs a base of operations from which to explore the unknown, and no body of knowledge can grow if only uncertainties exist. The poet and the artist also search for truth, for a glimpse of reality, the former through metaphor, the latter through the use of space, color, and line. Theirs, however, is a more personal vision of reality, to be accepted or rejected by the reader or viewer, but not subject to verification by test or measurement. There is, furthermore, a permanence to poetic and artistic truth that defies the passage of time; the response of human beings to what is good and true and beautiful is steadfast even while science and scientific knowledge are undergoing drastic alteration.

The aim and purpose of science is to gain an understanding of the natural world, living and nonliving. By "understanding" we really mean *the ability to predict* events or relationships in nature to a more or less correct degree. You readily *understand* that $1 + 1 = 2$, but in reality you have sufficient familiarity with numbers to *predict* that the same result will always be obtained. The more exact a science becomes, the more precise will be the power of prediction. We can, for example, predict the time of an eclipse of the Sun by the Moon to a fraction of a second, but you are equally aware that the science of weather forecasting is much less accurate.

Science, therefore, seeks to discover order in what often appears to be disorder. It does so on the assumption that such order exists, that what goes on in the universe is not a series of chaotic happenings, and that cause and effect relationships hold. Prediction would otherwise be impossible. Science also explores new ways of knowing, just as the arts explore new ways of seeing and expressing. The domain of science is the structure, behavior, and history of matter and energy. Unlike the arts, science deals best with those things that can be observed, described, measured, tested, and verified—things that we can directly or indirectly detect with our senses, and that enable us to project our thoughts beyond our senses. In the final analysis our senses are our *only* authority. They are our means of contact with the world around us. However, it is the mind that assembles these observations, putting them into meaningful relations to give us our ideas of an atom, a gene, or a universe (Fig. 1.10).

This does not mean that science must deal only with those phenomena which can be directly sensed; after all, we cannot see x-rays and we do not hear radio waves. Yet we know of their existence for we have instruments to detect them. These instruments become our extended senses: the telescope, microscope, and spectrometer are examples of such instruments. And in the hands of an imaginative scientist they can be used to enrich our lives every bit as much as the painter's brush or the writer's pen. The great scientist must be as much a creative artist as a Shakespeare or a Rembrandt. The great writings of science are permeated with imagination, enthusiasm, mystery, and humility.

We do not mean to imply that there is a clear and sharp separation between science and the arts. It is true that we generally think of science as dealing with phenomena lying outside

Courtesy of Mount Wilson and Palomar Observatories

Figure 1.10 Without the aid of instruments man's view of the universe is severely limited. The Hale telescope on Mount Wilson, California, enables us to probe outward into space and backward in time. In the photograph above, notice the observer in the focus cage and the reflecting surface of the 200-inch mirror ahead of him. Microscopes, like the student research microscope below, which can magnify about 1,200 times, enable us to view the microuniverse.

Courtesy of Leitz-Labolux

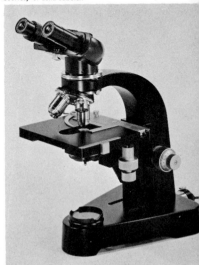

15

Figure 1.11 Above, the beautiful symmetry of a snow crystal, no two of which are exactly alike. Below, the equally symmetrical siliceous shell of a diatom, a single-celled alga.

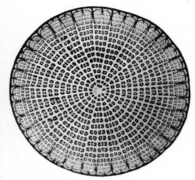

of us, phenomena that we can measure, test, and experiment with. But we must not forget that each of us is an organism, and our emotions, sensations, desires and impulses, thoughts, and purposes arise out of our animal attributes, and are then molded by the cultural environment in which we have our existence. In their totality, these behavioral phenomena comprise in each of us an inner, private world from which we view and interpret the outside, measurable world.

We can think of science in two ways. It is, first, *a body of knowledge organized by men and reflecting the order and disorder we see in nature.* Gradually we have come to sense that there is a rule of order instead of chaos, a pattern of law instead of anarchy. We can see this strikingly revealed in the symmetry of crystals and diatoms (Fig. 1.11). It can be just as beautifully demonstrated in the realm of ideas and discoveries. Our image of the origin of the universe, the Solar System, the Earth, and life itself has been laboriously and gradually built from ideas and discoveries of astronomers, geologists, physicists, chemists, and biologists over a period of many years. We can think of this approach—the gradual building up of a body of knowledge—as the cumulative and progressive side of science; but there is another and equally important side.

Science is also an attitude. When no answers are immediately forthcoming, the scientist devises means of wresting answers from the unknown. This is the aggressive side of science; it is what gives science its vigorous spirit of inquiry. It includes the priceless quality of wonder and curiosity that has understanding as its goal. In a way, the endless "what" questions of the wondering child are not so very different from the "what" and "how" questions of the scientist. This attitude also includes the restless, endless pursuit of knowledge, even though it is difficult to know in advance whether a bit of knowledge will be "useful" or enlightening. Understanding is also a means to power, that is, the power to control and use our environment to our advantage. Much of the financial support given to science is for the extension of this power: control of disease, use of space, improvement in the quality and quantity of food, and an understanding of ourselves are examples.

Why do we apply these terms to science? There are two reasons. The quest for new ways of knowing is an advance against the unknown. Here man finds his greatest challenge, and his sight is ever forward. His searching ground is at the boundary of the known and the unknown. The second reason is that science is the only field of human activity in which progress is certain and inevitable. Civilizations, cities, cultures, and empires rise, decay, and fall. But the high point of science is today, and tomorrow it will be still higher.

When we combine the several aspects of science, we find that science emerges as a process of growth similar to that taking place

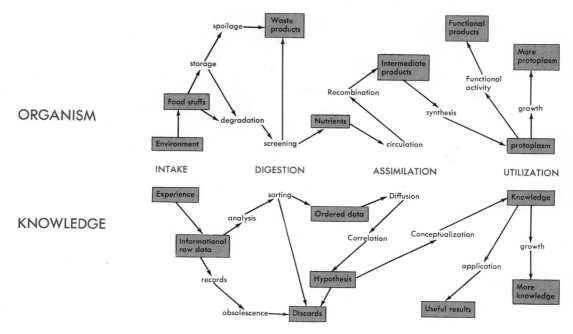

ORGANISM

KNOWLEDGE

Figure 1.12 An analogy, between the growth of an organism and the growth of knowledge. Part of the food an organism takes from its environment is converted to energy for the performance of activities; and part is excreted as waste products. Knowledge is "mental food"; it is generated from experience, absorbed, assimilated, and used. But some will never be used, and some will be incorrect. The organism—you—decides what will be used and what will be discarded.

in an organism. This has been depicted by Paul Weiss, biologist at the Rockefeller University. On the organism side of our illustration (Fig. 1.12) the key to the chain of events is the organism, which takes from the environment what it needs and refines, transforms, and incorporates matter and energy into more of itself. In this way the organism metabolizes, grows, and reproduces. On the side of knowledge, the key agent is not shown. It is the human mind, constantly observing, questioning, testing, and devising new tools and new techniques as it seeks new ways of knowing, and new ways of applying the knowledge gained. Along the way, the human mind constructs a new language to describe the reality it has discovered.

THE SCIENCE OF BIOLOGY *Biology is that part of science dealing with the matter and energy which is, or was, a part of a living system.* The word comes from the Greek *bios,* which means "life," and *logos,* which means "word" or "thought." In short, biology is a body of knowledge of living systems. We often add adjectives or the prefix or suffix *bios* to describe narrower segments of the larger subject of biology: molecular biology, developmental biology, biochemistry, symbiosis, and so on.

Biology is both an old and a vast science. Our earliest recorded biology comes from the ancient Greeks, but even prehis-

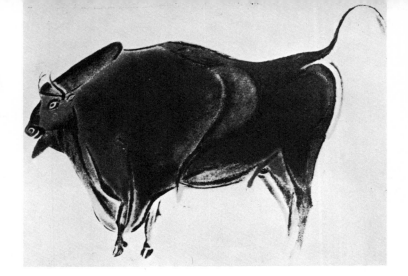

Figure 1.13 A portion of a cave painting—a bison—made by prehistoric man at Altamira, Spain. The bison belongs to an extinct species, and the history of this race of artists is not fully known.

toric man left behind him in his caves beautiful drawings of animals (Fig. 1.13). He seems to have had a sensitive awareness of proportion, anatomy, and motion. When man existed principally as a hunter, he practiced biology of a sort as he sought out his food, for he had to know the ways of the hunted animal, the ways of those animals that would prey upon him, and the sources of edible plants. When he turned from his nomad life to a more stable, agricultural existence, he had to have a greater knowledge of plants and animals before he could domesticate them sufficiently well to provide himself with a ready source of food. And as man domesticated these organisms, so did they domesticate him, for he had to adapt himself to their ways of life as well as adapting them to his own. It is from these early beginnings that biology had its start.

The vastness of biology today can be seen by the various subdivisions into which the science has been broken. A look at any natural habitat will impress you with the great diversity of living forms, which range from trees and large animals to organisms of microscopic size. Classification of organisms into convenient groups that show common similarities is **taxonomy.** Investigation of structure is **anatomy;** of function, **physiology;** of the details of cells, **cytology;** of inheritance from one generation to the next, **genetics;** of the relation of organisms to their environment, **ecology;** of disease, **pathology.** Biology, too, can be divided according to the particular group of organisms being studied: **entomology** (insects), **malacology** (shells), **mycology** (molds and mushrooms), **bacteriology, virology,** and **botany** (plants), and **zoology** (animals). The list can be greatly lengthened, depending on how narrowly one wishes to limit a given area of study.

Table 1.1 will give you some idea about subdivisions as one goes from the molecules of biological importance to the total

TABLE 1.1. The Science of Biology—Disciplines, Methods, Tools, Concepts

Level of Investigation	Disciplines	Methods	Tools—Equipment	Concepts
Molecule	Biochemistry Biochemical genetics Organic chemistry Immunology Enzymology Polymer chemistry	Biologically important molecules—isolation, purification, analysis, synthesis, function Use of radioactive tracers Determination of reaction rates	Analytical balance Centrifuges Spectrometers Chromatographs Electrophoresis Counting devices for radioactivity	Mechanism of enzyme action Source, storage, and transfer of energy Source and transfer of information Control and inheritance of biosynthesis Antibody–antigen relations Molecular structure and function Action of drugs
	Pharmacology	X-ray diffraction Electron microscopy Fluorescent microscopy	X-ray sources Electron microscope Fluorescent microscope	
Organelle	Cell biology Plant physiology Animal physiology	As above Cell fractionation Microdissection Staining procedures	As above Microscopes (various)	Organelles—formation, structure, function, interaction
Cell	Cell biology Embryology Protozoology Microbiology Bacteriology Pathology	Fixation, sectioning, staining of cells Cell culture techniques Autoradiography Microdissection	Microscopes Microtomes Incubators Tracer counters	Mitosis, meiosis Haploidy vs. diploidy Sex determination Differentiation, growth Host–parasite relations Cell death Classification
Organ	Anatomy Histology Physiology Pathology Embryology Pharmacology	Sectioning Staining Dissection Surgery Transplantation Transfusion Drug action	Microscopes Microtome Electrocardiograph Electroencephalograph Blood, bone banks Heart–lung machines	Comparative anatomy and physiology Action of drugs Action of hormones Replacement of organs Regeneration Problems of disease

TABLE 1.1. (Continued)

Level of Investigation	Disciplines	Methods	Tools—Equipment	Concepts
Individual	Embryology	As above	Psychological tests—	Homeostasis
	Physiology	Conditioning	IQ, aptitude,	Consciousness
	Pathology	Psychoanalysis	personality	Ideas of learning
	Psychology	Inheritance tests	Breeding plots	Individual differences
	Animal behavior	Psychological tests	Statistics	Gene action;
	Genetics	Diets	Twin studies	limits of variation
	Medicine (various)	Hypnosis		Genotype vs.
	Taxonomy	Learning		phenotype
	Physical anthropology			Goals, purposes,
				free will
				Aging
Species	As above	Methods of classification	Herbaria	Natural selection
	Evolution	Cytotaxonomy	Museums	Mutation rates
	Paleontology	Statistical methods	Microscopes	Breeding structure
	Cultural anthropology	Population sampling	Breeding plots	Species formation
	Climatology			Race concepts
	Geography			Ecotypes
				Fossil formation
Community	Ecology	As above	Control of environ-	Cultural evolution
	Population genetics	Sampling techniques	ment—physical	Social structure and
	Sociology	Field observations	and biological	evolution
	Social anthropology	Isolation techniques	Introduction of new	Group reaction
	Group psychology	Determination of	variables	Environmental
	Evolution	environmental		variables
	Environmental	variables		Population dynamics
	medicine			Species succession
	Epidemiology			
	Meteorology			
	Geology			

living world, and as one views the particular subject in terms of structure, function, or history. We can also look upon biology as being subdivided by ideas (concepts) or by methods (tools and techniques). Try doing this as you proceed through this book. It will help you to organize your thoughts and enable you to visualize the relation of one part of biology to all the other parts.

Many of the terms in the table are unfamiliar to you now, but by referring to them as your knowledge of biology grows, you will begin to see that structure always has a history or an evolution, and that structure and function are intimately related to each other.

How does a science grow from the humble beginnings of casual or studied observation stemming from a need to know something? Two things are required to develop a science—facts and ideas. Men of prehistoric times had to rely on their unaided senses. They witnessed the slow passages of the seasons and associated with them the migrations of birds, the coming of plants into flower and fruit, and the cycle of birth, growth, death, and decay. Men gradually recognized that the great variety of life about them exhibited similarities as well as differences. All these things had to be named and sorted out so that they could be handled conveniently and discussed intelligently. Imagine trying to keep track of your friends or enemies if they had no names. This is the **taxonomic,** or classification, stage of biology. No experimentation is needed here, for the purpose is to group similar things into similar categories to form broad generalizations or patterns that reveal basic similarities in a mass of different phenomena. This activity is based on the assumption that there is an orderliness and a pattern in nature.

At this stage the question generally asked is "What is it?" or "What is it like?" Having answered the "what" question, it is then natural to ask, "What is it made of?" or "How is it put together?" Sometimes the "how" of the situation requires instruments (the microscope is a good example), and at this point the **structural stage** of biology begins. Here again, experimentation is not a necessary part of the science, for good description is often enough.

Proceeding onward, the more curious will inevitably ask, "What do the parts do?" This is a question of function (for which the biological term is **physiology**), and we can say that we have arrived at an **analytical stage** of scientific inquiry. Description alone is inadequate at this stage, and some kind of experimentation becomes necessary. If we are keen-minded, we begin to notice that structure and function go hand in hand. It is also important to recognize that there can be no appeal to authority at this stage or at any other stage of science. We must find out for ourselves. The mind must be open, curious, and questioning, and the experimenter is judge for the moment. History and the test of time will be the final judge of what is good or bad observation or experimentation.

The final stage in the solution of a problem generally involves the question: "What is the relation of this information to other bits of knowledge?" **Synthesis** must follow analysis. Although description and experimentation are not involved, synthesis, which attempts to tie things together, depends on them for basic information. It is also the most difficult stage of science and the least certain to produce answers. But when an answer emerges, systhesis is a truly creative step. The theory of evolution put forth by Charles Darwin is an example of this aspect of scientific study. It is not fundamentally different from the struggles of the poet, who, in his way, also recreates experience. Synthesis is a tying together of information in order to answer questions such as "What are the functional requirements that demand this type of structure?" or "How is this structure related to its function?"

Probably science proceeded along such lines. Yet it would be wrong to suppose that each stage developed independently of the others, and that all stages developed in a 1, 2, 3 fashion. Furthermore, no stage is ever complete. Each generation tends to center its interests on some broad aspect of a science, and each generation reinterprets the past in order to make the past a useful part of the present. The Swedish botanist Carolus Linnaeus (1707–1787) established the rules of taxonomy in 1753 with his great book *Systema Naturae,* and taxonomy continued to be a most important part of biology throughout the 18th and 19th centuries. But taxonomy as a dynamic, rather than as a descriptive, part of biology had to wait for the 20th century, a time when the sciences of genetics, evolution, and biochemistry enriched it. The first half of the 1900s witnessed the beginnings of tissue and cell analysis, although the latter half of the 19th century was known as the "Golden Age of Cytology" because of the many discoveries made then; but, once again, biochemistry and the development of the electron microscope have changed our science of cellular biology (Fig. 1.14). By the beginning of this century, the organ systems of plants and animals were generally understood in a structural and functional way. At present biologists tend to concentrate their attention on the structure and function of the many parts making up a cell—**molecular biology** as it is popularly called.

THE SCIENTIFIC METHOD   Let us now turn the whole problem of biology around. Let us suppose that we wish to add to the present store of information. How do we do this? You have probably heard of something called the **scientific method.**

The term "scientific method" is like a battered football; it has been kicked around and abused by a good many people. Actually, the definition is simple: the scientific method is what

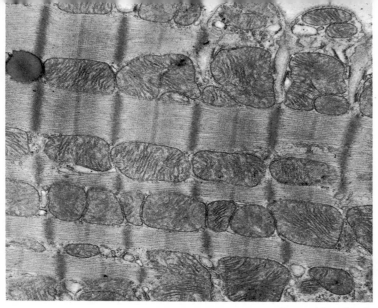

Figure 1.14 This electron micrograph of a section of heart muscle of a rabbit reveals the great amount of detail resolved by the electron microscope. Magnification is 50,000 times.

scientists use when they try to solve a problem. It can be a simple or a complex process; it varies with the scientist and with the problem. And its procedures are not rigidly set, despite attempts of many people to make it into a neat formula.

Let us consider a number of scientists and what they have done so that we may gain some idea of how they have contributed to our fund of knowledge. We shall find that these men have several things in common, but they followed no set formula in attempting to understand the world (Figs. 1.15, 1.16, and 1.17).

Figure 1.15 Left, Ptolemy (Claudius Ptolemaeus), Greek mathematician, astronomer, and geographer, whose idea of an Earth-centered universe, advanced in A.D. 127, was widely accepted until it was displaced by the idea of a Sun-centered Solar System, advanced in 1543 by Nicolaus Copernicus, a Polish astronomer, who is shown at the right.

Figure 1.16 Charles Robert Darwin (1809–1882), an English naturalist, developed the theory of evolution that is accepted today.

Figure 1.17 In 1953 James Watson (left), an American biologist, and Francis Crick (right), an English physical chemist, with Maurice Wilkins, an English crystallographer, discovered the structure of DNA, the crucial molecule of inheritance. A model of the molecule is in the background. These scientists were awarded the Nobel Prize for their discovery.

All recognized, however, that the isolated fact, unconnected to an idea, is a useless piece of information.

Today we accept the idea that the Sun is the center of the Solar System, and that the Earth, together with the other planets, revolves around the Sun. This hypothesis was put forth in 1543 by Copernicus, a Polish astronomer and mathematician, in a book published as he lay on his death bed. Certain Greek scientists had thought similarly, but a point of view, the Ptolemaic hypothesis, persisted and had dominated men's minds up to the time of Copernicus: namely, that the Earth was the center of the universe. To this was added the idea, from Judaic–Christian writings, that the Earth and all that lived on it were divinely created for man's use and delight. It was a tight, small, comfortable, and orderly world. Each planet was thought to revolve in a perfect circle, and the fixed stars were in crystalline heavens, the home of the many gods of the Greeks—Mars, Jupiter, Venus, and others—who oversaw all things. Some of the words in our language come from the world of the Greeks: *jovial* (Jove) and *saturnine* (Saturn) tell us something of the character of these gods. *Disaster* originally meant "against the stars," and *exorbitant* meant being "out of regular orbit."

However, some of the astronomical events that could be observed, and that were needed for navigation, the perfection of the calendar, and the prediction of the seasons, did not fit the neat and tidy picture of Ptolemy. Copernicus spent 30 years recalculating astronomical data and concluded that these data would equally well fit the idea that the Sun, not the Earth, was the center of things. Copernicus had not set out to change man's position in the scheme of things, but he did, and the world has not been the same since. In the process, man lost some of his dignity and divinity.

Charles Darwin was an equally tireless collector of information. An English country gentleman, he started out to be a doctor, then a clergyman, and finally became a naturalist. A trip on HMS *Beagle,* as ship's naturalist, took him around the world and brought him face to face with the enormous diversity of living things and their irregular patterns of distribution. He knew a good deal of biology and geology, and of the character and origin of fossil-bearing rocks. He also knew the practices of plant and animal breeders and the value of selection in changing the character of domesticated plants and animals. But all this diversity made sense to him only after reading an essay by Thomas Malthus on how human populations were controlled in number by food supply, disease, famine, and war. Darwin's theory of evolution through natural selection came, therefore, only after a long period of observations and the tedious collection of vast amounts of information. Alfred Wallace, an animal geographer, had virtually the same idea at the same time; but instead of putting his theory in the form of a book, as Darwin did, he sent

Darwin a short note that contained the essence of Darwin's theory.

Our world today is a safer one to live in because of our knowledge of antibiotics, but the discovery of these substances was a matter of chance. As mentioned earlier, some of the bacterial cultures of Sir Alexander Fleming, the English microbiologist, became contaminated with mold, and some of the bacteria were killed. But instead of just considering this bad luck or poor technique, Fleming investigated the matter further, and the science of antibiotics was born. He treasured his exceptions and made the most of them.

The most important biological discovery in the last 25 years has been the discovery of the structure of DNA, deoxyribose nucleic acid, the molecule of inheritance in the vast majority of organisms. This knowledge has changed the whole character of biology. The significance of DNA, which we will deal with later, was recognized in 1944. Through the efforts of three men working in Cambridge, England—J. D. Watson (an American virologist), Francis Crick (an English chemist), and Maurice Wilkins (an English crystallographer)—a model of the DNA molecule was painstakingly built, which corresponded to all of the information gathered by them and many other scientists. They were awarded the Nobel Prize for their brilliant efforts.

These men had several things in common: they knew their science and the tools of their trade thoroughly, they valued sound data, they trusted their observations, they were not afraid of new ideas, and they had the courage to believe in their convictions. Most importantly, they knew how to ask the right questions of nature, questions that were appropriate to their time and to the state of their science and that had some hope of being answered and accepted. The great idea that ties odds and ends of miscellaneous information together into a meaningful whole can come in a flash or only after a lifetime of work. Whenever it comes, it is the product of an alert and disciplined mind, a mind that can take advantage of ideas when they do appear.

The final step in the treatment of a problem is the process of making this information public. The scientist does this, usually, in writing, though he may on occasion simply present his ideas in lectures or discussions. The scientist has an obligation to communicate his findings; he cannot rightly shirk this responsibility, for he is, in a very real sense, a public servant. What is bad science is usually forgotten, but what is good becomes part of the great heritage of knowledge and understanding passed on to other generations.

Unfortunately, the more specialized the science the more technical and difficult the language, and the proportion of information that gets back into our common language is correspondingly small. Even so, you might be surprised at the number of technical words that have crept into common usage from the field of rocketry alone in recent years. How many can you think of? Where did you first hear of them? A word such as **feedback,** arising out of the fields of electronics and communications, often provides us with a new way of seeing a problem in our everyday experience. The extent to which such words and concepts find their way into common usage, thereby enriching our language, is one measure of public response to science.

BIOLOGY AND MAN As human animals, all of us, individually and collectively, are part of the world, related directly to, or in association with, all other things, living and nonliving. A study of biology is a study of ourselves and of certain of those associations. Its goal is to help us know ourselves and the world we live in: how we are processors of matter and energy at the cellular level; how we grow, differentiate, inherit, and evolve; how through the course of time we moved to a human status; how we must, from necessity, remain adapted in an environment that sustains us, but an environment that has limits to its resources of matter and energy. We also want to know where we stand in the stream of time and in the immensities of space. Some 250 years ago, Blaise Pascal, the great French philosopher, said:

> When I consider the short duration of my life, swallowed up in the eternity before and after, the little space which I fill, and even can see—engulfed in an infinite immensity of spaces of which I am ignorant, and which know me not, I am frightened, and am astonished at being here rather than there; for there is no reason why here rather than there, why now rather than then. The eternal silence of these infinite spaces frightens me.

But Pascal's despair was counterbalanced by a knowledge of his ability to think, and by a knowledge of his own limitations.

> It is not in space that I should look to find my dignity, but rather in the ordering of my thought. I would gain nothing by owning territories: in point of space the universe embraces me and swallows me up like a mere point; in thought, I embrace the universe.

You, too, can embrace the universe in your thoughts, and come to understand yourself as a dynamic system functioning in a changing world.

summary

Science is an attempt by men to understand, and thereby to predict, the structure and behavior of themselves and the physical, chemical, and biological world in which they live. Science does so through controlled and orderly observations, and through experiments designed to test our ideas about energy and matter— what they are, how they are put together, how they change, and how simple forms of matter can be built up into complex forms. When we study matter that is living, or has been part of a living organism, we are studying biology.

Biology is a vast subject, ranging from the study of molecules to the study of populations composed of millions of individuals. Biology has grown from simple descriptions of living things to the present-day emphasis on biochemistry, cell biology, genetics, growth and development, evolution, and population studies. This does not mean that other aspects of biology are unimportant. They are, and they will continue to be; but a concentration on specific aspects of biology reflects today's interests in biology.

for thought and discussion

1   Develop a problem that you think you are capable of solving; for example, your car does not start in the morning. What clues do you make use of? What hypotheses do you develop? How do you test these hypotheses? Extending this method to scientific problems in general, how do you recognize the existence of such a problem and how do you begin to get an intellectual grasp of a problem?

2   Think of an organism with which you are familiar. In scientific terms, how many ways can you think of to describe this organism? Into what subsciences of biolgy would these descriptions fit? What is the purpose of systems of classification?

3   Assume that a scientist has an idea or has made a discovery that can drastically alter our way of life if the knowledge is made public. When does consideration of this problem cease to be scientific and become ''nonscientific''? What is the obligation of the scientist in this matter? Should all knowledge be made public?

4   Why did the Copernican view of the universe alter our view of man?

5   Make a list of words common to your vocabulary which were not in use when your parents were as old as you are now. From what science did these words originate? Can you trace their origins and suggest why they were coined?

6   Consider an activity in which you are engaged, for example, riding a bicycle. Can you, in your present state of knowledge, describe the physical and biological structure and processes that make bicycle riding possible? When do you think you could say that you understand fully what bicycle riding involves? Are there different levels of understanding?

7   We have stated that any scientific discipline goes through several steps as it matures. Is this true for artistic or humanistic disciplines as well? If so, are the steps the same?

8   Each of us is unique in physical appearance. Why do you suppose there is uniqueness in our ability to perceive and discriminate?

9   Each of us is distinguished from all other human beings by our fingerprints, and many of us by our Social Security numbers. Are these comparable in scientific accuracy? Can you think of other systems of classification that would be equally discriminating and equally effective? What are the good and bad features of such classification systems?

selected readings

BRONOWSKI, J. *The Common Sense of Science.* New York: Random House, Inc. (Vintage Books), 1964. A description of science as a reasonable human endeavor to understand the world in which we live.

BRONOWSKI, J. *Science and Human Values.* New York: Harper & Row, Publishers (Torchbooks), 1965. A readable account of the relation of science to creativity, truth, and human dignity.

BUTTERFIELD, H. *The Origins of Modern Science* (rev. ed.). New York: Macmillan Publishing Co., Inc. (Collier Books), 1957. One of the best accounts of the history of science from 1300 to 1800 A.D.

CONANT, J. B. *Modern Science and Modern Man.* Garden City, N.Y.: Doubleday & Company, Inc. (Anchor Press), 1952. A popular account of the role of science in today's world.

DUBOS, R. *So Human an Animal.* New York: Charles Scribner's Sons, 1968. An excellent account of man as an animal and as a human being.

OPPENHEIMER, J. R. *Science and the Common Understanding.* New York: Simon and Schuster, 1953. A readable account by an eminent physicist of what science, particularly nuclear science, is all about.

YOUNG, J. Z. *Doubt and Certainty in Science.* New York: Oxford University Press, 1950. Deals with the manner in which man uses his brain to apprehend the world about him.

ZIMAN, J. *Public Knowledge: The Social Dimensions of Science.* Cambridge, Mass.: Cambridge University Press, 1968. One of the better expositions of how the scientist functions within the rules of the game.

the cell
chapter 2

The world around us consists of matter and energy in many different forms. To deal with them in an orderly fashion, we need to find those things that are common to all matter, and common to all energy. We need, in other words, to get down to *basic units,* and to reach some general agreement about what these units are.

When we deal with matter we say that the basic unit is the **atom,** and that atoms group together into **molecules.** Molecules, in turn, group together and form the objects and living things around us. Energy, too, has its own basic units. As our knowledge of such basic units increases, we can describe them more precisely. It is in this way that scientific language, common to all scientists, develops. Science makes use of two kinds of units. Those used to define time, mass, and distance are arbitrarily defined. Kilometers may be used instead of miles to measure distance, for example, or centimeters instead of inches. The United States has a variable system of use, employing a metric system for monetary values and nonmetric systems for measures of weight and distance. Other units, however, are worldwide in use. All scientists agree that atoms are the units of matter and of molecular construction even though no one has ever seen an atom. Mathematicians do not dispute the value of *pi* (3.1416). These units, like the others, have a physical reality, but they are not arbitrarily

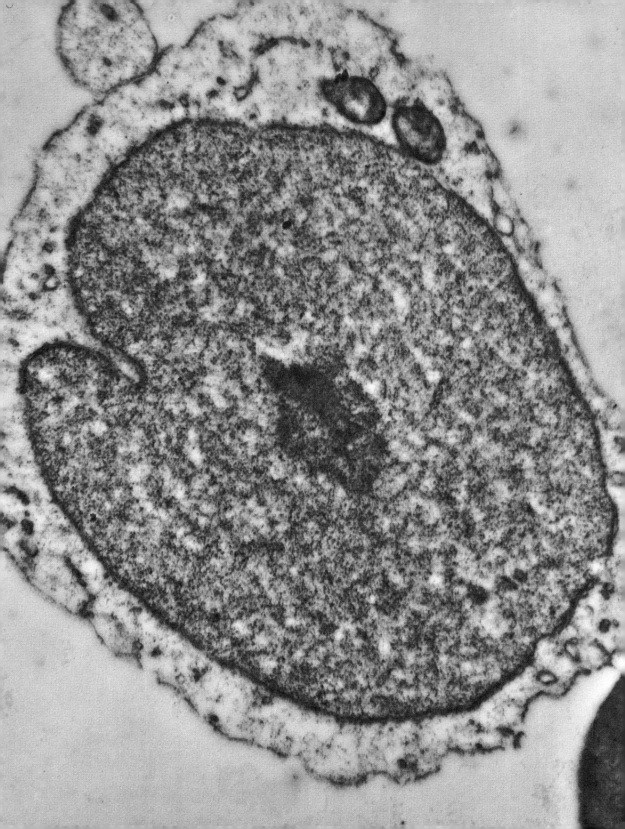

defined. Anyone having the proper instruments and knowledge can verify their existence, structure, and behavior.

The **cell** is such a unit (Fig. 2.1). It is the basic unit of all living things, except viruses, which may or may not be "living." We can break cells apart and extract fragments of them much as a physicist might break up an atom in a cyclotron. We find that these fragments, when separated from the cell, can carry on many of their activities for a time. They may consume oxygen, ferment sugars, and form new molecules. But these activities individually do not constitute "life" any more than a lone elec-tron or a proton typifies the behavior of a whole atom. A dis-rupted cell can no longer continue life indefinitely; so we con-clude that the cell is the simplest unit that can carry on life.

Compared to an atom or a molecule, a cell is very large and complex. It has a definite boundary within which chemical activity is going on constantly. In the area of **cytology** (the science of cells) one learns to recognize the many kinds of cells, and to understand their organization and structure in terms of their function. One also learns to visualize the cell not only as an individual unit, but also as a working member of a more elabo-rate organ or system of an entire plant or animal (Fig. 2.2).

HOW The now familiar idea that the
THE CELL CONCEPT cell is the basic unit of life is
WAS FORMED known as the **cell theory.** The the-
ory was formulated in 1839 by two German scientists, the botanist M. J. Schleiden and the zoologist Theodor Schwann. Today we regard the statement as one of the basic ideas of modern biology. We understand life in its more intimate physical and chemical details only to the extent that we understand the structure and function of cells. We also recognize that, when groups of cells form organs, organ systems, and organisms, other qualities of life make their appearance.

The general acceptance of a great truth such as the cell concept is a slow process. Every great thought has its origin in the mind of one man, but even when stated clearly, it is rarely accepted by everyone. The date 1839 and the names Schleiden and Schwann are not significant because these men discovered cells. They did not. Cells had been known since 1665, when the English scientist, Robert Hooke, first saw them in a piece of dead cork under his microscope (Fig. 2.3). It was Hooke who coined the word "cell" to describe the tiny structures he had observed.

Figure 2.1 The cell is the basic unit of all living things. This micrograph shows a human leucocyte, white blood cell. The large nucleus, with its smaller, dark nucleolus, occupies the middle of the cell, with the less structured cytoplasm surrounding it.

Courtesy of Imperial Cancer Research Foundation

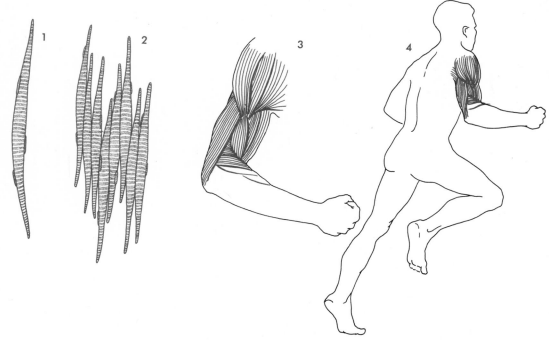

Figure 2.2 A single muscle cell (1) becomes grouped with other similar cells into a cellular mass (2), which is part of the muscle tissue of the arm (3), which in turn is but one of many parts of the entire organism (4).

He thought they looked like the cells of a monastery in which the monks lived. What he actually saw were really not cells at all, only the nonliving walls of cells that had died.

Also, Schleiden and Schwann were not the first biologists to believe in the idea that plants and animals were composed of cells and cell products. Many others had stated this clearly some years earlier. What they did was to associate a number of ideas and observations, and make a whole structure out of them. Theirs was an act of synthesis rather than of original discovery. By viewing the cell as both the structural and functional unit of organization, Schleiden and Schwann defined the basic unit of life. Because they were both prominent scientists of their time, they were more readily believed than were their predecessors.

Some 20 years later (1859), Rudolf Virchow, the great German physician, made another important statement: *cells come only from pre-existing cells.* When biologists further recognized that sperm and eggs are also cells which unite with each other in the act of fertilization, it became clear that life from its earliest beginnings was a continuous and uninterrupted succession of cells. From the time, some 3 billion years ago, when life first developed on Earth, there has been an unbroken chain of cells leading up to the organisms alive today.

Figure 2.3 With this primitive microscope and illuminating system, Robert Hooke observed cells in a plant tissue. Above the illuminating system is a portion of a page from his book *Micrographia*, in which he described his observations.

The Bettmann Archive, Inc.

Obferv. XVIII. *Of the* Schematifme *or* Texture *of* Cork, *and of the Cells and Pores of fome other fuch frothy Bodies.*

I Took a good clear piece of Cork, and with a Pen-knife fharpen'd as keen as a Razor, I cut a piece of it off, and thereby left the furface of it exceeding fmooth, then examining it very diligently with a *Micro-fcope,* me thought I could perceive it to appear a little porous; but I could not fo plainly diftinguifh them, as to be fure that they were pores, much lefs what Figure they were of: But judging from the lightnefs and yielding quality of the Cork, that certainly the texture could not be fo curious,

Courtesy of Burndy Library

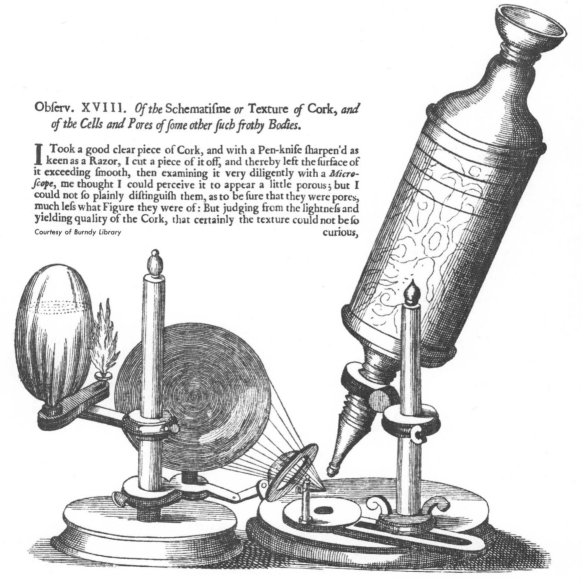

33

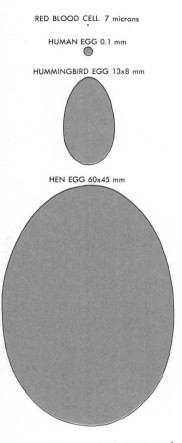

RED BLOOD CELL 7 microns

HUMAN EGG 0.1 mm

HUMMINGBIRD EGG 13x8 mm

HEN EGG 60x45 mm

Figure 2.4 Relative sizes of four cells: The eggs of the hummingbird and hen are about natural size; the human egg and red blood cell are enlarged.

The world of cells, except in a gross sense, lies beyond the limits of ordinary vision. Most cells are far too small to be seen with the naked eye. For example, the period at the end of this sentence is about 0.5 millimeter (mm) in diameter. It would cover about 25 to 50 cells of average diameter. Of those we can see, none of their details can be made out. We must, therefore, use instruments to extend our vision into the minute world of cells.

The 200-inch Hale telescope on Mount Palomar, California, reaches across billions upon billions of miles of space, bringing distant galaxies of the universe into view. The cell biologist, however, needs to overcome the problem of small size rather than that of great distance. Light microscopes and electron microscopes do this for him, thereby revealing otherwise invisible worlds. Magnification, therefore, is as much of a problem for the biologist as it is for the astronomer.

The problem of magnification is a problem of **resolving power.** An optical system has good resolving power if it enables us to distinguish objects—stars or parts of a cell—lying very close together. In reading these pages, for example, some people see each letter and word distinctly. Proper image formation is part of the problem, but others, with poorer resolving power, will see only a blur, and their resolving power and image formation have to be improved through the use of eyeglasses. Some people can see that the middle star in the handle of the Big Dipper is really a double star; others can see the two stars only with binoculars or a telescope. In a microscope, the resolving power of the **magnifying lens** is the critical factor. As Fig. 2.5 shows, the lens nearest the specimen being examined—the **objective**—is the key element of the compound microscope. The uppermost lens—the **ocular**—enlarges only what the objective lens has resolved.

The human eye is, of course, a crucial factor in making use of instruments. The unaided human eye with 20/20 vision has a resolving power of about 0.1 mm (Fig. 2.6); a human egg would be at the borderline of visibility. Lines closer than this appear as a single line. Objects that have a diameter smaller than about 0.1 mm are invisible or appear only as blurred images. Unlike the human eye, microscopes both resolve *and magnify,* but they cannot magnify that which they cannot resolve. We have the same problem in photography. A 35-mm negative can be enlarged into a huge picture, but if the negative is a blur because the lens did not resolve the object in question, the positive will also be a blur, no matter how we try to correct it.

The resolving power of a microscope is limited by the kind of illumination used. Objects that are less than one half the wavelength of light apart cannot be distinguished in a light microscope. This is a fundamental law of optical physics. Since light has an average wavelength of 5,500 **angstroms** (Å), even

34

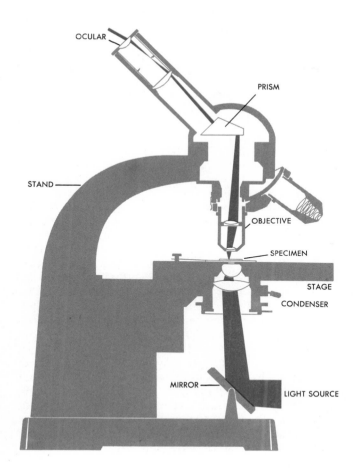

OCULAR

PRISM

STAND

OBJECTIVE

SPECIMEN

STAGE

CONDENSER

MIRROR

LIGHT SOURCE

Figure 2.5 The diagram at the left shows the optical system of a light microscope and the passage of light through the condenser, objective, prism, and ocular.

with the most perfectly ground lenses, the objective cannot resolve objects with a diameter less than 2,750 Å (which is 0.275 **micrometer** (μm), or 275 **nanometers** (nm)). Since many parts of biological systems are much smaller than 2,750 Å, we could not detect them unless a means of greater resolution were available.

The **electron microscope** (Fig. 2.7) provides increased resolving power by making use of "illumination" of a different sort. Highspeed electrons, which are parts of atoms, are used instead of light. As the electrons pass through a specimen being viewed, the more dense parts of the specimen deflect or absorb more electrons than less dense parts. This contrast forms an image of the specimen on an electron-sensitive photographic plate or fluorescent screen. The human eye, of course, is not stimulated by electrons, hence the need for plates or screens. The "optical" system of an electron microscope is similar to that in the light microscope, except that the "illumination" is focused by electromagnetic lenses instead of conventional glass lenses.

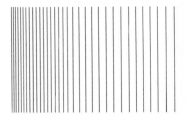

Figure 2.6 Anyone with good vision can resolve each of these into individual lines. If you have poor resolution, the lines at the left begin to merge and become indistinct.

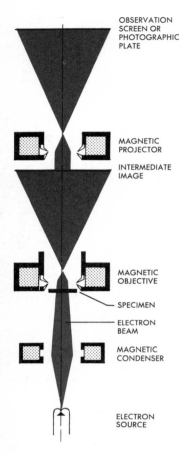

OBSERVATION
SCREEN OR
PHOTOGRAPHIC
PLATE

MAGNETIC
PROJECTOR

INTERMEDIATE
IMAGE

MAGNETIC
OBJECTIVE

SPECIMEN

ELECTRON
BEAM

MAGNETIC
CONDENSER

ELECTRON
SOURCE

Figure 2.7 In the optical system of the electron microscope, electrons instead of visible light provide a means of resolution and magnification.

When electrons are propelled through the microscope by a charge of 50,000 volts, they have a wavelength of about 0.05 Å. This is 100,000 times shorter than the average wavelength of light. An electron microscope can thus theoretically resolve objects with a diameter of one half of 0.05 Å, or 0.025 Å. This dimension is far less than the diameter of an atom (the smallest atom—hydrogen—has a diameter of 1.06 Å). However, due to difficulties in lens construction, the actual resolving power of a good modern instrument is about 10 Å. In approximate figures, then, the human eye can resolve down to 100 μm, the light microscopes to 0.2 μm, and the electron microscope to 0.001 μm. Or, to put it another way, if the human eye has a resolving power of 1, that of the light microscope is 500, and that of the electron microscope 100,000. The electron microscope has obviously opened up a whole new domain to the biologist.

Sometimes biologists need more than just clear resolution and high magnification. They also must be able to distinguish clearly between one small part of a cell and its immediate surroundings. The electron microscope can do this much better than the light microscope. To overcome this problem with the light microscope, we use killing agents **(fixatives)** and stains to bring out the parts we want to examine. Literally hundreds of fixing and staining procedures are known (Fig. 2.8).

Another way biologists study a living substance is to grind it up and examine it. This is done with a special mortar and pestle so as to burst the cells and release their contents in solution. This solution is then centrifuged at carefully regulated speeds. The heavier material settles out at lower speeds, the lighter material at higher speeds. Most of the individual parts of cells can be collected in this way. Once separated, each portion can be analyzed for chemical content or tested for activity, since in the test tube some parts continue to function for a while as they did in the intact cell.

Biologists, then, have several ways of making a cell give up its secrets. Any one tool or technique, however, is not enough by itself. Usually, several methods must be used before an answer to a particular problem can be found.

THE STRUCTURE OF CELLS UNDER THE LIGHT MICROSCOPE Let us now take a look at a cell under the light microscope so that we can become familiar with its parts (Fig. 2.8). Later we shall deal with the finer structure of the cell as revealed by the electron microscope (compare Fig. 2.1 with Fig. 2.9 for differences in detail). The cell we will describe now does not really exist, for there is no *typical cell*. Our cell is idealized for purposes of discussion (Fig. 2.9). We should remember that any cell in a living organism has a particular structure necessary for its activities, and it is, therefore, unique, not typical.

As the diagram shows, our idealized cell has two major components: a rounded central body, the **nucleus;** and a surrounding mass, the **cytoplasm.** Each plays an important role in the life of the cell.

The nucleus can best be seen in cells that have been stained. It is bounded on the outside by a double **nuclear membrane.** The membrane is invisible in the light microscope, but the electron microscope shows it. Within the nucleus is the **chromatin,** which appears as a network of fine threads, and the **nucleolus,** a more solid-appearing body. The nuclear membrane does two things: (1) it separates the nucleus from the cytoplasm so that each is partially independent of the other, and (2) it permits the passage of some materials in and out of the nucleus while keeping other substances out. Such a membrane is said to be **selectively permeable,** a term that can be applied to any living membrane wherever it is found. The nuclear membrane, unlike the **plasma membrane,** has pores, but there is some doubt that the pores are free passageways for materials in solution.

The chromatin of the nucleus is a very special substance made up of **proteins** and **nucleic acids.** These are very large molecules, which we shall come to know intimately, that are responsible for determining the unique structure and function of each cell. We shall return to the nucleus time and again in this book. At the moment we need remember only that the nucleus controls the cell, and that the chromatin and the nucleolus are its essential materials.

If we punctured a cell with a microneedle, we would find that the cytoplasm would leak out. This suggests two things: (1) cytoplasm is a watery substance; and (2) it is bounded on the outside by a membrane—this is the **cell membrane** (also called the **plasma membrane**), which separates the cell contents from the outside environment. Like the nuclear membrane, the cell membrane is selectively permeable.

The fluid nature of the cytoplasm is somewhat misleading. Although it contains water, it contains other materials as well. When properly stained, or when viewed in the living condition in a phase-contrast microscope, it presents a varied appearance. We can see many granules in the form of rods, long filaments, or spheres. These are the **mitochondria** (singular, **mitochondrion**). Bacteria, blue-green algae, and a few other primitive organisms are the only cellular organisms that lack them. No human cell, with the possible exception of the red circulating blood cell, is without them. A liver cell may have 1,000 or more mitochondria

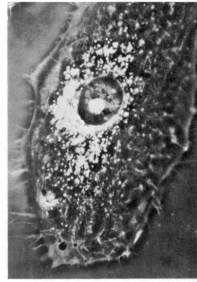

Courtesy Dr. George O. Gey

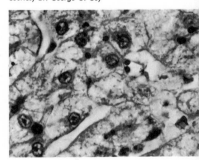

Courtesy E. Leitz

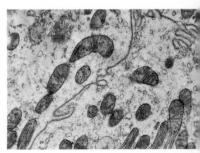

Figure 2.8 Top, a human cancer cell growing in tissue culture and photographed in a living state by means of a phase-contrast microscope, X100. Center, cells from the liver of a dog. The tissues were fixed, sectioned, and stained before being photographed—at about X1,000. Bottom, portions of three cells from the kidneys of a frog, revealing a richness of detail when photographed through an electron microscope, X38,000.

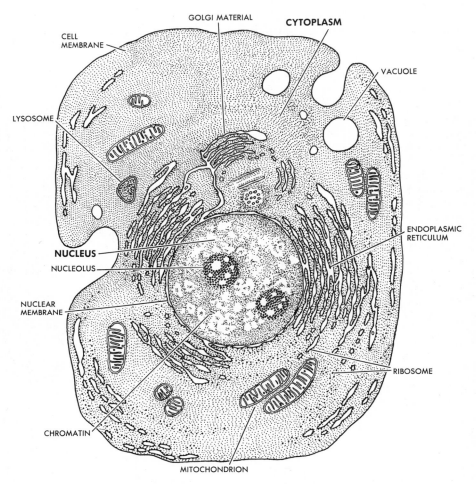

CELL MEMBRANE

GOLGI MATERIAL

CYTOPLASM

VACUOLE

LYSOSOME

ENDOPLASMIC RETICULUM

NUCLEUS

NUCLEOLUS

NUCLEAR MEMBRANE

RIBOSOME

CHROMATIN

MITOCHONDRION

Figure 2.9 This diagrammatic representation of an idealized cell shows all of the details that can be observed with the great resolving power of the electron microscope.

in its cytoplasm. The more active a cell is, the more mitochondria it is likely to have. This suggests that mitochondria play an important role in the life of the cell. This role, which we shall discuss in detail later, is concerned with the energy requirements of the cell.

If you wish to stain a cell to show the mitochondria particularly well, use **haematoxylin** on cells that have been previously fixed (killed in a fixative), or **Janus Green B** on living cells. Even with good staining techniques, however, the small size of the mitochondria does not permit us to see any intimate details of structure in the light microscope.

If a cell is stained with a silver-containing dye, a network of black substance becomes visible. This is the **Golgi material** (Fig. 2.10). Like mitochondria, Golgi materials (named after their discoverer) are found in most cells; it appears that they play a role in the formation, aggregation and secretion of cell products, often in membrane-bound vesicles.

Cells, and particularly plant cells, occasionally show large clear areas when they are stained. These are the **vacuoles.** As Fig. 3.4 shows, they force the cytoplasm into a thin layer against the cell membrane and keep the cell from collapsing. It appears that certain waste products are dumped into the vacuoles for indefinite storage. More recently they have been found to contain enzymes. A membrane enclosing the vacuolar material keeps the waste products away from the remainder of the cell.

Plant cells have two features that are generally lacking in animal cells. One is the rather thick **cell wall** that lies outside of the plasma membrane and serves as a supporting skeleton for the plant cell. The cell wall is made up largely of a substance called **cellulose,** which is of tremendous economic importance to man. It determines the strength and character of wood and paper. In the cotton plant, it provides us with the thread for our cotton fabrics, and it is the major ingredient in cellophane.

In the cytoplasm of green plants, and particularly in the cells of leaves, we find another structure—the **plastids.** They come in many shapes and sizes, each characteristic of a particular kind of plant. The number of plastids per cell can vary from one to a great many. Figure 2.11 (left) shows one kind of **chloroplast,** those plastids that contain the green pigment **chlorophyll.** Also, there are several kinds of storage plastids that may contain starch,

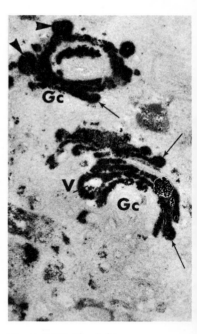

Figure 2.10 The Golgi material of an animal shows up clearly when stained with a silver-containing dye and photographed through a light microscope. (See also Fig. 2.9.)

Figure 2.11 At left is an electron micrograph of a chloroplast from a tobacco leaf cell that shows the stacked lamellae (layers) of the grana, and the more open spaces of the stroma. At right, the layered grana are shown at a higher magnification. (g = grana; ig = intergrana space; s = stroma; t = membrane surrounding a vacuole; cw = cell wall.)

Courtesy G. Schidofsky

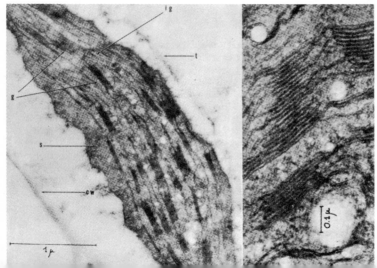

fat, or protein, and chromoplasts that contain pigments other than chlorophyll. The chloroplast is of particular importance as it is this structure that absorbs light energy from the Sun and converts it into chemical energy, which the cell can then use.

The rest of the cytoplasm seems to be without visible structure, but this is merely because the light microscope cannot resolve the fine structure of membranes and particles that are present. When a growing cell is stained with **basic dyes,** the cytoplasm often shows a deep, rich color, indicating the presence of definite substances. This organization remained unknown until the greater resolving power of the electron microscope revealed a wealth of details. We now know that the cytoplasm is the main synthesizing part, or manufacturing plant, of the cell. Its fine structure will be discussed in later chapters when we can relate the cellular structure to particular chemical activity.

POSSIBLE EXCEPTIONS TO THE CELL THEORY

For nearly every rule there is an exception. The cell theory, which is a rule of a sort, also has its exceptions. The fact is that the viruses do not fit into the cell theory scheme.

More than 300 different kinds of viruses are known. Many of them are the infective agents in such diseases as yellow fever, rabies, poliomyelitis, small pox, mumps, and measles in humans; and peach yellows and tobacco mosaic disease in plants. Some viruses appear to be the cause of some kinds of cancer. Those that infect plants tend to be elongated structures, or aggregated into crystal-like bodies. The viruses that infect animals tend to be spherical (Fig. 2.12). If we now apply our usual definition of a cell, viruses do not qualify as living organisms. They lack the internal organization normally considered indispensable to a functioning cell. A distinction cannot be made between nucleus and cytoplasm. Although they contain the nucleic acids characteristic of nuclei, they seem to be totally lacking in cytoplasm.

When they exist outside a living cell, viruses behave as in-

Figure 2.12 Two different kinds of viruses are shown here: at right, the sphere-shaped influenza virus; at left, the rod-shaped tobacco mosaic virus.

Courtesy Dr. Robley Williams Virus Lab., Univ. of Calif.

Courtesy Dr. Robley Williams Virus Lab., Univ. of Calif.

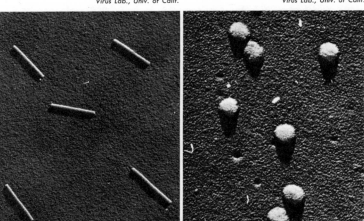

active molecules, although very elaborate and complex ones. Inside a cell, however, they act as parasites and display the usual characteristics of life: they multiply and produce exact replicas of themselves. They also have a type of inheritance not too different from our own. Viruses also contain the key molecules of protein and nucleic acid invariably found in every living organism.

Their doubtful nature has led biologists to describe them in various ways: as living chemicals, as cellular forms that have degenerated by adopting a life of parasitism, or as primitive organisms that have not reached a cellular state. Fortunately, we are not forced to decide whether a virus is or is not a cell, or even whether it is living or nonliving. We generally treat them as if they were individual cells. Their extreme simplicity of structure, when compared with a normal cell, makes them ideal objects for certain types of biological research.

Certain of the less complex forms of life, such as the protozoa, algae, and fungi are also difficult to fit into the plan that the cell is the basic unit of life. The protozoan *Paramecium* is seemingly a single cell, but it has a mouth, or gullet, vacuoles for the elimination of water and waste, other vacuoles for digestion, and many **cilia** (fine surface hairs) for moving about. Is *Paramecium* a true cell, or not? It is a difficult question to answer.

The same thing is true for certain algae, such as *Valonia,* or for fungi such as the black bread mold (Fig. 2.13). They are simply a mass of cytoplasm containing many nuclei, and are bound by a continuous outer retaining wall. It would be difficult to define the basic unit of such living bodies. Are they single-celled, or multicelled? These organisms, however, are related to more conventional cellular forms so we can speculate that they have simply lost the usual type of cellular organization and have acquired one that is mechanically and physiologically better suited to their mode of existence.

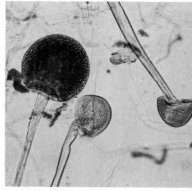

The Bergman Associates

Figure 2.13 There are no typical cells making up the body of this fungus, the black bread mold, *Aspergillus niger.*

Figure 2.14 Electron micrograph of a plasma cell which circulates in the blood stream and produces antibodies. The details revealed are much greater than can be resolved with the light microscope. The inset is a portion of the cell at a greater level of magnification.

THE STRUCTURE OF CELLS UNDER THE ELECTRON MICROSCOPE Let us now examine an animal cell with an electron microscope, remembering that with this instrument we can see details well below the limits of resolution of the light microscope (Fig. 2.9). The photograph of the animal cell (Fig. 2.14) shows such a cell greatly magnified. Our first impression is that of a great richness of structures. We shall want to examine each one, but before we do we need to become acquainted with the atoms and molecules of which these structures are composed.

The atoms are mainly carbon, oxygen, hydrogen, and nitrogen, along with sulfur and phosphorus. These are the principal atoms commonly found in living structures. The molecules formed from these atoms are of four major classes, and all are distinguished by the fact that they are **macromolecules.** That is,

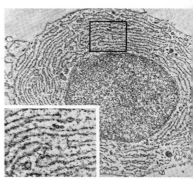

compared with ordinary small molecules such as water (H_2O), carbon dioxide (CO_2), and common table salt (NaCl), they are enormous structures. They are usually elongated molecules, although some are globular, and are composed of thousands of atoms hooked together, or **bonded,** in a particular way. The four molecules are as follows:

1. **Proteins,** each made up of amino acids strung together like beads on a chain.
2. **Polysaccharides,** constructed similarly, but of smaller sugar molecules.
3. **Lipids,** made up in large part of fatty acids.
4. **Nucleic acids,** constructed out of several different kinds of smaller molecules.

We need not be concerned now with their chemical structures, but we need to be familiar with their names and where they are found.

As we make our way from the outside of the cell toward its middle, the first structure we encounter is the cell membrane. This was invisible in the light microscope, but the electron microscope reveals its structure clearly. When we view it under high magnification, we can see three distinct layers. The outer and inner dark layers are separated by a light region in a sandwich-like arrangement. From a variety of experiments, we now believe that the dark layers are protein, while the central, light area consists of two layers of lipid molecules. Figure 2.15 shows how

Figure 2.15 At right is an electron micrograph of cell membranes at the juncture of three cells. At left is one interpretation of the cell membrane in molecular terms. The paired dark layers in the electron micrograph consist of proteins; the light layer sandwiched in between is made up of lipids (fats). The three layers together constitute a unit membrane. Neither layer, however, needs to be continuous.

Courtesy Dr. J. D. Robertson

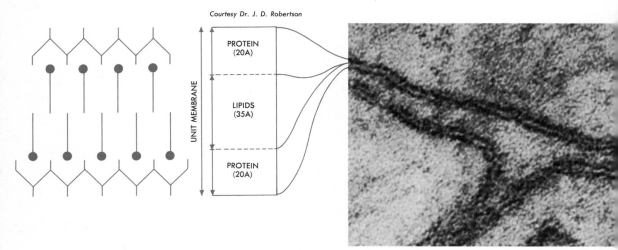

PROTEIN
(20A)

LIPIDS
(35A)

PROTEIN
(20A)

UNIT MEMBRANE

the layers are believed to be arranged. We need to remember, however, that the membrane is a living part of the cell, and that from one region to another the kinds of proteins and lipids can vary. The membrane, therefore, is a kind of dynamic mosaic of macromolecules, variously arranged.

Why do biologists believe that this is the structure of a cell membrane? If we suspend red blood cells in water, they swell and soon release their contents (hemoglobin) into solution. We can in this way obtain a mass of pure cell membranes and analyze them. They are called "ghosts." We can then separate the proteins and lipids. When digested, or broken down, the proteins yield amino acids, the lipids, fatty acids. There are small amounts of other molecules but they are unimportant to our present consideration.

But how do we know that there are two layers of lipids between the outer and inner protein coats? Suppose that we have a cell whose surface area on the outside is 100 μm^2. We now extract the lipids from the cell membrane and allow them to spread out on the surface of water in a dish. Like a drop of oil, these molecules will form a thin film one molecule thick. If there is one layer of lipids, an area 100 μm^2 will be covered. If there are two layers, 200 μm^2 of water will be covered; three layers, 300 μm^2, and so on. It turns out that the lipids from one cell will cover twice the area of the cell. Two layers of lipids are, therefore, sandwiched between the inner and outer protein layers.

We have spent this much time on the cell membrane because all membranes of the cell, and there are many kinds, are of a similar structural nature. Furthermore, knowledge of the cell membrane permits us to understand the passage of substances in and out of the cell, a topic of much importance to be discussed later.

As we continue our exploration of the cell, we come to the cytoplasm and find that it is not just a watery fluid. It is filled with membranes and particles. The long membranes running through the cytoplasm are conspicuous. In some cells they are quite numerous, filling the cell with many channels. In others, membranes are sparse, or even absent. These are the membranes of the **endoplasmic reticulum** (inside network). In some cells, these membranes are studded with small bodies called **ribosomes** (Fig. 2.18); in other cells the membranes are smooth.

The endoplasmic reticulum is believed to be the manufacturing, packaging, and transporting portion of the cytoplasm; it is the assembly line of the cell. The nature of the endoplasmic reticulum determines in part what the cell does. When ribosomes are present, as they are in virtually all cells, they are there for the purpose of making proteins. When the endoplasmic reticulum is free of ribosomes, some other substance is being made by the cell.

Another series of membranes, similar to the endoplasmic reticulum, but recognizably different, is the **Golgi material.** Com-

Figure 2.16 An electron micrograph of plant cells shows the heavy cell wall, the long membranes of the endoplasmic reticulum, the stacked membranes of the Golgi materials, and several mitochondria.

Courtesy Dr. G. Whaley

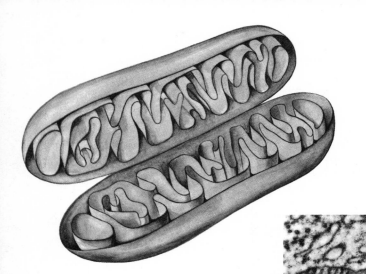

Courtesy Dr. B. L. Munger

Figure 2.17 In this electron micrograph of a mitochondrion the inner membranes, or cristae, are visible (pieces of the endoplasmic reticulum are also evident). At left, the drawing of a mitochondrion opened up reveals the arrangement as extensions of the inner membrane.

Figure 2.18 Many ribosomes, free in the cytoplasm, are revealed in this electron micrograph of part of a human white blood cell. Several rounded mitochondria are also visible.

Courtesy Imperial Cancer Research Fund

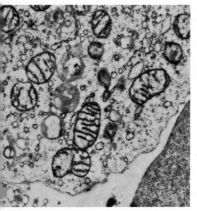

pare the photograph of Golgi material on page 39, taken with the light microscope, with Fig. 2.16, taken with the electron microscope. The difference in detail is striking. Actually, the Golgi material is like a group of balloons flattened together. They do not have ribosomes attached to them, and consequently are not concerned with protein formation. They seem, rather, to accumulate certain materials such as fats or mucus and then move them out of the cell. One might think of Golgi material as a place in the cell for packaging molecules for export.

The ribosomes, whether numerous or few in number, are about 250 Å in diameter. A bacterial cell of 1 μm diameter may have as many as 30,000 ribosomes. Often they are grouped in clusters of five to eight. In this state they are known as **polyribosomes,** or simply **polysomes.** Again, they are concerned with the synthesis of proteins.

The **mitrochondria** appear complex when viewed in the electron microscope. They have both an outer and inner membrane. The inner membrane is thrown into folds called **cristae,** which give a greatly increased surface area (Fig. 2.17). A great amount of research has been done on these structures, for the mitochondrion is of vital importance to the cell. It is the chief source of usable energy. Whenever the cell does work of any kind, the mitochondrion (and, in plant cells, the chloroplast) supplies it with the necessary energy.

A structure similar to the mitochondrion is the **lysosome.** It is different in appearance in that it does not have cristae. It

44

is also different in function. Its main task in the economy of the cell is to break, by means of a number of enzymes contained within the lysosome interior, large molecules of food materials into smaller ones, which are then passed on to the mitochondria to serve as energy sources. However, if the lysosomes are injured, their destructive action can cause digestion and death of the cell.

The nucleus is the last of the structures we shall view in our cell. Surrounding it is a double membrane, its two parts often separated from each other by a **perinuclear** space. The outer membrane often has pores and is connected with the endoplasmic reticulum (Fig. 2.20). Just as the cell membrane is the barrier separating the cell from the outside environment, so the nuclear membrane separates the nuclear contents from the cytoplasm. It differs from the plasma membrane, however, particularly in behavior, for it can disappear and reform during cell division. All substances passing in and out of the nucleus must pass through the nuclear membrane. Except for the nucleolus, the nuclear contents show very little defined structure. This is because the chromatin, the major nuclear component, is widely dispersed.

A plant cell differs from an animal cell in three major ways. First, there is usually a heavy wall just outside the cell membrane. This is a major supporting structure, and it is part of a cell that gives "character" to the different kinds of wood. Such heavy walls are lacking in animal cells. Second, the center of a plant cell is generally occupied by a water-filled sac, or **vacuole,** which is covered by a membrane called the **tonoplast.** The vacuole pushes the cytoplasm toward the outer edges of the cell where a ready exchange of gases can take place. The vacuole may be filled with colored pigments—for example, the red pigment, anthocyanin, of the beet cell. Third, the plant cell may contain structures known as **plastids.** These, as indicated, are of great importance, for they capture the energy of sunlight for use in the manufacture of sugars. The structure of such a plastid is very complex (Fig. 2.11). The many layers of its membranes are often stacked into **grana;** and the green pigment, chlorophyll, which traps the energy of sunlight, is layered on these membranes. Each plastid, therefore, has a tremendous amount of surface area, and is a most efficient energy trap. Other kinds of plastids are used for storage. Some store starch, others protein, and others fats. Each differs slightly from the other in appearance.

BACTERIAL CELLS   Bacteria of many kinds abound in the world. Some are disease-producing organisms that give rise to a variety of infections, such as tuberculosis. Others are useful and, indeed, necessary, such as those that break down waste organic materials or those that make vinegar. All bacteria are cellular organisms, but their structure is different from that of ordinary plant or animal cells. As Fig.

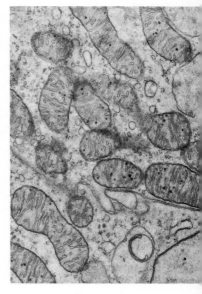

Figure 2.19 This electron micrograph of part of a kidney cell of a frog shows the elongated mitochondria characteristic of this kind of tissue.

Figure 2.20 The double membrane (with its pores) of the nucleus, the endoplasmic reticulum connected (connected to the membrane as well as free in the cytoplasm), the stacked membranes of the Golgi materials, and several round mitochondria are visible in this electron micrograph showing part of a plant cell.

Courtesy Dr. G. Whaley

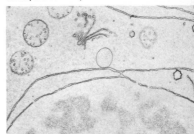

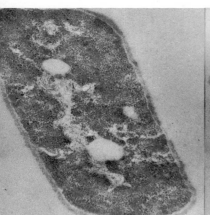
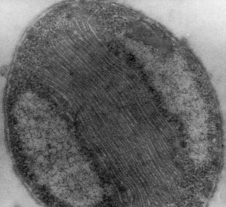
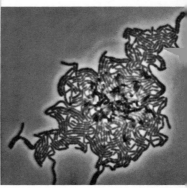

Courtesy Dr. C. Robinow   Courtesy Dr. S. Watson   E. Leitz,

Figure 2.21 Right: a low-power photomicrograph of a cluster of bacterial cells. Middle: electron micrograph showing the elaborate membrane system in *Nitrosocystis oceanus,* a marine bacterium. Left: a cell of *Bacillus subtilis,* showing its dense cytoplasm containing the bacterial ribosomes and the less dense nuclear area.

2.21 shows, a bacterial cell has a cell membrane, and a cell wall may lie outside the membrane. Although ribosomes are present, there are no plastids, no mitochondria, and no nuclear membrane. Some bacteria, however, have elaborate membrane systems. The functions carried on by these structures in more complex cells are not missing in the bacterial cell. They are performed in the cytoplasm, or at the surface of the plasma membrane, with less highly organized structures.

summary

Except for viruses, all organisms are composed of **cells,** or cell products. Cells arise only by the division of another cell. All organisms are, therefore, related to each other through **cellular descent.**

Cells are so small that we can see their details only with the aid of a microscope. With a light microscope, we can see that cells consist of three main parts: (1) a **nucleus;** (2) **cytoplasm;** and (3) a **cell membrane.**

The nucleus contains **chromatin** and a **nucleolus,** which are enclosed within a nuclear **membrane.** The cytoplasm contains **mitochondria, Golgi material,** and **vacuoles.** These individual parts may be seen well only when they are fixed and stained before being viewed through a microscope. An electron microscope reveals many more details than a light microscope does, particularly in the cytoplasm. Ribosomes, endoplasmic reticulum, and lysosomes show up well. Also, the detailed structure of membranes, the mitochondria, and Golgi material can be seen.

Plant cells differ from animal cells in that plant cells have a heavy cell wall and plastids in addition to the structures found in animal cells. Bacteria differ from both plant and animal cells by lacking a nuclear membrane, mitochondria, and plastids.

1   Imagine a spherical cell having a radius of 30 μm. What is the radius in millimeters? Nanometers? Angstroms? What is its surface area and volume in these units of measure?

2   Cells have often been compared to the bricks in a house. Do you think this is a reasonable comparison? What are the reasons for your conclusions?

3   A cell is a miniature factory: it either performs a task or makes a product. From what you know of a cell, and of a factory, draw up a list of comparisons for sources of energy, steps carried out by particular parts, and tasks performed or products made. Can you point out where you need more information to understand the functioning of a cell?

4   Animal cells, in contrast to plant cells, have no cell walls. How does an insect or crab or dog keep its shape and move in a purposeful manner? Why doesn't it collapse?

5   Two spherical cells, one having twice the diameter of the other, have the same metabolic rate (metabolism per unit mass). Let us assume that oxygen limits the rate of metabolism. If the concentration of oxygen outside of the smaller cell is 20 per cent, what must the concentration be outside the larger cell to maintain the same metabolic rate?

6   You want to examine two similar cells, one by light microscopy and the other by electron microscopy. Both are spherical and 30 μm in diameter, and both are to be sectioned for study, with a 5-μm thickness for light microscopy, and a 150-Å thickness for electron microscopy. How many sections of each do you need to include the entire cell?

DYSON, R. D. *Essentials of Cell Biology.* Rockleigh, N.J.: Allyn & Bacon, 1975. A recent and up-to-date text in cell biology, with an emphasis on physiology and biochemistry.

JENSEN, W. *The Plant Cell.* Belmont, Calif.: Wadsworth Publishing Company, Inc., 1964. A college-level treatment of the plant cell in terms of structure and function.

KENNEDY, D. (ed.). *The Living Cell.* San Francisco: W. H. Freeman and Company, 1965. A collection of *Scientific American* articles on cell structure and function, with the editor providing a preface to each article.

SWANSON, C. P. *The Cell* (3rd ed.). Englewood Cliffs, N.J.: Prentice-Hall, Inc., 1969. Chapters 1 through 6 deal with cell structure and function in a general way, and consider in some detail the problems of dimension in the light and electron microscopes.

WOLFE, S. L. *Biology of the Cell.* Belmont, California: Wadsworth, 1972. A comprehensive treatment of the cell with excellent photographs and extended discussions of structure and function.

general
cell
features
chapter 3

It must be apparent to you that all cells do not perform the same kinds of work. Muscle cells contract and relax, nerve cells send messages to various parts of the body, cells of the eye receive impressions of the outside world and transmit them to the brain, while cells of the stomach and intestine play a role in digestion. Cells become specialized and do different kinds of work, just as individual members of our society do particular jobs and so keep a community as a whole functioning in a coordinated way.

When an organism consists only of a single cell, that cell must carry on all the functions required by the organism to live, grow, and reproduce. We find such single-celled, or **unicellular,** organisms among the bacteria, protozoa, and algae. They differ from each other in appearance, fine structure, and behavior because each lives a somewhat different existence. Other organisms are multicellular, some consisting of relatively few cells. Others, such as man, have billions of cells. Each cell, or group of similar cells, has a special function and a special structure that enables it to perform its duties.

In this chapter we want to discuss general cellular features: size, shape, number, and the length of time each cell lives. We want also to determine, as well as we can, what it is that determines these features of the cell. As we shall see, this is not an easy task. Definite answers are not always readily available.

48

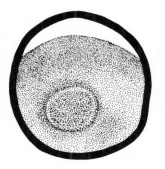
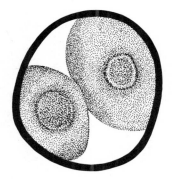
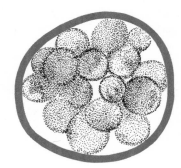

Figure 3.1 A single fertilized egg cell gives rise to countless billions of other cells, most of them specialized in the performance of certain functions. Shown here are five stages in the development of a human being: fertilized egg; two-cell stage; 16-cell stage; embryo with the three cell layers that give rise, respectively, to the brain, heart, and lungs. These three layers will also form all of the other tissues of the body.

CELL SIZE   By examining cells of plant and animal origin, we can see that their sizes vary widely. The smallest cells we know of belong to a group of **pathogens** that causes respiratory diseases. These belong to the genus *Mycoplasma*. Their diameter is about 0.1 μm, which means that they can be seen only with the aid of an electron microscope. Louis Pasteur, the great French bacteriologist, knew of them nearly a hundred years ago, but it is only recently that biologists have studied them intensively. Pasteur simply could not find them, even though he recognized their disease-causing properties. Their size places them among the viruses. But unlike the viruses, which can grow only within another cell, these cells grow readily in a test tube when given the proper nutrition. Their structure is simple, and not unlike that of the bacteria, to which they are probably closely related. Compare the photograph of the bacterial cell on page 46 with the drawing in Fig. 3.2. Within this tiny and simply constructed cell, all the functions of life are carried out.

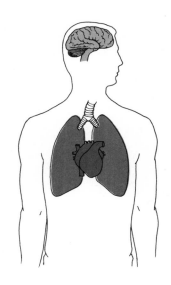

How small can a cell be? Clearly, it must have a diameter greater than 0.02 μm, or 200 Å, because the membrane surrounding the cell must have a width of at least 100 Å. But how much room is needed for the cytoplasm? (You will notice that *Mycoplasma*, like the bacteria, have no definitive nucleus.) We do not really know. About all we can say is that the smallest cell we know of is about 0.1 μm in diameter. Taking away 200 Å for the cell membrane, this leaves a sphere 800 Å in diameter for the remaining contents. Within the material filling this remarkably tiny space, life expresses itself. Here must be contained all the matter and forms of energy that we call "life." This minute

49

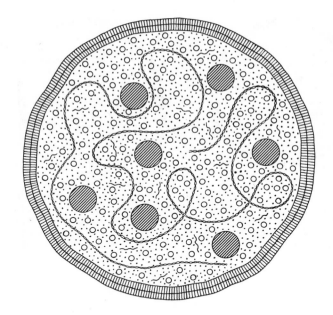

Figure 3.2 The smallest cell we know belongs to the genus *Mycoplasma*. A membrane surrounds the cell, which contains the hereditary material (long thread structure), ribosomes (large spheres), and molecules in solution (smaller structures in the cytoplasm).

amount of cellular material grows, reproduces more of its own kind, and undergoes variation or change.

The many forms of bacteria range in size from 0.2 to 5.0 μm in diameter. The smallest, therefore, are just at the limits of resolution of a good light microscope. The largest cell known is the ostrich egg, if we consider volume alone. It measures about 15 centimeters (cm) around the outside, and about 7.5 cm when the shell is removed. If we compare such a cell with the smallest bacterium, the ratio of linear dimensions would be about 75,000:1. The ratio of their volumes would be $75,000^3:1$. To put this in terms of easy reference, the order of difference is about the same as that between a sphere 1 inch in diameter and one that is more than 1 mile wide.

The range of cell size in the human body extends from the small leucocyte (white blood cell), which has a diameter of 5 to 6 μm to a nerve cell, which may be more than 1 meter in length. This is a difference of about 300,000:1. Such a comparison is misleading, however, for the *main body* of the nerve cell is not nearly so different in size (100 μm) from that of the small leucocyte. When computed on this basis, the ratio is only about 20:1. Even this comparison has little meaning without a consideration of other factors that govern size. A hen's egg, for instance, has outside dimensions of 60 × 45 mm. Like the ostrich egg, it is large because it contains food stored in the form of an enormous yolk, food that will be used for the growth of the developing embryo over a 3-week period. A human embryo, on the other hand, draws its nutrition from its mother during development, so the human egg (about 0.1 mm in diameter) need not be adapted for the storage of a large amount of food.

The different sizes of nerve, muscle, and blood cells reflect the particular task they do. With cells of similar function, we find that three things tend to govern size:

1. The ratio of the amount of nuclear material to the amount of cytoplasmic material.
2. The ratio of cell surface area to cell volume.
3. The rate at which the cell carries on its many chemical activities.

Although all three are related to each other, we shall consider them separately.

Let us first accept the fact, which we shall prove later, that the nucleus is the control center of the cell. In cooperation with the cytoplasm, the nucleus regulates the growth, development, and continued existence of the cell. Although a cell can function for a time without a nucleus, it eventually runs down and ceases to operate. The nucleus, however, cannot extend its control over too large an amount of cytoplasm. As the cell enlarges, the surface area of the nucleus, across which an interchange of materials must pass, increases only as the square of the nuclear radius (surface area = $4\pi r^2$), while the volume of the cell increases as the cube of the cell radius ($v = 4/3\pi r^3$). Too large an amount of cytoplasm would soon put some part of the cell far removed from nuclear control. The nucleus can increase its surface area by changing its shape, or by doubling its amount of chromatin, the main component of the nucleus. Adjustments can, therefore, be made. Most mature cells, however, seem to maintain a relatively constant nucleo-cytoplasmic ratio, and we find that growing cells do not vary greatly in size.

The amount of surface area tends also to limit a cell's size. As we saw earlier, chemical activity occurs continuously throughout the cell mass, and the various substances the cell needs to carry on this activity must pass through the surface membranes. Oxygen, for example, is required by most cells. If sufficient oxygen is to reach those parts of the cell where it is needed, its concentration outside the cell must be at or above a certain level. The particular concentration is related to the rate at which oxygen moves into the cell, the rate at which oxygen is being used, and the dimensions of the cell. A cell that is too large has difficulty in getting oxygen to its more inaccessible parts.

The surface–volume limitations can be overcome in a variety of ways. Cells can change their spherical shape by being flattened, folded, or elongated. These changes either increase the surface area without increasing the bulk of the cell, or they bring the cellular contents nearer to the membrane surfaces, as for vacuolated plant cells. If the surface area is increased, the flow of

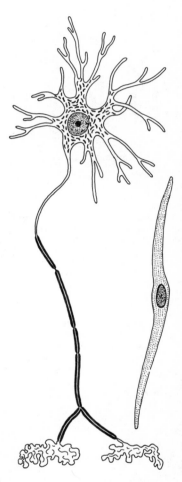

Figure 3.3 Two relatively long animal cells. Left: a nerve cell with its many-branched projections from the cell body, and the long axon terminating at points where it could make contact with a muscle. The axon may reach 1 meter in length. Right: a smooth muscle cell such as might be found in the wall of the intestine.

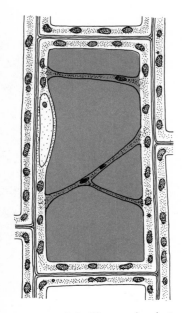

Figure 3.4 Diagram of a plant cell with the heavy cell wall, the nucleus and chloroplasts in the cytoplasm, and the center occupied by a large vacuole. Strands of cytoplasm cross the vacuole.

Figure 3.5 Here is a section of an intestinal wall with its convoluted surface, or *villi* (left), and with two of the absorbing cells at the right.

materials in and out of the cell is also increased, and the cell can consequently enlarge without changing the rate of its chemical activities, called **metabolism.** This enlargement can go on as long as the expansion does not become so great that the nucleus can no longer exert control over the cytoplasm.

The surface–volume problem appears over and over again in living organisms, and it is met and solved in various ways. A nerve cell, for example, may reach 1 meter in length, but the main body of the cell is not especially large. The **axons** and **dendrites** of the nerve cell are of such small diameter that an exchange of materials across their membranes can readily take place.

The individual plant cell, with its rather rigid cell wall, has solved the surface–volume problem in yet another way. It has a large liquid-filled vacuole in the center (Fig. 3.4). This pushes the cytoplasm to the outside so that a rapid exchange of materials is possible. In addition, the contents of the cytoplasm are kept in constant motion, thus preventing any portion of it from becoming stagnant through lack of oxygen or nutrients.

For cells that cannot exceed a certain size, an increase in number rather than size is the only solution. An organ of the body (as opposed to a single cell) has more elaborate means to meet the demands made on it. In mammals, for example, the digestive system is a long coiled tube, and its function is to digest and absorb the food that is eaten. To increase its ability to absorb, the lining of the intestine is composed of many folded surfaces arranged much like the piling of a bath towel (Fig. 3.5). Also, the surface of each of the absorbing cells is similarly folded so that its absorbing surface is greatly enlarged (Fig. 3.6). The inner surface of the lungs is also designed to make rapid exchange of oxygen and carbon dioxide possible. Each cell is in contact with a blood vessel on one side and air on the other. Oxygen can

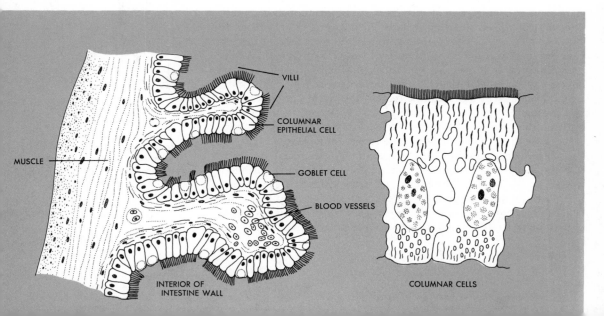

VILLI

COLUMNAR EPITHELIAL CELL

MUSCLE

GOBLET CELL

BLOOD VESSELS

INTERIOR OF INTESTINE WALL

COLUMNAR CELLS

readily pass into the blood, and carbon dioxide can easily move out.

The third condition that affects the size of a cell is the rate of chemical activity carried on inside it. It is tempting to say that the smaller a cell or an organism, the higher its rate of chemical activity. Although cell size is not absolutely correlated with the rate of metabolism, the rapidly metabolizing cells of such organisms as bacteria, hummingbirds, shrews, bees, flies, and mosquitoes are generally small. Those of the more slowly metabolizing animals such as man, elephants, amphibians (frogs, toads, and so on), and grasshoppers are considerably larger. The surface exchanges of materials in the cells of the smaller and more active animals must be more rapid because of their greater need for a constant supply of energy. Also, the cells must be smaller so that the amount of surface area relative to volume can be at a maximum. If this were not the case, the movement of substances into the interior of the cell would be insufficient to maintain the rate of metabolism; the metabolic processes would bog down. Also, heat is released from the breakdown of food substances, and the heat must escape. For example, if the elephant metabolized at the same rate as the hummingbird, it would roast itself. The tremendous amount of energy, some of it liberated as heat, could not escape. The same result would occur if a human cell metabolized at the same rate as a small bacterial cell.

Cell size can also be viewed as a structural problem. The cell contents need support of some kind so that they can be held together and function efficiently; hence, the cell membrane. Such an elastic but firm support would seem wasted on cells of minute size, but it is necessary as a boundary between the cell and its environment. On the other hand, oversized cells without the aid of a membrane would burst like punctured balloons because of internal pressure. A balance must be maintained between the need for support and the need for a proper surface–volume relationship.

What, then, is the ideal size at which a cell performs most efficiently? Obviously there is no simple answer. Since cells of various sizes exist, we must assume that each is efficient. If we exclude eggs from our consideration, since they really belong in a category by themselves, we find that maximum cell diameter is about 100 μm, while the minimum is approximately 0.1 μm, a ratio of 1,000:1. (We have also excluded viruses because, whether they are cells or not, their energy for metabolism comes from the cell they have parasitized.)

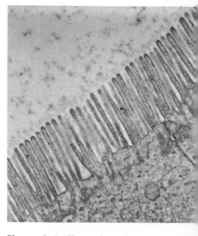

Figure 3.6 The surface of an absorbing cell (Fig. 3.5) has many projections, or *microvilli*, as shown by this electron micrograph. Each absorbing cell may have as many as 3,000 microvilli, so that its surface area is increased enormously.

CELL SHAPE Let us now look at organisms that are unicellular. When we remember that protoplasm is a somewhat fluid substance bounded by an elastic cell membrane, we might well assume that the shape

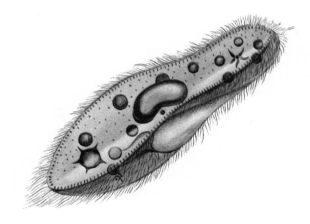

PARAMECIUM
Ciliate protozoan

Figure 3.7 Three species of
protozoans: *Paramecium* and
Ameba are commonly found in
waters all over the world.
Paramecia may range in length
from 200 to 350 *μ*m. Amebas
may be around 600 *μ*m in
diameter. Both are considered to
be rather large protozoans.

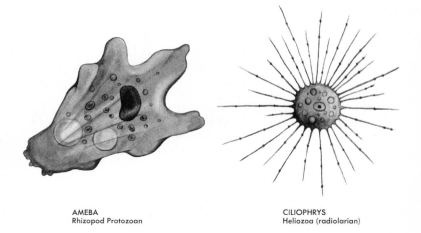

AMEBA
Rhizopod Protozoan

CILIOPHRYS
Heliozoa (radiolarian)

of these cells would be spherical. Their surface tension, particularly in those that are free-floating, should shape them in the
same way that surface tension shapes airborne soap bubbles or
rubber balloons. Many cells, indeed, do have a spherical shape:
the eggs of many marine animals, many yeasts and bacteria, and
a variety of unicellular algae. But from the different shapes
attained by other forms of unicellular life, it appears certain that
the organism itself, through its inheritance, determines its own
shape. Some bacteria are shaped like rods, spirals, or commas.
Among the algae, the diatoms, desmids, and dinoflagellates, with
their unusual contours and outer skeletons, take on a bizarre
appearance. Even the ameba, familiar to most of you, is not
normally a sphere. Generally flattened because it rests on a surface, it has no particular shape, but rather is a fluid glob of

protoplasm that can flow this way or that. Only at rest or in death does it become spherical.

One of the most remarkable single-celled organisms, and one widely used in biological research, is *Acetabularia,* an alga found in warm marine waters (Fig. 3.8). Some species are 9 to 10 cm in height. A distinctive cap, characteristic of each species, tops the whole structure. Until it begins its fruiting stage, it is essentially a single cell with the nucleus located at the base of the stalk, and just above the root-like structure **(rhizoid).**

When we turn to a consideration of the cell shapes found in multicellular organisms, we find that mechanical forces still determine shape to some extent, but that the function a cell performs can also help determine the shaping process. As pointed out above, free-floating cells with thin membranes tend to be spherical. This is the most economical (that is, the most compact) shape a given mass of protoplasm can assume. A physicist would say that the cell is obeying the "law of minimal surfaces." When spherical cells are packed together, however, they tend to become faceted as they come in contact with neighboring cells, much as the sides of soap bubbles become flattened when the bubbles are jammed together in a confining space. In animals this can be seen in the early stages of embryo development (Fig. 3.10). For a brief period the cell mass still retains the rounded shape and size of the original egg, but in adjusting to the available space the cells shape themselves accordingly. Similar arrangements of plant cells can be seen in the growing tips of roots or stems.

Cells are not always packed in the same way. Some are

General Biological Supply House, Inc., Chicago

Figure 3.8 *Acetabularia* is a stalked one-cell plant living in warm marine waters. The plant may be 1 cm or more in height, but it remains as one cell until the nucleus in the rhizoid (base) rises to the cap and divides, thus forming many cells.

Figure 3.9 Four species of protozoans: three are free-swimming and one lives fastened by a stalk and is colonial in habit. *Noctiluca* is the protozoan responsible for the luminescence in warm marine waters.

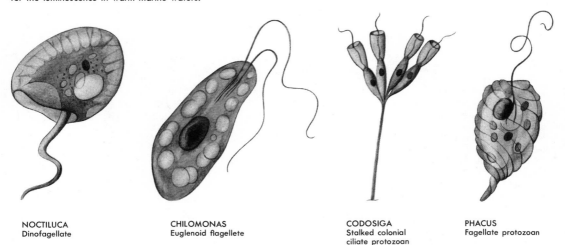

NOCTILUCA
Dinofagellate

CHILOMONAS
Euglenoid flagellete

CODOSIGA
Stalked colonial
ciliate protozoan

PHACUS
Fagellate protozoan

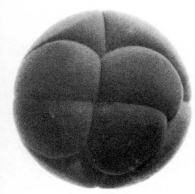

Figure 3.10 Above: the cells of an amphibian embryo become faceted as the number of cells (eight here) increases through division. With additional divisions they become more and more crowded together. Below: layered cells, such as those found in the skin or in the walls of arteries. They become flattened, and are greater in width and length than they are in depth.

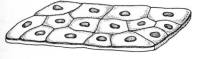

Figure 3.11 Left: red blood cells appear rounded in face view, but dumbbell-shaped in profile. This shape prevents clogging in the capillaries and allows for an easy and rapid exchange of gases. Right: the melanocyte, containing the black pigment melanin, changes its shape as it matures. These cells from a human being do not alter their shape once they are mature, but those in the flounder (Fig. 3.12) can expand or contract, depending on the environment in which the flounder finds itself.

layered in flat sheets, as in the linings of blood vessels, or skin. Such cells tend to be longer and wider than they are thick (Fig. 3.10). Presumably, they are forced into this shape by tension. However, it would be a mistake to overemphasize the role of tension in shaping cells. The function a cell performs is also related to its shape, but we cannot say that function *determines* shape, or that shape *determines* function. In any particular organism, the two go together. Let us consider the human red blood cell as an example.

Viewed face on, a red blood cell from a vertebrate has a circular shape. From the side, it may appear flattened and concave (Fig. 3.11). Its function is to carry oxygen from the lungs to various tissues of the body, and to carry carbon dioxide from the tissues to the lungs. While its flattened shape allows for the exchange of gases, its rounded contours and small size permit it to slide easily through the smallest blood vessels **(capillaries)** without clogging them. A spherical cell of the same size would be inefficient, because the rate of gas exchange between the exterior of the cell and its center would be very slow in proportion to its size.

The shape of muscle and nerve cells further emphasizes the relation of shape to function. Both kinds of cells are elongated.

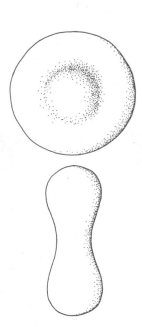

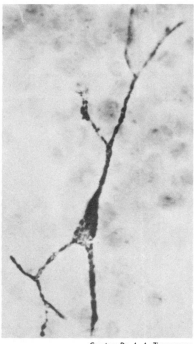

56

The muscle cell can alternately contract and relax. The nerve cell is part of the communication system of the body. Why would these cells perform their function less well if they had a spherical or a flattened shape?

Some cells can alter their shape quite rapidly. An example of this kind of cell is the **melanocyte,** so called because it contains a pigment named **melanin,** which is responsible for the dark color in the skin of animals. As these melanocytes mature in man, they go through a series of changes that transform them from a spherical into an elongated and branched shape. Once they reach maturity, they do not change shape. However, in the flounder or the chameleon, the shape varies with the background of the environment in which the animal finds itself. When the melanocytes are contracted, the animal appears light. When expanded, with cell branches extended, the animal is darker (Fig. 3.12). The melanocyte, therefore, not only gives color to the skin but also provides protective coloration: it enables the animal to blend into either a light or a dark background. No other cell shape could perform this function so well.

Figure 3.13 shows a number of cell shapes from various parts of a plant, each type having a special function. The particular shape adopted reflects the particular position and function of each cell. We find, therefore, that the shapes of cells are related both to the body plan of the organism and to the varied activities the organism performs.

Plants and animals differ in their structure, mobility, and mode of nutrition. Except for certain free-floating plants, plants generally have devices for anchorage and for the absorption of water and mineral salts (roots), for the conduction of substances through the system (stem and veins of leaves), for the manufacture of food through photosynthesis (leaves), and for reproduction (flowers or other reproductive organs). Animals, however, must search for their food, and must have a bodily construction de-

Figure 3.12 These photographs show a flounder on rocky (left) and sandy (right) bottoms. Its melancytes are expanded above and contracted below so that it can blend better with the environment, and so escape detection.

R. H. Noailles   *R. H. Noailles*

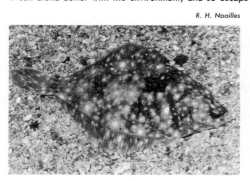 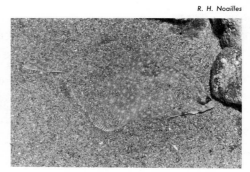

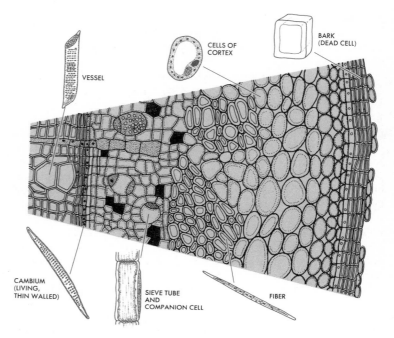

VESSEL

CELLS OF
CORTEX

BARK
(DEAD CELL)

CAMBIUM
(LIVING,
THIN WALLED)

SIEVE TUBE
AND
COMPANION CELL

FIBER

Figure 3.13 A cross section of a woody stem shows different kinds of cells: vessels for water conduction; cambial cells for the production of new cells through division; sieve tubes and companion cells for the conduction of food materials; fibers for protection and tensile strength; cortical cells for photosynthesis; and cells of the bark for protection.

signed for this purpose. They need structures for support (bones, cartilage, or **exoskeleton,** as in the lobster), for movement (muscles), for communication (nerves), and for digestion, excretion, secretion, circulation, and reproduction. Since the cell is the unit of construction in all these structures, we must conclude that it is an extraordinarily flexible unit. Each cell in a body has the same inherited constitution, yet it is capable of adapting to the situation it finds itself in and to the function it performs.

CELL NUMBER   Let us consider the *number* of cells found within any given organism. This number may seem impossible to calculate in so large an organism as man, but since cells are roughly the same size, simple arithmetic can give us a fairly good estimate. The difference between a human dwarf and a giant is primarily a matter of cell number rather than of cell size. Generally speaking, within any one species the larger the organism the greater the number of cells in its body.

Among unicellular forms, the cell and the organism are one

Ewing Galloway

Figure 3.14 The difference in size of a giant and a dwarf is due to numbers of cells, not to a difference in cell size.

and the same. Most multicellular organisms consist of an indefinite number of cells, but in a few the number is fixed. The green alga, *Pandorina morum* (Fig. 3.15), found in freshwater ponds is a plate made up of 8 or 16 cells. A similar and related species, *P. charkowiensis,* has either 16 or 32 cells. Intermediate numbers of cells are not found in these species. When new colonies are formed in *P. morum,* each of the 16 cells divides four times, producing 16 new colonies, each with 16 cells apiece. Such definite cell numbers are found in only a few of the animal forms.

Most organsims have an indefinite number of cells. The number is limited only by the ultimate size the organism reaches at maturity. Take man as an example. During the first 285 days of prenatal life, a human being grows from a single fertilized egg that divides again and again, forming a newborn infant weighing about 7 pounds. The egg cell originally weighed about one-millionth of an ounce, but at birth the infant consists of approximately 2,000,000,000,000 (2×10^{12}) cells. At maturity the average human male, weighing about 160 pounds, consists of about 60 thousand billion (6×10^{13}) cells. If all cells divided simultaneously and at the same rate, how many cell divisions would have occurred by the time of birth? How many by the time of maturity?

Not all cells, however, divide at the same rate and for the same duration of time. Some reach maturity and stop dividing long before others; some continue to divide through old age and until death. The development of the human being from a single fertilized egg involves many things; an increase in cell number is but one of them. For example, at birth most or all of the nerve and the great majority of muscle cells are formed; growth after

Figure 3.15 *Pandorina morum* is an alga that has a definite number of cells (16) per colony. When new colonies are to be formed, each cell in a colony undergoes four divisions, forming 16 new 16-celled colonies.

Courtesy of Annette Coleman

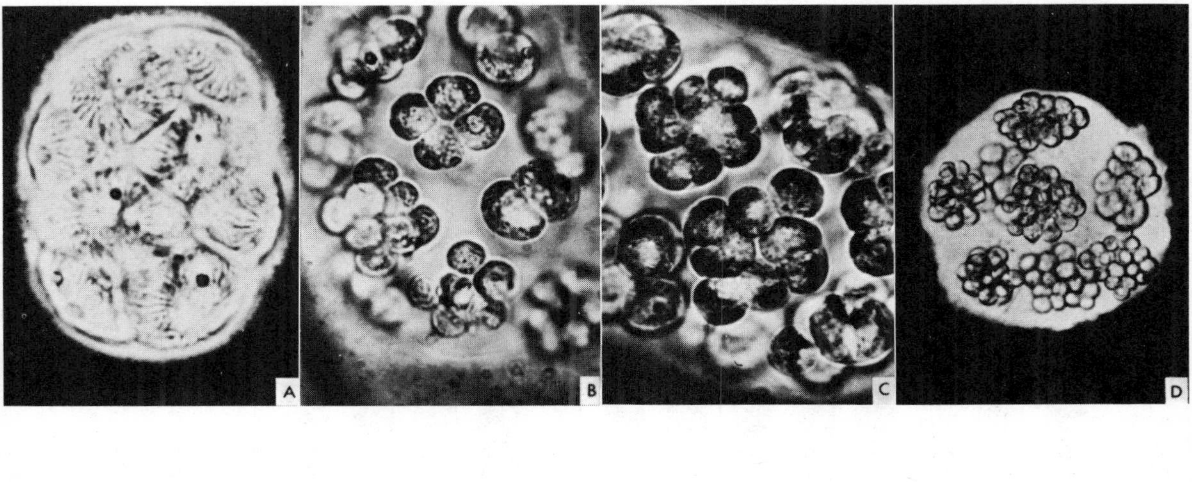

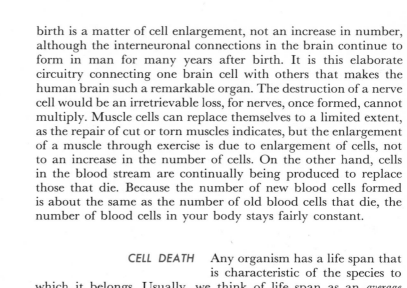

birth is a matter of cell enlargement, not an increase in number, although the interneuronal connections in the brain continue to form in man for many years after birth. It is this elaborate circuitry connecting one brain cell with others that makes the human brain such a remarkable organ. The destruction of a nerve cell would be an irretrievable loss, for nerves, once formed, cannot multiply. Muscle cells can replace themselves to a limited extent, as the repair of cut or torn muscles indicates, but the enlargement of a muscle through exercise is due to enlargement of cells, not to an increase in the number of cells. On the other hand, cells in the blood stream are continually being produced to replace those that die. Because the number of new blood cells formed is about the same as the number of old blood cells that die, the number of blood cells in your body stays fairly constant.

CELL DEATH　　Any organism has a life span that is characteristic of the species to which it belongs. Usually, we think of life span as an *average* figure: a few days for certain insects, a few months for annual plants, 60 to 70 years for man, 250 to 300 years for an oak tree. The sequoia (redwood) of our West Coast and the bristle-cone pine of California's White Mountains are probably the longest-lived organisms. Some of these trees reach an age of 3,000 to 4,000 years.

Cells, too, have a life span that is characteristically long or short. Yet some cells can be considered to be immortal. When a unicellular organism divides, the life of the single mother cell becomes part of the life of two new daughter cells. So long as a unicellular species continues to live, so too does the life of the original cells that began the species sometime in the long-distant past. Among sexually reproducing organisms, only the cells of

Figure 3.16 The maximum life spans of a number of different organisms are shown in this chart. Sequoias may live for several thousands of years, and their full life span would extend well beyond the limits of the page.

MAXIMUM LIFE SPANS FOR A VARIETY OF ORGANISMS

LIFE SPAN IN YEARS

the germ line (those cells producing eggs or sperm) can lay claim to immortality. They are the only ones that span the generations, thus keeping the species alive. But among the cells of the body, death is a very necessary process. If its role were altered, the functioning of the organism would be drastically affected.

Some parts of your body are continuously renewed as new cells replace old ones that die. These parts are characterized by rapid rates of cell division. Other tissues cannot renew themselves. The original cells live on and on; they die only through disease, accident, or when the individual dies. Cell division does not occur in such parts of your body. As pointed out above, all nerve cells are formed by the time of birth. They continue to mature and function as long as the individual lives (barring injury, of course). Yet nerve cells have been estimated by pathologists to die at a rate of 12,000 to 14,000 per day. Using a brain capacity of 1,350 cm^3 and a nerve cell diameter of 100 μm, can you calculate how many brain cells you had at birth, and what percentage you will have at age 50 or 100 years? How long would you have to live to have no brain at all?

The fact that an organ remains constant in size tells us nothing about its rate of cell replacement. The constancy of size merely indicates that there is no *net* gain or loss of cells. The death of the old cells can be equaled by the production of new cells.

Some biologists estimate that 1 to 2 percent of the cells of the human body die each day. But if a person's weight remains constant, these old cells must be replaced by new ones—by the billions every day. Since muscles and nerve tissue to not produce new cells, there must be some very active centers of death and replacement elsewhere. These include the protective layers of skin, the blood-forming centers, the lining of the digestive tract, and the reproductive system. The other organs of the body have much slower replacement rates. A cell in the liver, for example, is thought to have an average life span of about 18 months. Consequently, if we look at a liver slice under the microscope, we expect to find very few cells in division.

The outer surface of the human body is covered with a protective layer of skin. This includes the cornea of the eye and such modified skin derivatives as nails and hair. The cells of these structures are constantly being lost through death. You can easily scrape the inside of your mouth and pick up living cells that can be viewed in the microscope. The skin constantly sloughs off, and the growing nails and hair are composed of cells that have died. The process of replacement, then, must be a relatively rapid one. The underlying cells are constantly dividing, and are pushed outward toward the skin surface, while the outermost cells harden, or become **cornified,** as they die (Fig. 3.17). It takes about 12 to 14 days for a cell in the skin of the forearm to move from the dividing to the outermost layer of the skin. Callouses on the hands are thickened areas of dead cells.

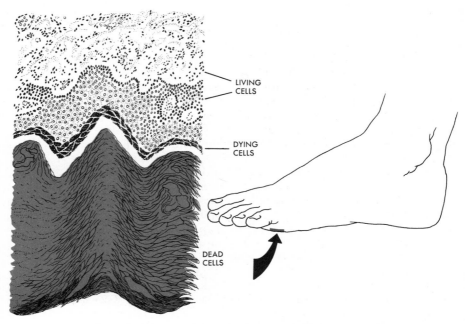

LIVING CELLS

DYING CELLS

DEAD CELLS

Figure 3.17 A section through the skin of the human foot would show that cells are continually dying at the surface but are continually being replaced by cells beneath them.

The cornea of the eye is a special type of skin in which the rate of cell death and replacement is high. The cornea, in fact, is an excellent type of tissue to examine for active cell division since it is only a few cell layers thick. When it is fixed, stained, and mounted intact on a microscope slide, the dying cells can be seen at the outer surface, while the cells underneath can be seen in active division.

The cells of the blood are not formed in the blood. The red blood cells are formed in the marrow of the long bones; the white cells form in the lymph nodes, spleen, and thymus gland. Together, these cells and a clear fluid, the **plasma,** constitute the blood, which has an average ratio of 1 white cell to 400–500 red cells. The blood-forming areas usually manage to maintain this ratio.

Let us consider the red blood cell. Its life span is about 120 days. Mature red blood cells do not have a nucleus, which means that the damage done to them as they pass through the vessels cannot be repaired. The cells grow fragile and eventually burst. Certain types of illness may shorten the life span of red blood cells. In a person afflicted with pernicious anemia, the life span is reduced to about 85 days; with sickle-cell anemia, to 42 days. The rate of replacement in persons afflicted with these diseases cannot keep pace with the loss of cells, and the red-cell count falls below normal and results in an anemic state. The cause of the shortened life span of these abnormal cells is not known, but it is probably determined by the character of the cell membrane rather than the cell contents. The digestive system consists of some organs in which the cell death rate is very high. Figures for the human are not known, but it has been estimated that the cells lining the intestine of the rat are replaced every 38 hours.

In the plant kingdom, we find that the lower plants—the algae and fungi in particular—have a rather low loss of cells through death. In the higher plants however, the rate is enormous. In herbaceous plants, all the cells above ground are lost every season. But consider a large tree. The annual loss of cells in the leaves, flowers, and fruits is high enough, but when all the cells that form dead wood and bark are added, the loss of cells through death in animals is small by comparison. A high rate of cell death, therefore, is as much a pattern of existence as the continuation of living cells. Unfortunately, we know very little about why cells die. Until we advance our knowledge in this field we cannot hope to understand the whole pattern of aging.

summary

Cells from various organisms and from the several parts of a single multicellular organism differ greatly in size, shape, length of life, rate of metabolic activity, and function performed or product produced. What a cell does and what kind of morphology it possesses are determined by several things: its heredity, which is largely located in the nucleus; the environment in which the cell finds itself and which determines its nutrition, metabolic rate, and imposing stresses; the part of the body in which it is located; and the function that the cell performs. All cells have a common group of organelles, and some have special structures, but the organelles shared by most cells can vary in kind and number. We must, therefore, view the cell as a very flexible structure, capable of doing many things in a variety of environments. Every cell is an efficient unit; otherwise, it would not persist.

for thought and discussion

1 We have about 5 liters of blood in our body, and there are 5,400,000 red blood cells per milliliter in the average human male (females have about 600,000 fewer). These cells have an average life span of 120 days. How many RBC's must die and be formed each day to keep the number of RBC's constant?

2   Wood of various kinds consists only of dead cells. How do you suppose the different kinds of wood—pine, oak, cherry, walnut, balsa—get their distinctive characteristics?

3   Plan a menu consisting of soup, salad, fish course, meat course, vegetables, dessert, and something to drink. What cells and cell products do you consume?

4   *Metabolic needs are proportional to the volume of cells; exchanges are proportional to the surface area.* What does this statement mean? What does it have to do with the limitation of the size of cells? If the metabolic rate increases, what can the cell do to meet the increased needs (assuming that the volume does not change)?

selected readings

All the books listed in the previous chapter can serve also for this chapter.

GERARD, R. *Unresting Cells.* New York: Harper & Row, Publishers, 1940. One of the finest books available on cells. Although somewhat out of date, it should be read by all students of biology.

HOFFMAN, J. *The Life and Death of Cells.* New York: Hanover House, 1957. A nontechnical account of the world of cells, examined in the light of modern theories of matter and energy.

cell
division
chapter 4

Cells increase their numbers by **division.** In unicellular species, such as an ameba or a bacterium, division results in an increase in the number of individuals in a population. In multicellular forms, such as a human being, the cells do not separate after division. This means that an increase in cell number is part of the processes of growth and maintenance of the individual. As we saw in the last chapter, an adult man has about 6×10^{13} cells. Among these are cells of many kinds, each kind having a different structure, function, and life span. The cells that die must be constantly replaced if the body is to maintain itself. Within any organism that grows and requires routine repair, the process of cell division must produce new cells at varying rates and for varying periods of time (Figs. 4.1 and 4.2).

The process of cell division is essentially the same in all organisms. If we describe this process as it takes place in one or two kinds of cells, we gain a reasonably clear picture of how it operates in all organisms, including man. Most of the activity of cell division is centered in the nucleus. However, the cytoplasm also undergoes a significant series of changes, so the whole process of division involves both nucleus and cytoplasm.

William Bateson, the great English biologist of an earlier generation, once wrote: "When I look at a dividing cell I feel as an astronomer might do if he beheld the formation of a double

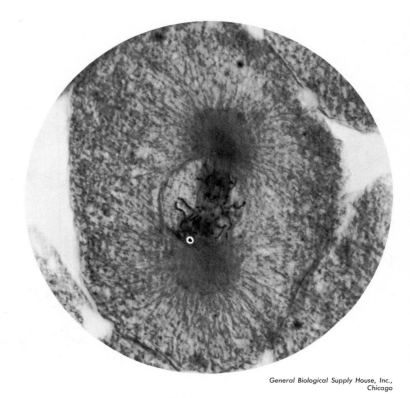

Figure 4.1 This cell, part of a whitefish embryo, is dividing. The nuclear membrane is still intact, the spindle is being formed between the poles, and the chromosomes are readily visible.

star: that an original act of creation is taking place before me." Other biologists who have also watched a cell divide, particularly through motion pictures, have been as fascinated as Bateson was.

Each of us developed from a single cell. This cell came from preceding sex cells, and so on back to the beginnings of cellular life. Our most important heritage from the past, and our most precious gift to the future, is the unbroken chain of perfectly formed, individual cells. They are "individual" in the sense that eggs and sperm are individual cells, and the fertilized egg is an individual being, even if not yet fully formed. They are "perfect" in the sense that they are able to live, reproduce, and give rise to new cells, and hence to new individuals. Every time you wash your hands thousands of dead cells making up the outer layers of skin are removed. So far as these cells are concerned, the chain of cellular life has been broken. Death is the inevitable and inescapable fate of most cells, just as it is for us as individuals. Among sexually reproducing organisms, only the eggs and sperm maintain the unbroken chain that links one generation to another.

Ever since biologists came to understand the cellular nature of organisms, they have been investigating and debating the ways in which new cells originate. Not until the middle of the 19th century was it generally accepted that cells originate through the division of pre-existing cells. We attribute this idea to the German, Rudolf Virchow, who stated in 1858:

> Where a cell exists there must have been a pre-existing cell just as the animal arises only from an animal and the plant only from a plant. The principle is thus established, even though the strict proof has not yet been produced for every detail, that through the whole series of living forms, whether entire animal or plant organisms or the component parts, there are rules of eternal law and continuous development, that is, of continuous reproduction.

Although we cannot reconstruct the beginnings of life from what we know today, presumably it did not begin in the form of the cells that we now see under our microscopes. They are the products of ages of evolution. Nor can life be originated anew, at least with the techniques and knowledge we possess today. It can come only from pre-existing life. This theory, known as **biogenesis,** we credit to the French biologist Louis Pasteur, although he was not the first scientist to demonstrate this fact. Pasteur showed that life could not arise anew under the conditions existing in his laboratory. He took two flasks of broth, a good culture medium for micro-organisms, and boiled the broth in such a way as to kill all the organisms in it. One flask he left open; the other he made airtight by sealing it. Within a few days the open flask contained "germs" of various sorts—bacteria, yeast, or molds—which, as we now know, were present in the air. The sealed flask, however, contained no life whatsoever, and could not acquire any until air (a carrier of germs) was once more admitted.

ROOTTIP CELLS IN DIVISION The growing roottips of plants have long provided a good source of actively dividing cells. Because the cells do not all divide at the same time, a single roottip has cells in all the stages of division. Here we shall examine the process as it takes place in the roottips of the broad bean, *Vicia faba,* and the onion, *Allium cepa,* although almost any kind of actively growing roottip will serve for this purpose.

Roottip cells present a variable appearance, depending on how the cells are prepared and stained. Study the low-magnification photograph of an onion roottip, sectioned in a longitudinal direction and stained with haematoxylin (Fig. 4.3). Nuclei are

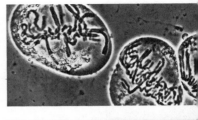
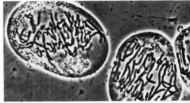
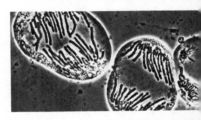
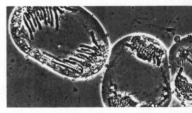
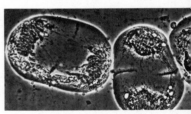
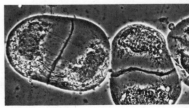

Figure 4.2 These living plant cells are also undergoing division, as shown in the series of time-lapse photographs. The difference in time from the top view to that at the bottom is 81 minutes.

Courtesy W. T. Jackson

67

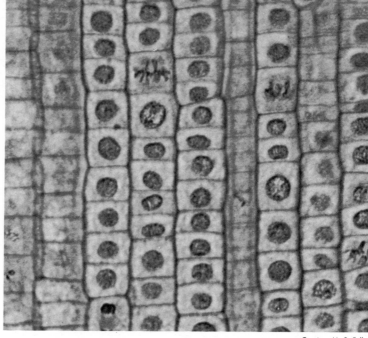

Figure 4.3 A number of different stages of cell division are visible in this low-power view of a longitudinal section through the roottip of an onion.

in various stages of division, and nucleoli are prominent. Now compare these cells with those of the broad bean (Fig. 4.5), stained with Feulgen and then squashed on a slide. Feulgen stains only the chromatin, leaving all other cellular structures virtually invisible; it is a highly specific stain. Haematoxylin, on the other hand, is a general stain. When we use it we can see other structures in addition to the chromatin.

Our knowledge of why a cell divides at a particular time is not very complete. However, the cell goes through an active stage of synthesis prior to division, doubling the amount of nucleic acid in the nucleus. This activity occurs during **interphase** (Fig. 4.4), a stage in which the nucleus shows little definable structure, except for the lightly stained chromatin and the more deeply stained nucleoli. The **chromosomes,** into which the chromatin will form itself, are not individually distinguishable during interphase.

Prophase begins when the chromosomes first become visible as long, slender threads that are longitudinally double. Each longitudinal half is called a **chromatid,** and the two chromatids of each chromosome are twisted around each other like two wires of an electrical cord. During prophase the two chromatids shorten by becoming coiled. The progress of coiling is shown in Fig. 4.4. The mechanism of coiling is not yet known, but it is obviously a means of converting a long and unmanageable strand into a more compact, maneuverable one.

During prophase the nucleoli are large at first, but gradually

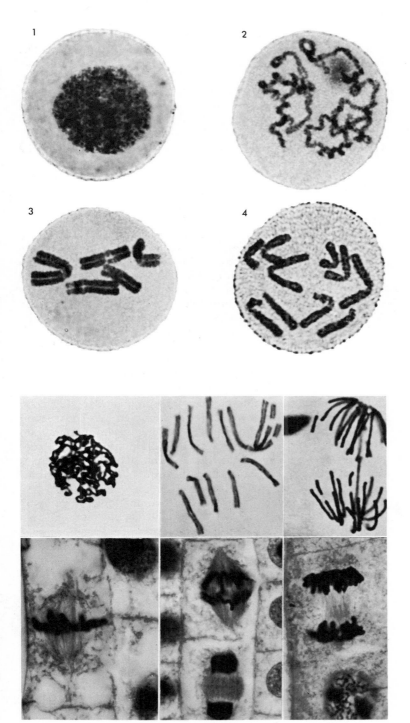

1

2

3

4

5

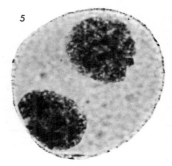

Figure 4.4 Cell division in the microspores (immature pollen grains) of the wake robin, *Trillium erectum:* (1) interphase; (2) mid-prophase, during which the chromatids shorten by becoming coiled; (3) metaphase, the spindle is not stained, hence does not show; (4) anaphase; and (5) telophase.

Figure 4.5 Top: Feulgen stain was used to show the prophase, metaphase, and anaphase stages of division in the roottip of the broad bean, *Vicia faba*. The stain reveals only the structure of the chromosomes, leaving the spindle, cytoplasm, and cell walls unstained. Bottom: The spindle is clearly revealed when iron haematoxylin stain is used on dividing cells of the onion roottip.

Figure 4.6 Here is the spindle of a dividing animal cell with the chromosomes on the metaphase plate. The fibers that stretch from chromosomes to poles, and from pole to pole, are actually fine tubules. The poles show astral rays radiating outwardly into the cytoplasm.

Courtesy S. Inouyé

they become smaller. They also free themselves from the chromosomes to which they are attached, and finally disappear. The nuclear membrane breaks down and disappears in late prophase, and the chromosomes become attached to a new structure called the **spindle** (Fig. 4.6). The cell is now in **metaphase.** At this stage, it is evident that the number of chromosomes per cell is constant, and that each chromosome has a particular shape and size which is maintained from cell to cell. It becomes possible, therefore, to identify the chromosomes individually. In the broad bean, the shapes and sizes of the chromosomes are of two major classes; in the human, they are grouped differently because of variable sizes among the 46 chromosomes.

The spindle consists of tubules of protein molecules arranged longitudinally between the two poles of the spindle. These proteins are formed in the cytoplasm during interphase and prophase. The mechanism that causes the proteins to aggregate into a spindle structure is unknown. However, by placing roottips in a solution containing the drug colchicine, spindle formation can be prevented. The chromosomes shorten normally, but they fail to aggregate. They lie free in the cytoplasm, where they are easy to count and where their shapes and sizes can be seen clearly.

When the spindle appears, the chromosomes move to a position midway between the poles. Here they become attached to the spindle at their **centromere** regions. The centromere of the chromosome is its organelle of movement. Without it, the chromosome fails to orient properly on the spindle, and later will fail to separate its two chromatids into the new daughter cells. The position of the centromere is clearly marked by a constriction in the chromosome. It consequently divides the chromosome into two arms of varying length. Very few chromosomes have centromeres at their ends. Since the position of the centromere is constant, this serves as an additional feature to aid in the identification of particular chromosomes.

The two chromatids of each chromosome now move apart from each other and migrate to the poles of the spindle. This period of movement is called the **anaphase** stage. Protein tubules running from the poles to the centromere of each chromosome shorten during anaphase and bring about the movement. Anaphase ends when the two groups of chromatids (which can now be called chromosomes) reach the poles.

Telophase begins when chromosome movement has stopped. At this stage a new nuclear membrane forms, the nucleus enlarges, the spindle disappears, the chromosomes uncoil and become long and slender, and nucleoli appear. These events are essentially the reverse of what took place in prophase. Across the middle of the spindle a new cell membrane is formed. Eventually, the **cell plate,** as it is called, grows outward until it reaches the side walls and cuts the cell in half. Two new cells are now fully formed.

cell division

CELL DIVISION IN ANIMAL CELLS Cells from the embryo of the whitefish illustrate very beautifully the division process. They also reveal the differences that distinguish animal from plant cell division. The behavior of the chromosomes is the same as in plant cells, although the chromosomes are so small and numerous in the whitefish that they cannot be individually identified. The most immediate difference is in the process of spindle formation. As Fig. 4.1 shows, the whitefish cell in prophase has a radiating structure adjacent to the nuclear membrane. This is the **centrosome,** with **astral rays** radiating from it, and with a central body, or **centriole,** within it (but not visible in the illustration). In early prophase, the centrioles, which had previously divided, migrate along the nuclear membrane until they lie opposite each other. As the centrioles migrate, they organize the spindle between them. The nuclear membrane disappears and the chromosomes line up in the middle of the spindle (Fig. 4.7, metaphase stage). Anaphase movement and the telophase reorganization of the two new nuclei take place as in plant cells.

The division of the cell into two daughter cells is another point of difference. A process of *furrowing,* which begins at the outer edges of the cell, cleaves the cell in two. Plant cells, with their rigid cell walls, cannot do this, but cell-plate formation accomplishes the same thing.

Figure 4.7 shows the complete sequence of events that take place during the course of cell division. The several processes and structures involved must be coordinated in time and place if the cell is to divide successfully. One mistake, and the daughter cells will be abnormal. They may die.

THE TIME SEQUENCE OF CELL DIVISION The time required to complete an entire cycle of cell division varies quite widely. Bacterial cells, for example, accomplish a cycle in 15 to 20 minutes. Roottip cells in the broad bean require about 20 hours at room temperature.

Figure 4.7 Stages of cell division in the whitefish embryo. Left to right: prophase, metaphase, anaphase, and telophase.

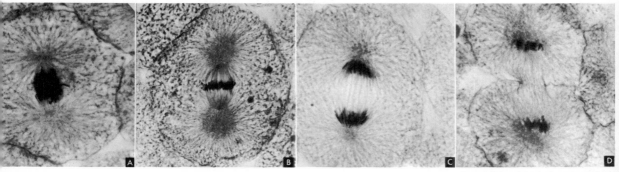

General Biological Supply House, Inc., Chicago

Higher temperatures can shorten the time span; lower temperatures lengthen it.

The overall time for cell division can be readily determined by finding the time necessary for doubling the cell number. This can be done with cells in tissue culture, or bacterial cells in a test tube; one simply counts them. Such knowledge, however, does not tell us much about the timing of the various stages of division. Let us consider this in terms of a human cell, easily grown in tissue culture. At 37°C, the temperature of the human body, a division cycle takes about 18 hours. Yet the period from the beginning of prophase to the end of telophase is only 45 minutes. More than 17 hours is spent in interphase, preparing for division. The dramatic events of cell division take place, therefore, in a rather explosive fashion.

Figure 4.8 shows the cycle of division of a human cell. It is possible to show that interphase has three distinct stages: the G_1, S, and G_2 stages (S stands for synthesis, and G for the interphase gaps before and after synthesis). The important stage is the S period when the nucleic acids and some of the proteins of the chromosome are being synthesized. Some 6 to 7 hours are required for the process, and the end result is that the chromosomes become longitudinally double. Some of the chromosomes

Figure 4.8 The cycle of cell division: interphase is the longest stage because during this time the chromosomes must replicate and the cell prepare itself for division. Metaphase and anaphase, the most dramatic of the stages when the chromosomes are clearly visible, are of shortest duration.

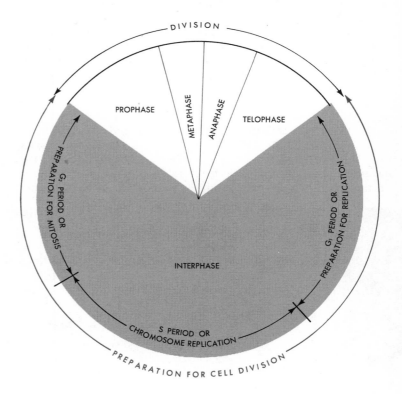

double, or replicate, early in the S period, others later. **Histone,** a protein component of the chromosome, is also synthesized at this time. The stages from prophase to telophase occur in rapid fashion, with metaphase and anaphase being the shortest stages.

Cell division is an act of survival—survival of life itself, survival of a species, and survival of each individual organism. A cell that does not, or cannot, divide will ultimately die. Cell division is also part of the process of growth and maintenance, providing more cells for the developing organism and replacing cells that die. As we saw earlier, the control center of a cell is the nucleus. It is here that chemical "decisions" of great importance are made, decisions which not only affect all division, but the day to day general well-being of the cell. In the next chapter we shall look at this control center in detail.

summary

Cell division is the source of new cells. Each cell, as it divides, goes through a number of recognizable stages: **interphase,** when the chromosomes in the nucleus replicate; **prophase,** during which the chromosomes shorten and the chromatids become visible, the nuclear membrane breaks down, and the nucleoli disappear; **metaphase,** when the spindle forms and the chromosomes orient on the spindle; **anaphase,** when the chromatids separate and move to the poles; and **telophase,** when the nuclei reorganize and the cytoplasm is divided for the formation of two new cells.

The time required for division can vary from several minutes to many hours. Although plant and animal cells differ somewhat in the way their cytoplasm divides, cell division is basically similar in all organisms.

for thought and discussion

1 Construct a chromosome out of pipe cleaners or wire so that the centromere and nucleolar organizer are visible features. Show how this chromosome changes as it goes through the several stages of division from interphase to telophase.

2 Why do you suppose that it is necessary for the chromosome to shorten before the two chromatids separate at anaphase?

3 Every daughter cell is like the mother cell from which it originated. How does the behavior of chromosomes during division ensure that this will be so? What would happen if it were not so?

4 Why would it be difficult for a plant cell to divide its cytoplasm in the same manner as an animal cell?

5 Asexually reproducing organisms are essentially immortal. Why is this so? The introduction of sexual reproduction into the plant and

animal kingdoms made sexually reproducing individuals mortal; each individual must eventually die. What does this statement mean? What keeps the species going if the individuals die?

6 What advantage is there to an organism being made up of many cells instead of just one large cell?

selected readings

KENNEDY, D. (ed.). *The Living Cell.* San Francisco: W. H. Freeman and Company, 1965. A collection of articles from *Scientific American* covering most aspects of cell structure, behavior, division, and metabolism.

MAZIA, D. *Cell Division.* D. C. Heath & Company, 1960. A small pamphlet written as a supplement to high school biology texts, and containing a clear account of the processes of cell division.

MCLEISH, J., and B. SNOAD. *Looking at Chromosomes.* New York: St. Martin's Press, Inc., 1958. This little book contains one of the finest collections of photographs illustrating mitosis and meiosis.

SWANSON, C. P. *The Cell* (3rd ed.). Englewood Cliffs, N.J.: Prentice-Hall, Inc., 1969. Chapters 7 and 8 in particular give a clear description of the processes of cell division in both plants and animals.

the nucleus—
control center
of the cell
chapter 5

We have mentioned several times that the nucleus exercises a control over the activities of the cell. What we are saying is that a cell carries out its functions because it has a nucleus that in some way provides it with a blueprint for action. In stating this, we do not mean that the nucleus controls every immediate action of the cell. With a change in temperature or nutrition, for example, the cell may change its metabolic action or even its appearance. It is also possible for a cell without a nucleus to function for a fair length of time. The human red blood cell, for example, loses its nucleus at about the time it enters the blood stream; it then operates very well for about 4 months without a nucleus. It may perhaps be more accurate to say that the nucleus in all kinds of cells provides a kind of long-range guidance over the affairs of the cytoplasm. But before accepting this statement, let us see what evidence can be found to support it.

First, observation tells us that the usual state of cellular affairs is one nucleus per cell. A departure from this rule is rather unusual in the cells of higher organisms, although among some bacteria and fungi, multinucleate cells may be the usual state of affairs. Living material, therefore, is generally so subdivided as to provide a single nucleus for a given amount of cytoplasm. There must be a good reason for this. From many observations, we also know that the character of the cytoplasm can change, the cell functions may differ, and the size and shape of the cell can be altered drastically. Yet the nucleus remains stable in what is often a changing cellular picture. This is not proof of a nuclear control, but it suggests that we should probably expect a control center to have a degree of stability such as that observed.

Another fact that suggests nuclear stability is the remarkably constant number and kind of chromosomes in each nucleus. The illustrations in Fig. 5.2 show the chromosomes of several organisms, including man. If you compare parts A and B, you will notice that B shows 47 chromosomes instead of the usual 46, the number characteristic of normal human individuals. Notice that the extra chromosome in part B is one of the very smallest, an extra chromosome 21. Nevertheless, its addition is sufficient to alter a normal child into one decidedly abnormal. This is a dramatic, if tragic, example of the nuclear influence of one extra chromosome on the character of the individual possessing it.

We can, therefore, adopt the hypothesis that, for normal development and behavior, a normal number of chromosomes (23 pairs in man) is necessary. Since the nucleus consists very largely of chromatin that, during division, changes into distinct chromosomes, we add evidence to our idea that the nucleus is the control center of the cell.

We still, however, lack definite proof. There are many kinds of proof, but let us use that provided by an ingenious set of experiments carried out by the German biologist, Max Hämmerling, on the single-celled alga, *Acetabularia.* The technique and the results are illustrated (Fig. 5.3). The two species of *Acetabularia,* *A. mediterranea* and *A. crenata,* differ principally in the shape of their caps. If the cap of either is cut off, it will form again as before and without change in shape. But it is also possible to

Figure 5.1 The dark objects in this electron micrograph are chromosomes in a dividing cell at the metaphase stage. The chromosomes are so compacted at this stage, and so dense to the electrons, that internal structure cannot be seen.

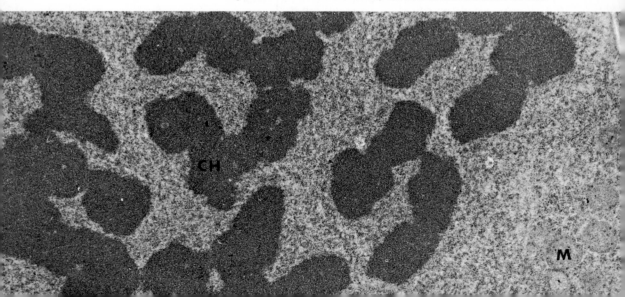

A

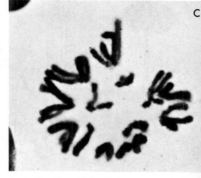

1	2	3	4	5

6	7	8	9	10	11	12

13	14	15		16	17	18

19	20	21	22	X Y

B

1	2	3	4	5

6	7	8	9	10	11	12

13	14	15		16	17	18

19	20	21	22	X Y

Courtesy V. McKusick

Figure 5.2 Chromosomes of a variety of organisms are shown on this page. (A) Normal human male, with the chromosomes arranged as 22 pairs plus an X- and a Y-chromosome. (B) Chromosomes from a human male characterized by *Down's syndrome* (Mongolian idiocy), which is caused by the individual having three instead of two chromosome 21's.

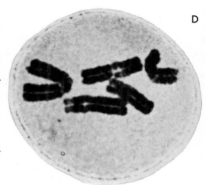

Courtesy Brookhaven National Laboratory

(C) Metaphase stage from the testes of a salamander (normal body cells would show 28 individual chromosomes instead of 14 pairs).

remove the cap and graft a portion of the stalk of one species onto the nucleus-containing rhizoid of the other species. This means then that the new stalk contains both *med.* and *cren.* cytoplasm and cell wall but has either a *med.* or a *cren.* nucleus in the rhizoid. When the cap forms again, the cap is always characteristic of the species contributing the nucleus and not that of the grafted stalk.

The proof can be made even more secure. It is possible to graft two rhizoids together, one containing a *med.* nucleus, the other a *cren.* nucleus. Again, a cap will form, but in this instance the cap will be intermediate in shape, reflecting the influence of both nuclei. We can now say with a good deal more assurance that it is the nucleus that instructs the cell to do as it commands. In other words, the nucleus contains the hereditary elements of the cell. This is an important fact to remember when one recalls that, when a sperm fertilizes an egg, the major contribution of the sperm in the act of fertilization is a nucleus, but virtually no cytoplasm. Yet the contributions of the egg and sperm to the inheritance of the newly developing offspring are equal.

Courtesy Brookhaven National Laboratory

(D) The five chromosomes of *Trillium erectum*, each one different in shape and in size.

THE CONTROLLING ELEMENT IN THE NUCLEUS We have now reached the point where we can say with some degree of scientific support that the chromosomes are the principal controlling elements of the cell. There is, in fact, very little else in the nucleus except chromosomes. It is possible to extract the chromosomes from burst cells and to analyze them chemically to find out what kinds of molecules they are made of.

Figure 5.3 Hämmerling's grafting experiments demonstrated that the character of the cap of *Acetabularia* is determined by the kind of nucleus present. (See text for explanation.)

Before looking into this, however, we want to consider a series of experiments that proved in a most conclusive way the nature of the *molecular basis of heredity. Pneumococcus,* the bacterium responsible for causing some forms of pneumonia, is a small, somewhat elongated cell covered with a thick coating, or capsule, of **polysaccharide,** a complex sugar. When plated out on a solid agar medium, this cell multiplies rapidly and forms a smooth, shiny colony. Occasionally, a colony is rough and dull appearing. The bacteria in the latter colonies are **avirulent** (Fig. 5.4); they

Figure 5.4 Griffiths' experiment showed that heat-killed virulent bacteria could contribute ''something'' to avirulent strains and cause them to be changed to a virulent form. Avery and his colleagues eventually showed that this ''something'' was DNA (see Fig. 5.5).

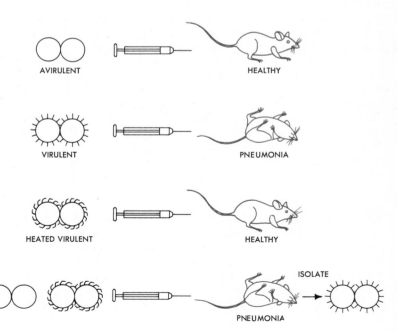

AVIRULENT · HEALTHY

VIRULENT · PNEUMONIA

HEATED VIRULENT · HEALTHY

ISOLATE

PNEUMONIA

are incapable of inducing pneumonia if injected into mice. They produce rough colonies because they lack the polysaccharide capsule, and they are unable to manufacture it. The rough trait, therefore, is heritable. Frederick Griffiths, an English bacteriologist, injected into mice a mixture of two strains of pneumococci. One strain was living, but rough and avirulent. This was called Type II, and it would not induce infection. The second strain, Type III, was virulent, but its virulence had been destroyed by killing the bacteria with heat. One would naturally assume that the mice should not have been bothered by these injections. Strangely enough, however, the mice died of pneumonia, and from them a living virulent Type III pneumococcus was recovered.

The results of the experiment suggest that some interaction took place between the two strains of bacteria in such a manner that the heat-killed cells changed a living, rough, avirulent Type II strain into a virulent Type III strain. This is basically a change in a heritable trait, because the cells remained Type III after repeated divisions. The phenomenon is called **bacterial transformation**. It was later demonstrated that the same phenomenon could take place in a test tube as well as in mice. Further, it was also shown that smashed, dead Type III cells could do the same thing as intact, but heat-killed, cells. The most likely explanation was that some molecule in the Type III cells was incorporated into the avirulent Type II cells, and subsequently changed its heritable characteristics.

Oswald Avery and his colleagues at the Rockefeller Institute set out to track down this *transforming principle*. By bursting bacterial cells open, and by laboriously isolating one kind of molecule after another for testing, Avery finally demonstrated that the critical molecule was **deoxyribose nucleic acid (DNA)** (Fig. 5.5). By adding purified DNA, extracted from virulent Type III cells, to a living culture of avirulent Type II cells, the latter can be permanently transformed into Type III cells. The frequency of transformation is not high—about one in every 1 million cells was transformed at first—but the fact that it could be done at all was an immensely important discovery.

Since Avery's work in 1944, scientists have shown that a

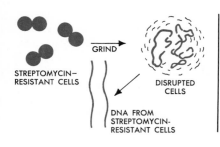

STREPTOMYCIN-
RESISTANT CELLS

GRIND

DISRUPTED
CELLS

DNA FROM
STREPTOMYCIN-
RESISTANT CELLS

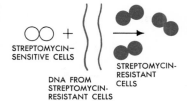

STREPTOMYCIN-
SENSITIVE CELLS

DNA FROM
STREPTOMYCIN-
RESISTANT CELLS

STREPTOMYCIN-
RESISTANT
CELLS

Figure 5.5 Avery's experiment demonstrated that DNA from one strain of bacteria possessing a known inherited trait could enter another strain lacking the trait, be incorporated, and thereby alter the host bacterium.

number of other heritable traits—penicillin resistance, ability to ferment sugars, and so on—can be similarly transformed.

THE CHEMISTRY
OF CHROMOSOMES

The importance of Avery's discovery, when considered in relation to our hypothesis that the nucleus is the control center of the cell, becomes obvious: DNA is found *only* in structures capable of self-replication—mitochondria, plastids, and, most importantly, in chromosomes. DNA is not, therefore, a unique nuclear substance, although most of it is found in the nucleus, and specifically in the chromosomes. In bacterial transformation, DNA from killed Type III cells must be incorporated into the DNA of Type II cells, thereby changing them into virulent Type III cells. Interestingly enough, however, DNA was discovered nearly 100 years ago. From that time it has been known to be a nuclear constituent. One can see that it often takes a long time for all the facts to fall into a pattern that makes scientific sense.

Avery's discovery prompted a closer scrutiny of the chemical nature of chromosomes. From a mass analysis of isolated chromosomes it appears that they contain from 26 to 40 per cent DNA; a small amount of **ribose nucleic acid (RNA),** which is related to but different from DNA; an amount of **histone** (a low-molecular-weight protein) which is bound closely to the DNA; and a complex protein called *nonhistone chromosomal* **protein** which varies in amount from one kind of tissue to another. Some calcium, magnesium, and **lipids** (fats) are also present, but their function is not known.

The important question that has been raised is whether molecules other than DNA occupy the same central role in governing cellular activity. The answer at the moment appears to be that DNA is the master molecule—the chemical basis of heredity—and that other molecules play supporting roles.

In this regard, however, it is important to point out that some viruses—tobacco mosaic virus (TMV) is an example—contain *only* RNA and protein, but no DNA. It is possible to separate the TMV protein from the RNA, and then to test each for its ability to produce the mosaic disease. The RNA is infective, the protein is not. The RNA of the virus is its hereditary material, but the answer is not quite so clear as in the bacterial transformation experiment. It is known that the TMV requires the cooperation of the DNA of the tobacco cell for the formation of additional viral particles.

But let us return to the chemistry of chromosomes. If we postulate that the DNA is the hereditary molecule of the cell, we would also expect it to be exceedingly stable in character and amount. This turns out to be true. By "tagging" a DNA molecule with radioactive atoms, we can show that, once formed, it remains intact and does not break down and reform. The histone is also quite stable and seems always associated with DNA. But only DNA has the capacity for transformation. On the other hand, RNA and nonhistone chromosomal protein in the nucleus and in most molecules in the cytoplasm continually break down and reform. They do not have the stability we would expect of the critical molecule of heredity even though they are important in the regulatory activities of cells.

The amount of DNA in nondividing cells is constant. In dividing cells it goes through a regular cycle that is coordinated with the stages of division. At anaphase, when the chromosomes move to the poles, the amount of DNA in each daughter nucleus is one half that of the original mother cell. During interphase, DNA is again synthesized, and the normal amount is soon attained (Fig. 5.6).

The amount of DNA per human cell is approximately 5.6×10^{-12} gram, or about two hundred billionths (2×10^{-11}) of an ounce. This is the amount contained in the fertilized egg that developed into you, and that was responsible for determining that you are what you are. This remarkable molecule somehow has built into itself the exceedingly complex store of information needed to direct the growth and activities of the simplest and

Figure 5.6 This diagram shows the sequence of cellular events and their relation to the amount of DNA per cell. The amount of DNA in an egg or sperm equals C. When these cells unite at the time of fertilization and their nuclei fuse, the DNA value is 2C. (In a human cell, the 23 chromosomes equal C; the 46 chromosomes, 2C.) When a cell prepares to divide, the DNA is doubled to a 4C value at interphase, and then reduced to a 2C value at anaphase, when the chromosomes separate into their chromatids. Successive cell divisions continue the 4C–2C–4C–2C alternation of DNA change.

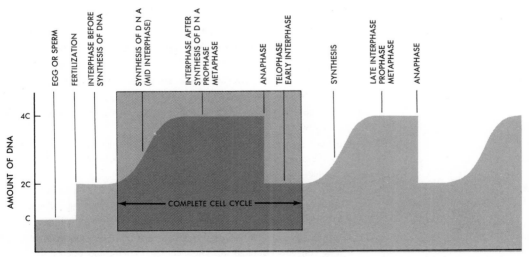

SEQUENCE OF CELLULAR EVENTS

most complex organisms. To understand how this is accomplished, we need to know something of the chemistry of DNA—what kind of a molecule it is, how it can store and transmit information, and how it can form more DNA. The next few chapters will consider the chemistry of cells, after which we shall return to DNA and its role in cellular behavior.

summary

Usually, cells consist of a single nucleus surrounded by cytoplasm. Through grafting experiments performed on the alga *Acetabularia,* it can be shown that the character of a cell is determined by the kind of nucleus it contains. It can also be shown that a particular species is characterized by a given number of chromosomes, and if this number is changed, the individual organism is usually abnormal in some way. The nucleus, therefore, is the control center of the cell, and the chromosomes are the key structures governing cellular morphology and function.

Chemical analyses of chromosomes show that they contain four kinds of macromolecules: deoxyribose nucleic acid (DNA), ribose nucleic acid (RNA), and two kinds of protein. Transformation studies in bacteria have shown that DNA is the crucial molecule governing heredity. DNA is also stable in amount from one cell to another in the same organism, as we would expect of a controlling element. We can, therefore, trace the control of cellular characteristics to the nucleus, then to the chromosomes, and finally to the DNA molecule within the chromosome.

for thought and discussion

1 What advantage is there in having only one nucleus per cell? Why not have an indefinite number?

2 What does the expression "the molecular basis of heredity" mean to you?

3 Why do you think stability in the amount of DNA is a necessary condition for exercising cellular control?

4 Why do you think it important that we know the chemical basis of heredity?

5 It was once thought that our destiny, as individuals, was controlled by the stars. How would you classify such a belief? How does science help to alter our thought structure?

6 DNA extracted from a human cell has not been shown to be capable of transforming another human cell, even in cell cultures. What do you see as difficulties in this kind of experiment?

7 Without knowing in advance the structure of DNA as a molecule, how can you visualize this chemical as a controlling agent? Can you think of any analogies that would help you in your thinking?

ASIMOV, I. *The Genetic Code.* New York: New American Library, Inc. (Signet Science Library), 1962. An elementary discussion of how decoding the gene has aided us in understanding the meaning of inheritance.

BARRY, J. M. and E. M. BARRY. *Molecular Biology: An Introduction to Chemical Genetics.* Englewood Cliffs, N.J.: Prentice-Hall, Inc., 1973. An excellent discussion of how we have come to understand the gene from Mendel to today.

MADDOX, J. *Revolution in Biology.* New York: Macmillan Publishing Co., Inc., 1964. A nontechnical discussion of how our understanding of the genetic code has changed the entire field of biology.

WATSON, J. D. *Molecular Biology of the Gene.* Menlo Park, Calif.: W. A. Benjamin, Inc., 1965. The most up-to-date treatment of the structure and function of genes.

atoms
and molecules
chapter 6

Nature can be defined as "matter in motion," and includes living and nonliving things. In view of the highly organized nature of cells, we can define living things as organized systems (of matter) in motion. One attribute of all organisms, plant and animal alike, is that life impresses special kinds of organization on all the matter that becomes part of a living system. As atoms and molecules are brought into a cell, they enter into organized chemical reactions and become part of an organized structure.

To remain alive, grow, and reproduce, all living things, from microorganisms to man, require certain substances and undergo certain chemical changes. A cell, therefore, must be able to take up certain substances from the surrounding fluid. It must also contain the necessary machinery for making use of the substances in one way or another. Some of the absorbed **nutrients,** as these substances are called, become part of the cell itself. But this building process requires energy. The cell obtains energy by breaking down the molecules it takes in, or by manufacturing new molecules and then breaking them down.

Basically, the primary purpose of the machinery of a cell is to convert nutrients into useful energy, and at the same time to make new chemical compounds which the cell needs for a variety of functions. We speak of these chemical transformations as **metabolism.** For the past 75 years research into the metabolism

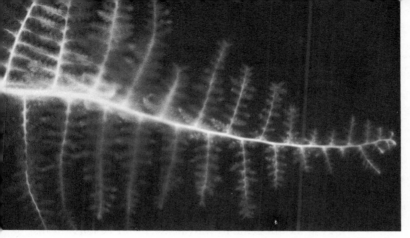

Courtesy Brookhaven National Laboratory

Figure 6.1 Radioactive elements can be used to trace the distribution of nutrients taken up by plants. In this case sulfur-35 is fed to a fern frond. The sulfur accompanies the food to various parts of the plant. The frond is then placed against a photographic film. The radiation emitted by the radioactive sulfur leaves light traces against the dark outline of the frond.

of cells has been a primary activity in the field of biochemistry. One of our purposes of including chemistry chapters in this book is to pave the way for a detailed discussion of cell metabolism later. Our task now, however, is to examine the composition and structure of a variety of important substances of biological origin.

All matter in the universe is made up of specific combinations of a limited number of substances called **elements.** Today we know of 103 elements—among them oxygen, gold, hydrogen, uranium, and so on. An element is matter composed of identical atoms. We classify the atoms of different elements by the differences in atomic mass, that is, the quantity of matter in, say, a hydrogen atom compared with the quantity of matter in a uranium atom. The smallest amount of an element that can be involved in a chemical change is an atom. But there are many subatomic particles, building blocks of the atoms themselves. More than 30 have been discovered so far; however, only three need concern us in this book.

THE STRUCTURE OF ATOMS Atoms are made up of three primary particles: **electrons, protons,** and **neutrons.** The electron was the first particle to be identified as a constituent of all atoms; it is a negatively charged particle that moves about a positively charged central mass, called the **nucleus.** (The term "nucleus" is the Latin word for "nut" or "kernel." Just as a nucleus is the central part of a cell, so is a nucleus the central part of an atom.) The nucleus is usually composed of two different kinds of particles—**protons** and **neutrons.** A proton is a positively charged particle with a mass 1,845 times that of the electron. Neutrons have about the same mass as protons, but are neutral; that is, they do not have an electric charge. Except in certain cases, which we shall discuss later, atoms are electrically neutral. This means that each must contain an equal number of negative electrons and positive protons. The electrons are the particles directly involved in chemical reactions.

The hydrogen atom, which has one electron and one proton, is the simplest of all the atoms. The proton accounts for most of the mass of a hydrogen atom. As a unit of measure, we use the mass of the proton (or neutron), calling it **one mass unit,** or **one atomic mass unit.** Accordingly, the weight of hydrogen is approximately 1 because of the single proton forming its nucleus. The electron mass is so small (1/1,845 of a proton) that we consider its mass to be negligible. Helium, the next largest atom, has a nucleus composed of two protons and two neutrons (giving it an atomic mass of 4) plus two electrons. As we work our way up the atomic scale toward heavier and heavier atoms, we find some, such as uranium, with dozens of protons. One form of uranium, in fact, has 92 protons and 146 neutrons, hence an atomic weight of 238.

In general, electrons travel in specific **orbitals,** or shells, around the nucleus of an atom. Although an atom can have a large number of orbitals, all atoms seem to have at least seven in which electrons can be located. They are called the **K, L, M, N, O, P,** and **Q** levels. There is a maximum number of electrons that each orbital can accommodate. This does not mean, however, that an electron is bound to any one orbital. The electrons circling the nucleus of an atom are free to jump back and forth from one orbital, or energy level, to another.

However, energy uptake or release is necessary for this movement. For example, when an intense high-energy ultraviolet light ("black light") is directed at certain chemicals, the electrons in the outer energy orbital are "pushed" into a higher energy orbital for a brief period of time. This is usually an unstable situation and subsequently the electrons drop back spontaneously to the

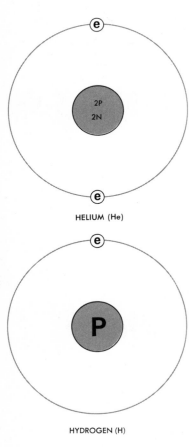

HELIUM (He)

HYDROGEN (H)

Figure 6.2 Diagrammatic sketch of an atom (right). This is not a picture of the atom but rather a working model. The electrons move about the dense nucleus at varying distances and at high velocity so that one speaks of a "cloud" or "haze" of electrons about the nucleus.

The representations of a hydrogen and a helium atom are shown above.

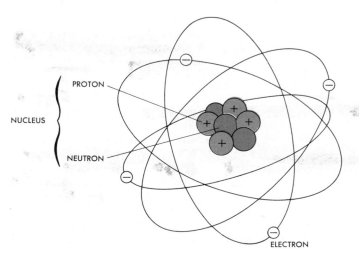

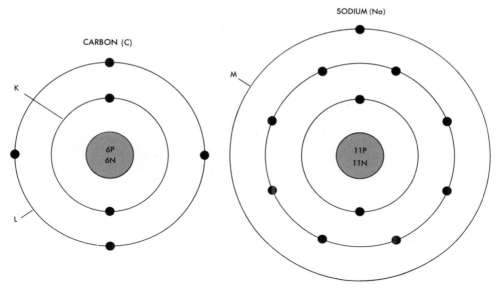

Figure 6.3 A sodium and a carbon atom are represented here. Notice the distribution of electrons at different energy levels.

lower energy level. Energy is released in the process and may appear as visible light. The shining of safety badges and of a television screen are examples of this process. We shall discuss how light can move electrons to higher energy levels when we consider how green plants convert sunlight into useful chemical energy.

The distribution of electrons in the various orbitals gives each element its particular chemical property. The electron distribution of several atoms is shown in Table 6.1. Notice that the outermost orbital, or energy level, of an atom never contains more than eight electrons. This is the stable electron pattern; atoms containing this arrangement are usually inactive, or **inert.** Neon and argon are two examples of inert gaseous elements.

HOW ATOMS COMBINE The number of electrons in the outermost energy level of an atom is what determines the ability of one atom to combine with another and form a molecule. These outer orbital electrons are called **valence** electrons. Understanding how atoms combine and are held together by a **chemical bond** is important to an understanding of the chemical changes that take place in living organisms. When two atoms collide or come close to one another, their electron clouds may overlap. The overlapping may lead to a sharing of electrons by the two atoms, or it may lead to an

TABLE 6.1. Energy Levels and Orbitals

Atomic Number	Element	Energy Levels						
		K	L	M	N	O	P	Q
1	Hydrogen	1						
2	Helium	2						
3	Lithium	2	1					
4	Beryllium	2	2					
5	Boron	2	3					
6	Carbon	2	4					
7	Nitrogen	2	5					
8	Oxygen	2	6					
9	Fluorine	2	7					
10	Neon	2	8					
11	Sodium	2	8	1				
12	Magnesium	2	8	2				
13	Aluminum	2	8	3				
14	Silicon	2	8	4				
15	Phosphorus	2	8	5				
16	Sulfur	2	8	6				
17	Chlorine	2	8	7				
18	Argon	2	8	8				
19	Potassium	2	8	8	1			
20	Calcium	2	8	8	2			
21	Scandium	2	8	9	2			
22	Titanium	2	8	10	2			
23	Vanadium	2	8	11	2			
24	Chromium	2	8	13	1			
25	Manganese	2	8	13	2			
26	Iron	2	8	14	2			
27	Cobalt	2	8	15	2			
28	Nickel	2	8	16	2			
29	Copper	2	8	18	1			
30	Zinc	2	8	18	2			
36	Krypton	2	8	18	8			
47	Silver	2	8	18	18	1		
53	Iodine	2	8	18	18	7		
56	Barium	2	8	18	18	8	2	
79	Gold	2	8	18	32	18	1	
92	Uranium	2	8	18	32	21	9	2

actual transfer of an electron from one atom to the other. If an electron is transferred, the atom losing it will be less negatively charged; that is, it becomes positive. The atom gaining the electron will be more negatively charged. This creates a situation in which the two atoms are held together because of their opposite charge. In such cases, **electromagnetic attraction** is the chemical bond. Thus, the chemical bond is an energy relationship between atoms; so in any chemical reaction we can expect energy changes. When a chemical bond is broken, potential energy is converted into **kinetic** energy, or energy of motion.

Gases such as neon and argon are chemically inert because they have eight electrons in their outermost shell. Atoms that are **reactive,** or combine with other atoms, do so because they do not have a full stable of eight outer-level electrons. In general, atoms with fewer than four electrons in the outer orbital tend to give up electrons; those with more than four tend to gain electrons.

Let us see how this works by examining a common kitchen chemical—sodium chloride, or table salt. During the process in which sodium (Na) and chlorine (Cl) atoms combine and form the compound sodium chloride (NaCl), one electron seems to be transferred from the sodium to the chlorine atom (Fig. 6.4). Sodium has 11 protons and 12 neutrons in its nucleus, giving it an atomic mass of 23. Because of its 11 protons, it has 11 electrons in the various orbitals. Two of these electrons are at the K energy level, eight at the L energy level, and one in the M. The nucleus of chlorine, on the other hand, has 17 protons and 18 neutrons, giving it an atomic mass of 35. Two electrons are at the K level, eight at the L level, and seven at the M level. Neither atom is stable (eight electrons in the outer orbital are required for stability or chemical inertness).

When sodium and chlorine react, the sodium atom acquires a net positive charge by losing an electron; it is designated Na^+. The chlorine atom acquires a net negative charge by gaining an

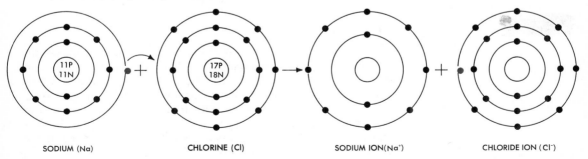

Figure 6.4 This diagram shows the formation of sodium and chloride ions by the transfer of an electron from sodium to chlorine.

SODIUM (Na) CHLORINE (Cl) SODIUM ION(Na⁺) CHLORIDE ION (Cl⁻)

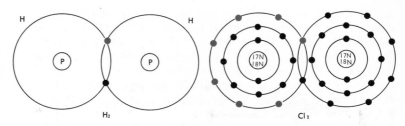

Figure 6.5 A shared pair of electrons is called a *covalent* bond. Atoms of a number of elements form covalent bonds. Hydrogen and chlorine are shown here.

electron and is designated Cl^- (Fig. 6.4). Both atoms are now **ions.** An ion is an atom with an unbalanced electrostatic charge; or, putting it another way, an ion is a charged atom. For this reason, we say that sodium and chlorine are joined by **ionic** bonding. A crystal of NaCl is held together, therefore, by the electrostatic forces that act between oppositely charged ions.

Another type of bonding between atoms is one in which electrons are not transferred, but shared. In effect, a shared electron fills an orbital for each atom, giving stability to each atom by completing an electron pair. A shared pair of electrons is called a **covalent** bond. Atoms of a number of elements form covalent bonds, and can also occur as two-atom, or **diatomic,** molecules. Hydrogen and chlorine gases are good examples. The two nuclei of chlorine atoms are held together because both electrons of the electron pair are attracted to both positively charged nuclei. Knowing that chlorine exists as a diatomic molecule, and that sodium does not, we can now write a chemical equation showing that the reaction of chlorine with sodium requires two molecules of sodium, but only one of chlorine gas:

$$2Na + Cl_2 \rightarrow 2Na^+ + 2Cl^- \rightarrow 2NaCl$$

A number of elements of biological interest combine by forming covalent bonds. Oxygen and hydrogen are two. For all organisms, the most important compound these two elements form is water. In the next chapter we shall take a detailed look at this substance and other compounds that are of particular importance to life processes.

summary

All matter in the universe is made up of specific combinations of substances called **elements,** such as hydrogen, oxygen, and nitrogen. The basic unit of an element is called an **atom.** Atoms are composed of three primary particles: (1) heavy positively charged **protons;** (2) **neutrons,** which are electrically neutral and which are packed with the protons in the nucleus; and (3) **electrons,** negatively charged particles, which surround and circle the nucleus in definite energy levels (orbits).

The number of electrons in a given atom always equals the number of protons in the nucleus, so that the atom is electrically neutral. Atoms combine to form specific chemicals by sharing electrons **(covalent bond),** or by transferring electrons **(electromagnetic attraction,** or **ionic bonding).** When table salt, NaCl, is dissolved in water, the sodium and chlorine exist as ions in solution (Na^+, Cl^-).

for thought and discussion

1   Look at the table of elements and determine the number of electrons, protons, and neutrons in some of the common elements that are familiar to you.

2   How many of the following terms do you understand?

atoms	nucleus	ionic bonds
electrons	atomic mass	covalent bonds
neutrons	orbitals	electromagnetic attraction
protons	valence electrons	ions

selected readings

BAKER, J. J. W., and G. E. ALLEN. *Matter, Energy and Life.* Reading, Mass.: Addison-Wesley Publishing Company Inc., 1965. An excellent small paperback written to provide students with a background in the chemistry and physics essential to understand modern biology.

BUSH, G. L. and A. A. SILUIDI. *The Atom: A Simplified Description.* New York: Barnes & Noble, Publishers, 1961. An elementary description of atom structure.

DRUMMOND, A. H. *Atoms, Crystals and Molecules* (Part 2). Columbus, Ohio: Xerox Education Publications, 1964.

WHITE, E. *Chemical Background for the Biological Sciences* (2nd ed.). Englewood Cliffs, N.J.: Prentice-Hall, 1970. A general elementary discussion of the chemistry of compounds of biological origin.

the chemistry
of
biological
compounds
chapter 7

Life as we know it on this planet is intimately associated with a water environment. Even those organisms that live in deserts are made up of large quantities of water. It is only in this aqueous environment that cells can function and maintain normal life. When cells or organisms lose too much water, all their life processes cease. Yet certain cells, particularly among the microorganisms, can be completely dehydrated and later restored to activity by submersion in water. To gain some appreciation of the importance of water to metabolic function, let us now look into the structure and properties of this most important of all biological compounds.

WHAT IS WATER? The union of two hydrogen atoms and one oxygen atom produces the remarkable molecule we call water. As you saw in the previous chapter, a hydrogen atom consists of one proton and one electron. An oxygen atom, on the other hand, has eight protons and eight neutrons in its nucleus. This makes it 16 times as massive as a hydrogen atom. The eight protons give the nucleus of the atom a positive charge of eight. An oxygen atom also has eight electrons, two of which are in the K orbital close to the nucleus, while the remaining six are in the outer (L) orbital. As we said earlier, atoms with more than four electrons in the outer orbital are electron gainers. Oxygen atoms tend to gain two electrons from some other source, thus filling the outer orbital (to a total of eight electrons) and becoming a stable electronic structure. One source of two electrons is two hydrogen atoms. When one oxygen atom and two hydrogen atoms combine by covalent bonding, a stable molecular union results. So much for the composition of a water molecule. What about its structure?

Because of their like (positive) charges, the two hydrogen nuclei tend to repel each other. This results in their taking the position shown in Fig. 7.2. The fact that the electrons of hydrogen are shared with the oxygen atom in a covalent bond means that each hydrogen atom electron no longer spends as much time on one side of its nucleus as on the other side (see Fig. 7.2). As a consequence, the hydrogen "end" of a water molecule has a positive charge, while the oxygen "end" has a negative charge. Such a molecule is called a **polar** molecule.

The polarity of a water molecule makes the molecule an electron seeker. It will attach itself to any other atom or molecule having an available electron. This results in a **hydrogen** bond, a very important bond in biochemical reactions. Thus the hydrogen bond results from the tendency of a hydrogen atom to share electrons with two other atoms, usually oxygen atoms—for example, with other water molecules. Formic acid is another molecule that tends to form hydrogen bonds with itself (Fig. 7.4).

Figure 7.1 To function and maintain normal life, the cells of all organisms require a water environment. How does the desert cactus plant obtain water?

Figure 7.2 The formation and geometry of the water molecule.

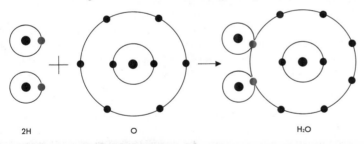

2H   O   H₂O

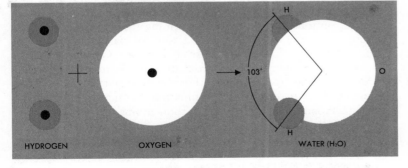

HYDROGEN   OXYGEN   WATER (H₂O)

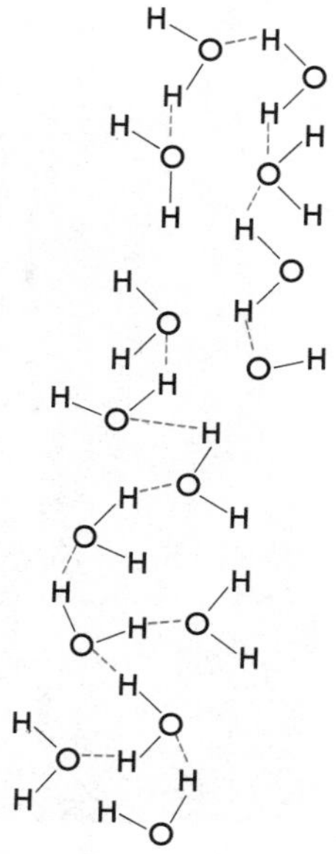

Figure 7.3 Hydrogen bonding between water molecules is indicated by the colored lines. The tendency of hydrogen atoms in one water molecule to combine with oxygen in a second water molecule gives water an organized structure.

Although hydrogen bonds are weak compared with covalent bonds, they are strong enough to give rise to some of the unique properties of water. Water, for example, is the only substance that is commonly present in all three states, as a solid, liquid, and gas, in the range of temperatures found at the Earth's surface. It is true that various other atmospheric gases can also be converted into a liquid or solid, but only in the laboratory where we can make use of extreme pressures or temperatures.

If we compare the boiling and freezing points of water with other chemicals similar to water, the variations we see in other compounds are surprising. For example, hydrogen sulfide (H_2S) has a freezing point at $-83°C$ and a boiling point at $-62°C$. Water, on the other hand, has a freezing temperature that is 83 centigrade degrees higher ($0°C$), and a boiling temperature that is $162°$ higher ($100°C$). This difference is accounted for if we consider the relatively greater thermal energy required to break the hydrogen bonds that join water molecules. Thus water molecules tend to attract one another into an organized structure because of the hydrogen bonding. Energy is required to break these water molecules apart before the individual molecules can escape from solution as a vapor. This is what we mean by "boiling"—the escape of individual water molecules.

The great stability in the water molecule structure was important for the origin and evolution of life as we know it on this planet. At room temperature, the thermal energy in a water molecule is so great that hydrogen bonds between molecules are being broken continuously and are reforming continuously. When water flows, we can visualize the molecules as tumbling over each other, breaking and reestablishing hydrogen bonds as they go. When the thermal energy of water molecules drops below the level necessary to break these bonds, ice forms. Later we shall return to the role water and hydrogen bonds play in relation to the function of enzymes.

CARBON COMPOUNDS The carbon atom is of special interest to students of chemistry, biochemistry, and biology. It plays a key role in the structure of molecules that are essential to living things. A carbon atom has six protons and six neutrons forming the nucleus, giving the atom a mass of 12. It also has a total of six electrons, two at the K level and four at the L level (Fig. 7.5). Carbon is neither strongly electronegative (electron attracting), nor strongly electropositive (electron repelling). For the most part, therefore, carbon enters into chemical combination by sharing electrons—covalent bonding—with other carbon atoms, or with atoms of other elements.

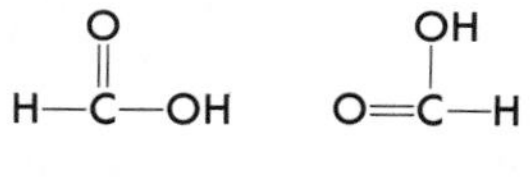

From a biological standpoint, the most outstanding property of carbon is the fact that it can share electrons with other carbon atoms and form long, straight or branched chains to which atoms

Figure 7.4 Formic acid is another molecule tending to form hydrogen bonds with itself.

94

of other elements can attach themselves. There are so many different carbon compounds (more than half a million) that a special branch of chemistry—**organic** chemistry—is devoted to their study.

When carbon atoms combine and form a chain, and when all the leftover electrons are shared with hydrogen atoms, substances known as **hydrocarbons** are formed. Figure 7.6 shows some of the simplest hydrocarbons: methane, CH_4; ethane, C_2H_6; and propane, C_3H_8. Notice that the electronic structure is shown for methane. The open dots represent the carbon electrons originally in the L orbital; the solid dots represent the hydrogen electrons in the K orbital. Neither atom has *lost* its electrons. Instead, they *share* their electrons with each other, and the result is a stable configuration of eight electrons in the outer orbital.

The shared electrons form a moving cloud and may circle the nucleus of any atom in the molecule. This makes a very stable covalent link between carbon and hydrogen. Since hydrogen has only one electron to share, and carbon needs four to complete the stable configuration, it should be clear why four hydrogens combine with one carbon, forming methane. Because carbon will accept four electrons from another source, and because it can share its own four electrons with another atom, we say that carbon has a valence of four.

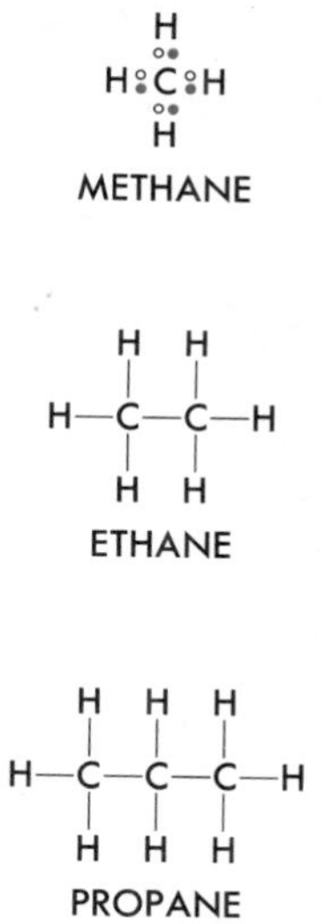

Earlier you saw that two hydrogen atoms each contribute one electron when they combine with oxygen and form water. Oxygen, then, has a valence of two. Oxygen also combines with carbon, forming carbon dioxide (CO_2). However, in this case two oxygen atoms are involved, each sharing a pair of electrons with the carbon atom. This type of covalent bond is called a **double bond,** and is represented by the double bar in Fig. 7.7. Both the electronic and bond formula's are shown. Note that the sharing of four electrons by each oxygen atom and the carbon atom helps complete the stable electronic configuration of both atoms.

As in the case of carbon dioxide, one carbon atom can form a double bond with another carbon atom, and the remaining valence electron can combine with hydrogen to form the **unsaturated** hydrocarbons. They are called "unsaturated" because the *carbon* atoms are sharing the electrons, rather than some other atom, such as hydrogen. Some examples are shown in Fig. 7.8.

IMPORTANT CHEMICAL GROUPS   In addition to a straight-line arrangement, carbon atoms can link together and form ring structures. Because many of these **ring compounds** have rather fragrant odors, they are called **aromatic.** The parent compound of all these aromatic substances is **benzene,** the structure of which is shown in Fig. 7.10. Many important biological compounds have a ring structure.

If we replace one of the hydrogen atoms of a hydrocarbon

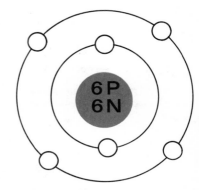

Figure 7.5 Schematic representation of the carbon atom.

H
H ꞉C꞉ H
H
METHANE

H   H
| |
H—C—C—H
| |
H   H
ETHANE

H   H   H
| | |
H—C—C—C—H
| | |
H   H   H
PROPANE

Figure 7.6 Saturated hydrocarbons.

Figure 7.7 The atoms forming carbon dioxide are held together by a type of covalent bond called a *double bond.*

O꞉꞉C꞉꞉O   O=C=O

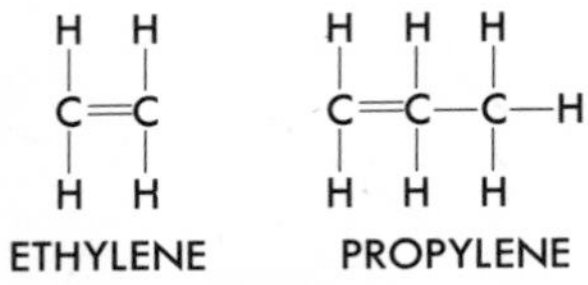

Figure 7.8 Unsaturated hydrocarbons are shown above.

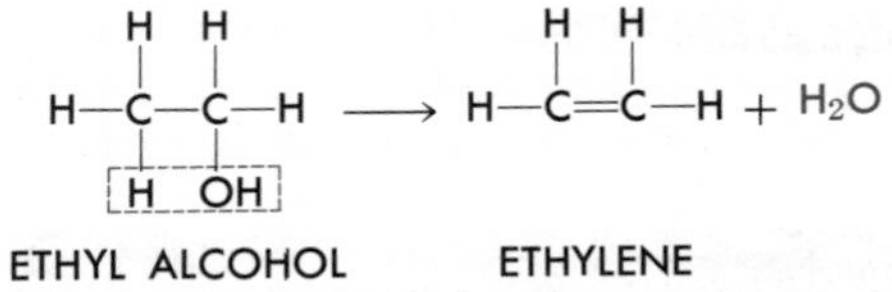

Figure 7.9 The removal of water from alcohol results in the formation of a double bond in the ethylene.

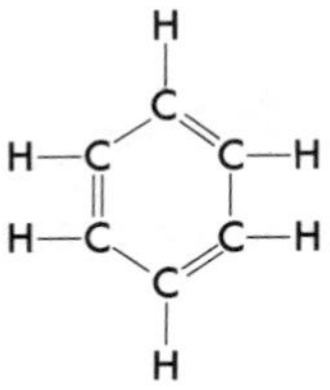

Figure 7.10 The benzene ring is shown above.

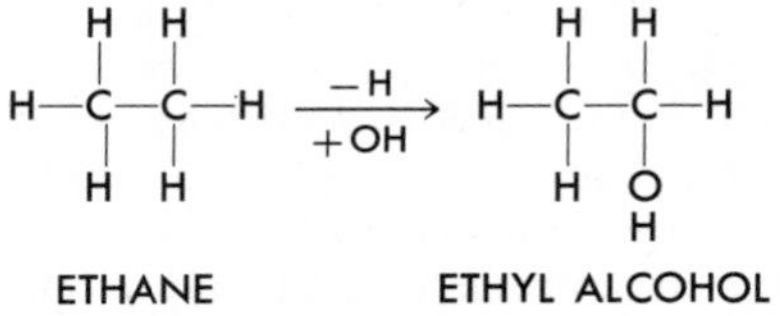

Figure 7.11 The formation of an alcohol can be brought about by adding an OH (hydroxyl) group to a hydrocarbon.

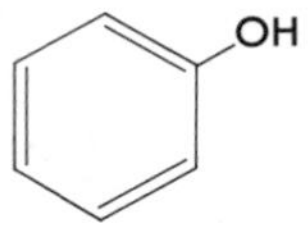

Figure 7.12 Phenol.

with a unit of oxygen and hydrogen (OH), called a **hydroxyl group,** we produce a compound belonging to the class of **alcohols.** If the hydroxyl group replaces a hydrogen of an aromatic ring, the compound formed is known as a **phenol.** The chemistry of alcohols and phenols is primarily that involving the properties of a hydroxyl group attached to a carbon atom.

When an alcohol such as ethyl alcohol is **oxidized** (Fig. 7.13), that is, when it loses electrons, it is converted to an **aldehyde.** In the example illustrated, it is **acet**aldehyde. Compounds containing a **carbonyl** group (C=O) are known as aldehydes or **ketones.** If one of the valences of the carbonyl compound is used in bonding with a hydrogen atom, the compounds are aldehydes. The remaining valences may bond with the hydrogen atom or a number of other groups. If one is a **methyl** group (CH$_3$), as in Fig. 7.13, then the compound becomes acetaldehyde. If both valences of the carbonyl carbon are used to bind other carbon atoms, then a ketone is formed. **Acetone** is a good example.

The process of **oxidation and reduction** is very important in biological systems. It is the primary mechanism for the liberation of useful energy and is essential for the synthesis of cellular com-

Figure 7.13 The oxidation (removal of hydrogen) of an alcohol produces an aldehyde.

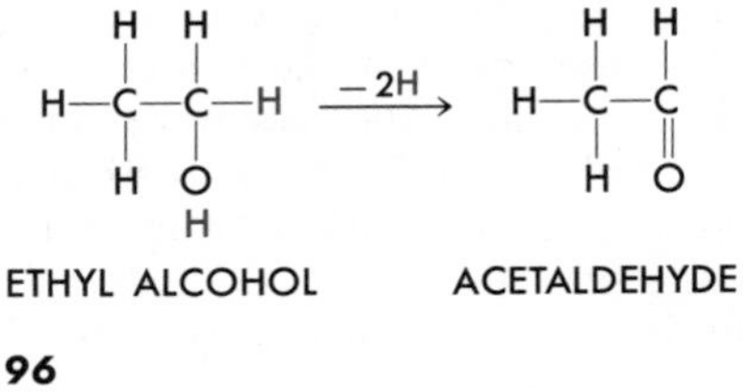

96

ponents. We shall consider some of these reactions in detail later, but for the moment it is essential that we understand the general process. As illustrated in Fig. 7.13, oxidation can be defined as the removal of hydrogen (and reduction the addition of hydrogen). The removal of hydrogen implies the removal of both a proton (H^+) and an electron (e^-) at the same time. In some cases oxidation (and reduction) proceeds in two steps. First, there is the removal of the proton, followed by the removal of an electron, as illustrated in the following reaction:

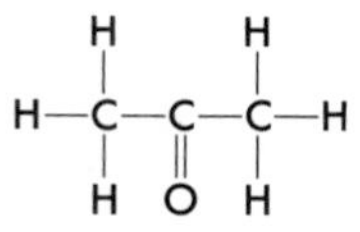

$$AH \rightarrow H^+ + A^-$$
$$A^- \rightarrow A + e^- \quad \text{(oxidation)}$$

The removal of the electron in this case is called the **oxidative step.**

Another group of organic substances of biological importance is comprised of those compounds containing the **carboxyl** group. This is a name coming from "**carb**onyl" and "hyd**roxyl**." The structure of a carboxyl group is shown in Fig. 7.15. Compounds of this general class are known as **carboxylic acids.** In the ordinary straight chain series, those compounds containing the carboxyl group are known as **fatty acids,** two of which are shown in Fig. 7.15. There are many other general classes of acids that contain the carboxyl group.

CHEMICAL REACTIONS   When a substance such as hydrochloric acid (HCl) is dissolved in water, the hydrogen separates from the chlorine. When this happens, however, the electron normally associated with the hydrogen atom remains with the chlorine. This dissociation in water, therefore, leads to the formation of ions—a hydrogen ion (a proton with a net charge of $+1$), and a chloride ion with a net charge of -1. The ionization of HCl in water can be written as follows:

$$HCl \rightleftharpoons H^+ + Cl^-$$

Normally, H^+ and Cl^- would immediately recombine and form HCl again; however, in water this charge separation is partly maintained by water molecules located between the two

Figure 7.14 Acetone.

Figure 7.15 The R group shown attached to the carbon atom of the carboxyl group represents any number of groups. If R=H, we have the structure of formic acid. If R=CH$_3$, we have acetic acid.

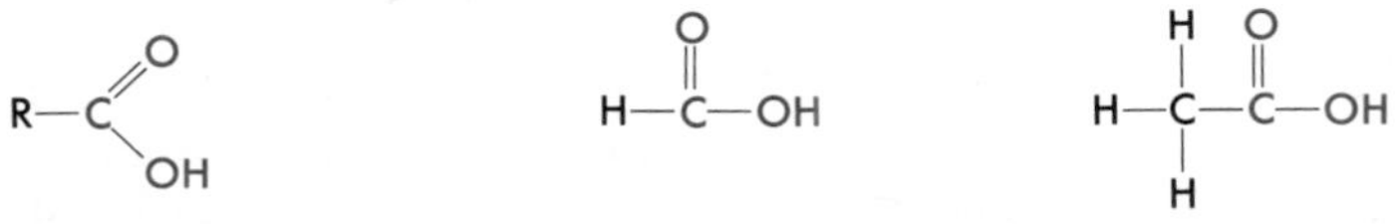

CARBOXYL GROUP   FORMIC ACID   ACETIC ACID

97

ions. Not all compounds ionize when they are dissolved in water. Molecules held together by ionic bonds, such as NaCl, tend to dissociate. Those held together by covalent bonds do not. In the case of the ionic bond, one atom has *given up* an electron to another, so the only force holding the two atoms together is an electromagnetic attraction, which water can overcome. In covalent bonds, the electrons in the outer energy levels are *shared* by the atoms involved; this means that more energy is needed to separate them.

We discuss this problem of charge on ions because of its importance to the combination of atoms in chemical reactions. In all chemical reactions, atoms combine in very definite proportions. Knowing whether an atom tends to give up or accept electrons, and knowing how many, helps us to predict the nature of the reaction. For example, NaCl dissociates into Na^+ and Cl^-. When calcium reacts with chlorine, it forms calcium chloride ($CaCl_2$). Calcium has two electrons in the N energy level, and chlorine needs only one electron to complete the octet at the M level. Thus for each calcium atom we need two chlorine atoms, as follows:

$$Ca + Cl_2 \rightarrow CaCl_2$$

If we then add water to the calcium chloride, the $CaCl_2$ ionizes and forms two chloride ions for each calcium ion. The calcium ion carries a $+2$ net charge:

$$CaCl_2 \rightarrow Ca^{2+} + 2Cl^-$$

OXIDATION–REDUCTION When an atom loses an electron, we say that it is **oxidized;** when an atom gains an electron, we say it is **reduced.** We emphasize electron transfer at this time for the following reason: *when electrons move to different energy levels (orbitals), there is a release of energy that can be used by organisms.* When sodium and chlorine combine as sodium chloride, there is a transfer of an electron from the sodium to the chlorine. The sodium is oxidized; the chlorine is reduced. Another way of saying the same thing is that sodium is a **reducing agent,** and chlorine is an **oxidizing agent.** When oxidation-reduction reactions occur in a cell, we need to know if *useful* energy is released, and, if so, whether it is available for use by the cell. As we shall see later, the oxidation of carbohydrates, fats, and other foodstuffs is the primary source of energy for cell function.

ACIDS AND BASES To describe chemical reactions that take place in cells and organisms, we must know something about the chemical nature of **acids** and **bases.** There are many types of acidic and basic substances in organisms. Some play major roles in cell structure; others are important in metabolism. Acids are substances that can donate a proton (hydrogen ion). The concentration of the hydrogen ions determines the degree of acidity of a solution. Bases are substances that combine with hydrogen ions. In aqueous solution, for example, hydrochloric acid dissociates into hydrogen ions and chloride ions. The chloride ions comprise a base (because they can combine with hydrogen ions).

$$HCl \rightleftharpoons H^+ + Cl^-$$

Acid   Hydrogen   Chloride ion
        ion   (a base)

Sodium hydroxide (NaOH), on the other hand, is a base. It dissociates into a sodium ion (Na^+) and a base (OH^-).

$$NaOH \rightarrow Na^+ + OH^-$$

Base   Sodium   Hydrogen ion
        base   (a base)

Consequently, NaOh can neutralize the acid HCl, forming water and NaCl.

$$HCl + NaOH \rightarrow H_2O + NaCl$$

The carboxyl groups of fatty acids, such as acetic acid, can also donate protons.

$$CH_3-C\overset{O}{\underset{OH}{\big<}} \rightleftharpoons H^+ + CH_3-C\overset{O}{\underset{O-}{\big<}}$$

Acetic acid   Acetate ion
                   (a base)

Notice that when an acetate ion takes up a proton it is converted to an acid. Thus, when a base accepts a proton, it is converted into a molecule that acts as an acid. Hydrogen ion concentration, then, is very important in determining the net charge on molecules inside cells. The negative and positive charges on very large molecules are important in determining their biological reactivity.

CONCENTRATION OF H⁺   One way of expressing the hydrogen ion concentration of a substance is to speak of the **pH** of the substance. On the pH scale, 7 is the neutral point. Any substance with a pH of 7 is neither acid nor base; it is neutral. Water, for example, is a substance that is neutral, so its pH value is 7. A small amount of HCl in water, on the other hand, has a much higher concentration of hydrogen ions; its pH value is less than 7.0. NaOH in water gives

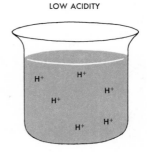

LOW ACIDITY

HIGH ACIDITY

Figure 7.16 The hydrogen ion concentration determines the acidity of a solution: high concentration produces high acidity, low concentration, low acidity.

pH values much greater than 7.0. (The definition of pH is "the negative log of the hydrogen ion concentration." Thus, if the hydrogen ion concentration is 10^{-3} molar, the pH is 3; if the concentration is 10^{-8} molar, the pH is 8, and so on.)

acidic   basic
←————————————————Neutral————————————————→
0   1   2   3   4   5   6   7   8   9   10   11   12   13   14

The pH is very important in cell metabolism. Big changes in the acidity of the cells of an organism can greatly affect metabolism. The pH of blood in mammals, for example, is kept very close to 7.35. A shift as small as 0.2 can result in death. Organisms have evolved chemical systems called **buffers,** which tend to keep the pH relatively constant. Buffer systems resist changes in pH when acids or bases are added. There are many buffers in the human body, but in principle they all work the same way. We can use acetic acid (CH_3COOH) as an example. When the acid is dissolved in water, it dissociates into the negative acetate ion and hydrogen ion to give a pH of about 4.0.

$$CH_3\overset{\displaystyle O}{\overset{\|}{C}}\!\!-\!\!OH \rightleftharpoons CH_3\!\!-\!\!\overset{\displaystyle O}{\overset{\|}{C}}\!\!-\!\!O- \; + \; H^+$$

If we add a small amount of base, such as NaOH, to this acetic acid solution, OH^- ions neutralize the H^+ and water is formed; however, additional acetic acid will dissociate and contribute more H^+ to the solution, making it acid again. In other words, acetic acid acts as a reservoir for H^+ so that the pH of the solution tends to remain constant. The buffer role of acetic acid is shown in Fig. 7.17.

You have now seen how certain compounds vital to cells are built up from elements. In Chapter 8 we turn our attention to several major compounds present in cells.

summary

Life on this planet is intimately associated with a water environment. Water molecules tend to associate with each other because of their **hydrogen bonding** properties. The tendency of hydrogen in one water molecule to associate with oxygen in a second water molecule leads to the formation of a structure involving multiple combinations of water molecules. Energy is required to break these water molecules away from one another, which suggests why water has a high boiling point.

The carbon atom is of special interest because it is part of so many compounds that are essential to life. Carbon atoms can combine with each other and form long-chain hydrocarbons. They can also form ring structures, such as benzene.

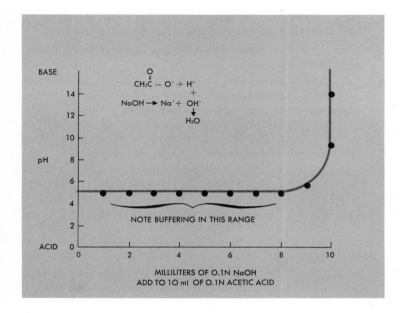

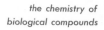

Figure 7.17 Titration of acetic acid with the base, sodium hydroxide.

There are a number of important chemical groups such as the **hydroxyl** (OH), **aldehyde**

$$\left(-C{\overset{\displaystyle O}{\underset{\displaystyle H}{}}} \right)$$

and **carboxyl**

$$\left(-C{\overset{\displaystyle O}{-OH}} \right)$$

groups. Substances that contain the carboxyl group are **acids** because they can dissociate, which enables them to contribute a hydrogen ion (H^+). The concentration of hydrogen ions determines the degree of acidity of a solution. **Bases** are substances that combine with hydrogen ions. We use **pH** to express the concentration of hydrogen ions, or acidity, of a solution:—pH 7.0 is neutral; values less than 7.0 are acid; those greater than 7.0 are alkaline, or basic.

1   When ice is thawing and forming liquid water can you visualize the state of the water molecules? Why can H_2O exist as a solid *and* a liquid? Remember hydrogen bonding and thermal energy.

*for thought
and discussion*

2   What are **oxidation** and **reduction?** Can you oxidize something without reducing something in an ordinary chemical reaction?

3   What is a **hydrocarbon, hydroxyl group, carbonyl group, acid, and base?** What is *pH?*

4   Can you think why the *pH*, or hydrogen ion concentration (H^+), is important in determining the electronic charge of chemical substances? Write down an example and explain.

suggested readings

See the books listed at the end of the previous chapter.

major
compounds
of cells
chapter 8

Carbohydrates, fats, proteins, and nucleic acids are the major large molecules that occur inside cells. These molecules, in turn, can associate with each other and make even larger cellular structures.

the role of carbohydrates

Carbohydrates play a key role in the energy requirements of organisms. They are the principal products formed when a green plant captures light energy in the process of **photosynthesis.** In this process, the energy of sunlight is used to convert CO_2 and H_2O to the energy-rich bonds of carbohydrates:

$$CO_2 + H_2O \xrightarrow[\text{energy}]{\text{light}} C(H_2O) + O_2$$

Carbon
dioxide   Water   Carbo-
                                hydrate

Sugars and starches are the principal sources of energy in the ordinary diet of most organisms. The carbohydrates, although used primarily as an energy source, also supply important carbon "skeletons"—carbon atoms linked together and associated with

Figure 8.1 The sunlight reaching these green plants underwater provides energy needed by the plants in the production of carbohydrates.

Richard F. Trump

hydrogen and oxygen—that are necessary in the manufacture of the basic components of the living material of all cells.

The basic units in carbohydrates are carbon, hydrogen, and oxygen. The term "carbohydrate" means *hydrate of carbon*. This name was used because carbohydrates include many compounds that contain atoms of hydrogen and oxygen in the same proportion as occur in water—two hydrogens to one oxygen. So a carbohydrate can be described by the general formula $C(H_2O)$. Carbohydrates range from relatively simple molecules called sugars to the complex molecules of starches and cellulose. The simplest class of sugars is the **monosaccharides.** They are further classified according to the lengths of their carbon chains: three-carbon sugars **(trioses);** four-carbon sugars **(tetroses);** five-carbon sugars **(pentoses);** six-carbon sugars **(hexoses);** and so on to the 10-carbon sugars (Fig. 8.2).

Glucose and **fructose** are the two hexoses that serve as the

Figure 8.2 Carbohydrates with different numbers of carbon atoms. Names of specific examples are given in parentheses.

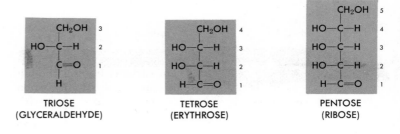

TRIOSE
(GLYCERALDEHYDE)

TETROSE
(ERYTHROSE)

PENTOSE
(RIBOSE)

principal source of energy and building material for most cells. Both can be represented by the same formula, $C_6H_{12}O_6$. Glucose contains five hydroxyl groups, each attached to a different carbon atom (Fig. 8.3). But many sugars can have the same chemical composition. The differences among them depend on the positioning of the hydrogen and hydroxyl groups around the carbon. Different groupings produce different chemical properties. These sugars of the same chemical composition, but different chemical properties, are called **isomers.** Cells are able to distinguish one isomer from another because of the different chemical groupings of the H and OH around the carbon atoms. When we talk about isomers, therefore, we must consider their molecular geometry.

Since cells are able to discriminate between isomers, because of the high degree of specificity of reaction, the empirical formula of a compound does not give us information of any great value. To understand cellular metabolism, we must know the shape of the compounds; thus we must consider molecular geometry when we discuss the reactions of all classes of compounds. (Note the cyclic form that ribose can take, as indicated below.)

The open-chain form of sugars occurs in aqueous solution but is in equilibrium with the ring form. The **ring structure** is the usual form that one encounters for longer-chain carbohydrates. The rings are formed from the chain form by the reaction of the hydroxyl group in the 4 or 5 position with the carbonyl group. The result, as indicated, is the formation of an oxygen bridge and a ring containing five to six elements.

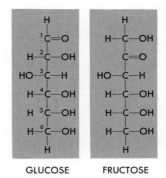

GLUCOSE   FRUCTOSE

Figure 8.3 Structure of the hexose sugars, glucose and fructose. The carbon atoms of glucose are numbered as shown. The positioning of the H and OH groups is important. For example, if we exchange the positions of H and OH groups on carbon 4 of glucose, we have the formula for the sugar galactose.

To show the correct **steric** (spatial) relations of the various groups, the ring structures are drawn as shown next for ribose and glucose. The ring is imagined as being looked at obliquely from above with the three thickened edges nearest to the observer.

Ribose

Glucose

105

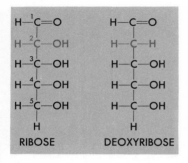

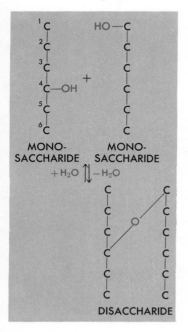

Figure 8.4 Ribose and deoxyribose. Notice that carbon atom number 2 of deoxyribose lacks an OH group. It has been replaced by an H.

Figure 8.5 The formation of a disaccharide from two monosaccharides by the loss of water. For simplicity, only the carbon skeleton and the two OH groups are shown.

Although we shall make use of straight-chain formulas for sugars at various times, it should be kept in mind that the ring structure and the stereochemical specificity of enzymes (biological catalysts) are very important in understanding biochemical processes.

Other important monosaccharides are the two pentoses, **ribose** and **deoxyribose.** They are components of the nucleic acids comprising RNA and DNA, which are discussed in Chapters 12 and 13. Figure 8.4 shows the difference between ribose and deoxyribose. Notice that carbon atom 2 of deoxyribose lacks an OH group. The OH has been replaced with a hydrogen.

Monosaccharides can be linked to form larger units. The **disaccharides,** those complex sugars containing two monosaccharides, are the most common. Disaccharides are formed by combining two monosaccharides. Energy is required to do this—an OH from one sugar combines with a H from a second sugar and forms water. This leads to the formation of a carbon–oxygen–carbon bond between the two sugars (Fig. 8.5). When two glucose units are linked through the carbons 1 and 4, the disaccharide **maltose** is formed.

One of the most important disaccharides is **sucrose,** or cane sugar. Sucrose is made up of equal quantities of glucose and fructose. Very complex molecules, such as starch and glycogen, are called **polysaccharides** because they are made up of a large number of monosaccharides joined in a long chain. **Starch** is the reserve carbohydrate in most plants and is formed by green plants in the process of photosynthesis. Starch is made up of glucose units linked in much the same way as maltose. **Glycogen** is the reserve carbohydrate of animals. It is found in high concentration in liver and muscles. Disaccharides and polysaccharides are readily broken down to monosaccharides by splitting the C—O—C bond with water.

Cellulose is another polysaccharide of great importance. It is the chief ingredient of the cell walls of plants. Cellulose is the major polysaccharide occurring in nature and is probably the most abundant organic compound found on our planet. It comprises at least 50 per cent of all the carbon in the plant world. Cellulose is the main constituent of cotton, wood pulp, linen, straw, and many other substances of plant origin.

Cellulose is insoluble in water and most organic solvents. Yet it is made up of glucose units linked together in very much the way maltose is, but with a slightly different configuration. These differences in shape are very important in stabilizing the molecule and making it resistant to biochemical attack by organisms. For example, the digestive systems of man and most animals do not contain the necessary chemical agents to break down cellulose. This is in contrast to their ability to break down compounds such as maltose, which has a very similar linkage. However, many microorganisms can decompose cellulose and use it as a food. Although termites eat wood, they cannot themselves break the

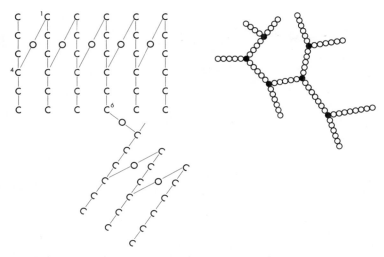

Figure 8.6 The glycogen molecule is made up of many units of glucose. In Fig. 8.3 the basic chemical unit (glucose) is shown. Glucose units are held together by a bond between the number 1 carbon of one glucose molecule and the number 4 carbon of a second glucose molecule. At intervals a glucose molecule is attached to the number 6 carbon. This leads to a branching point. Both arms of the branch grow by the addition of glucose making use of the 1–4 bond. A limited part of a glycogen molecule is shown schematically at the right in this diagram. The circles represent glucose molecules.

cellulose down, or digest it, as a source of nourishment. However, microorganisms living in the intestines of termites do the job of digesting. They are able to split the cellulose into the simple monosaccharides, which the termites can use. This is an example of a **symbiotic** relationship. The microorganisms live in a sheltered environment and in turn provide the host (termite) with sugar.

Thus, a very subtle difference in the chemical makeup of the compound can determine whether it can be metabolized by an organism, and hence be useful or useless as a source of food.

lipids

The **lipids,** or fats, are a group of organic substances originating in the living cell. They are classified into several subdivisions on the basis of their chemical and physical properties. All fats, such as butter, the fat on meat, olive oil, are built up from carbon, hydrogen, and oxygen atoms, and consist of two major components—fatty acids and **glycerol** (Fig. 8.7).

In general, animal fats are in a solid or semisolid state at room temperature; vegetable fats or oils are in a liquid state, largely because their fatty acid components have a greater num-

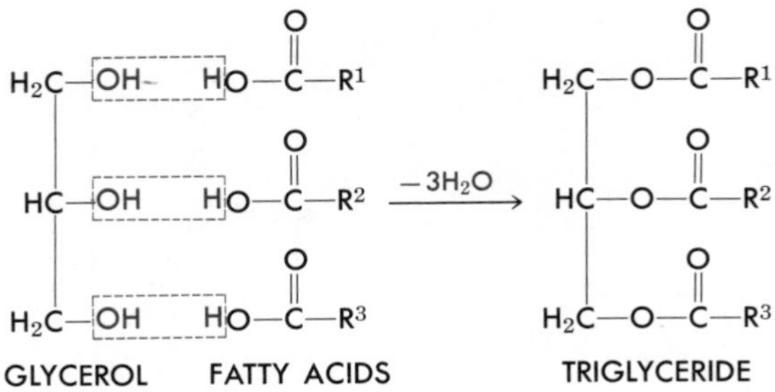

Figure 8.7 The synthesis of a triglyceride. The fats contain a mixture of what are called *triglycerides*, which are links between glycerol and fatty acids. As shown in this figure, the hydroxyl group (OH) of glycerol reacts with the carboxyl group

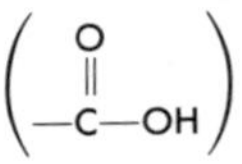

$$\left(\begin{array}{c} O \\ \parallel \\ -C-OH \end{array} \right)$$

forming the bond. The loss of water and the formation of this new bond requires metabolic energy. The R^1, R^2, and R^3 compounds shown in the figure can be identical long-chain fatty acids or they can be a mixture. The hydrophobic ("water-repelling") region of the molecule is due to the long-chain fatty acids while the hydrophilic ("water-loving") region is due to attached compounds that have OH, COOH, H_3PO_4, or other such water-soluble groups. See the structure of lecithin, for example (Fig. 8.8).

ber of unsaturated bonds (double bonds). Waxes, such as bees wax and the cuticle waxes of fruits, leaves, and flower petals, are principally esters of a fatty acid with a long-chain alcohol rather than with glycerol. Bees wax, for example, is an ester of palmitic acid (a 16-carbon saturated fatty acid) and myricyl alcohol (a 30-carbon-chain saturated alcohol).

In the simple lipid, one of the fatty acids may be replaced by compounds containing phosphorus and nitrogen. When this happens, **lecithin** and **cephalin** are formed. Called **phosphotides,** they frequently represent the major portion of cellular lipids. These two compounds are soluble in both water and fats and, therefore, serve a vital role in the cell by binding water-soluble compounds, such as proteins, to lipid-soluble compounds. Lecithin is a key structural material in the cell membrane. It maintains continuity between the aqueous and lipid faces of the inside and the outside of the cell. In Fig. 8.8, showing the structure of lecithin, the R groups represent a long-chain fatty acid.

Lipids have a number of important functions:

1. The storage of intracellular lipids, which can be used later for metabolic fuel.
2. As structural components for cellular membranes.
3. As biologically active hormones, such as the sex hormones, adrenal cortical hormones (steroids), and the relatively newly

Figure 8.8 Structure of lecithin.

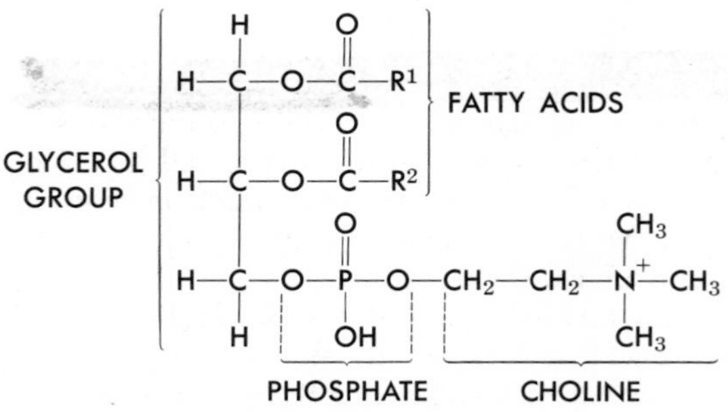

108

discovered prostaglandins (complex unsaturated fatty acids), which affect a wide array of physiological processes such as secretions in the stomach wall and contraction of the uterus.

In addition there are a number of vitamins that are classified as lipids since they are soluble in organic solvents (chloroform, ether, and the like) and insoluble in water. Fat-soluble vitamins include vitamins A, E, K, and D.

proteins

The structure, function, and metabolic activity of a cell or tissue all depend on a class of molecules called **proteins.** Proteins make up a significant portion of the protoplasm of plant and animal cells alike. Proteins also are an important part of the structure of the chromosomes, the nucleoplasm, and the nuclear membrane. All the structures in the cytoplasm—including the mitochondria, the ribosomes, the spindle fibers, and the flagella structures that are used for motion—are made up in part of protein molecules.

In addition, there are various proteins that operate as **catalysts**—agents that speed up chemical reactions. These catalytic proteins are called **enzymes.** Enzymes are intimately associated with all chemical processes occurring in the cell—muscle contraction, nerve conduction, excretion, absorption, and general metabolic reactions. Many proteins, such as the **keratin** of the skin, fingernails, wool, and hair, and the **collagen** of connective tissue and bone, serve in a structural capacity. **Antibodies,** which combine with disease-producing agents and destroy them, are also proteins of a very specific kind. There are important protein hormones that regulate cellular and tissue functions. Insulin is one. The amount of insulin in the blood determines whether we are normal or diabetic. **Contractile proteins** in muscle are essential for movement.

Figure 8.9 Electron micrograph of a fragment of subcutaneous connective tissue, air dried and shadowed with chromium. The photograph probably represents the normal arrangement of the fibers in the tissue.

The Upjohn Company—Jerome Gross, M.D.

THE STRUCTURE OF PROTEINS Proteins are molecules of gigantic size, sometimes containing tens of thousands of atoms. They are tremendously complex and have no competitors in the diversity of roles they play. In addition to containing carbon, hydrogen, and oxygen, proteins also contain nitrogen. They range in molecular weight from about 5,000 (insulin) to 40,000,000 (tobacco mosaic virus protein). Although we have much more to learn about the chemical and physical properties of protein molecules, we do know that their complex and diverse properties are related to at least three major aspects of their structure.

Primary Structure. When proteins are broken down in the

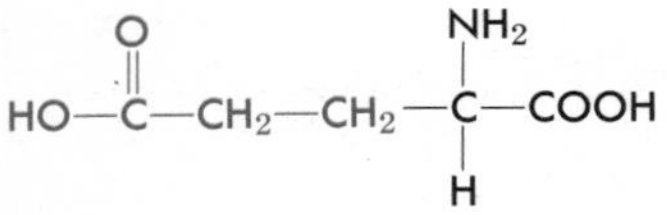

Figure 8.10 Structure of glutamic acid.

presence of water **(hydrolyzed),** they yield a mixture of simple organic molecules that contain nitrogen. These molecules are called **amino acids** (Fig. 8.10). The —NH$_2$ group is the amino group while —COOH is the carboxyl or acidic group, since it can donate hydrogen ions to the solution. All amino acids contain these two groups. Thus, in general, we can write the structure of an amino acid as follows:

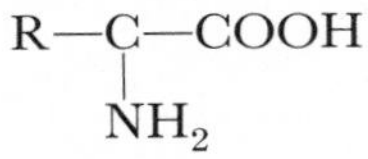

$$R-\underset{\underset{NH_2}{|}}{C}-COOH$$

The chemical nature of the R group in amino acids can vary considerably. In fact there are about 20 different R groups in nature; therefore, there are about 20 different amino acids. For instance, the glutamic acid shown here has an R group that is made up of two CH$_2$ groups and one carboxyl group.

In a protein molecule the amino acids are joined by a carbon–nitrogen bond between the carboxyl group of one amino acid and the amino group of another by the removal of water. The reaction is shown in Fig. 8.11. This bond between two amino acids

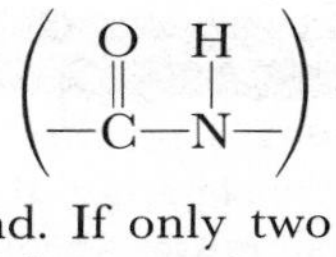

$$\left(-\underset{}{\overset{O}{\underset{}{C}}}-\overset{H}{\underset{}{N}}-\right)$$

is called the **peptide** bond. If only two amino acids are linked, it is called a **di**peptide; if three, a **tri**peptide; and so on. If a large number are connected, it is called a **poly**peptide (Fig. 8.12).

Proteins are made up of long polypeptide chains consisting of hundreds upon hundreds of amino acid units connected to one another by peptide bonds. Since there are approximately 20 different amino acids in nature, it follows that the amino acids in the long polypeptide chains of proteins must be present in many different combinations. The possible number of different combinations, in fact, is enormous. The thousands of different proteins are in part ascribed to an almost countless variety of possible combinations. Thus what we call the *primary* structure of the protein is determined by the number, kind, and sequence of the amino acids in the polypeptide.

Hydrogen bonding, which we discussed earlier, is important in maintaining the structure of proteins. In the protein molecule,

Figure 8.11 Synthesis of a peptide bond by joining two amino acids.

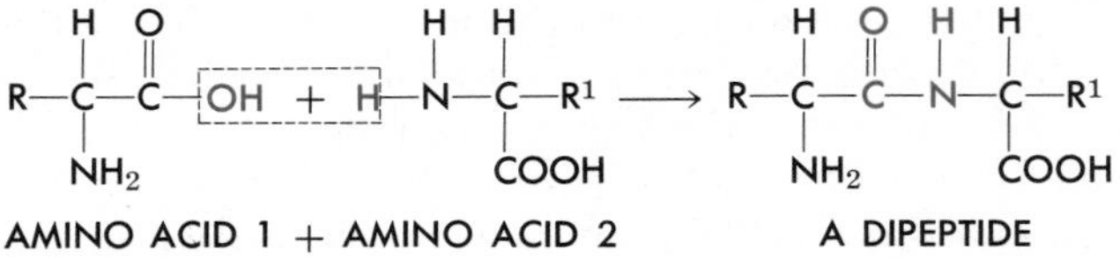

AMINO ACID 1 + AMINO ACID 2   A DIPEPTIDE

the hydrogen bonds generally occur between the oxygen of one peptide group and the amino nitrogen of a neighboring group. Although the forces of hydrogen bonding are relatively weak compared to most covalent bonds, the fact that a large number of hydrogen bonds can occur in a polypeptide makes them collectively a major force in protein structure. However, hydrogen bonds are weak enough so they can form and break with relative ease, thus giving the protein molecule a mobile structure.

Amino acids behave both as weak acids and weak bases, since they each contain at least one carboxyl group (—COOH) and at least one amino group (—NH$_2$). Such compounds are called **ampholytes.** An amino acid such as glycine, therefore, can carry a positive and/or negative charge, depending on the pH of the solution (Fig. 8.13). The addition of hydrogen ions (H$^+$) to a solution of glycine suppresses the ionization of the carboxyl group, and the molecule acquires a net positive charge. On the other hand, adding a base (OH$^-$) removes a proton from the ammonium group, resulting in a net negative charge. At a certain pH, the molecule is electrically neutral (the number of + charges equals the number of − charges); this pH, at which the dipolar ion will not migrate either to the positive or negative pole in an electric field, is called the **isoelectric point.** This point can be determined by titration with acid and base or by electrophoresis, as discussed below. The pH of the isoelectric point depends on the dissociation constants of the basic and acidic groups.

Proteins behave as ampholytes because in the formation of a peptide bond the ampholytic character of the amino acids is preserved. A tripeptide is shown in Fig. 8.14. Note that on the right side we have a free acid group and on the left a free amino group; R^1, R^2, and R^3 refer to the remainder of the carbon skeletons of the three amino acids. The number of acidic and basic groups in proteins depends on the number and types of amino acids present. Obviously, if we linked 100 glycine molecules together by peptide bond formation, we would end up with a large polypeptide with one free carboxyl and one free amino group.

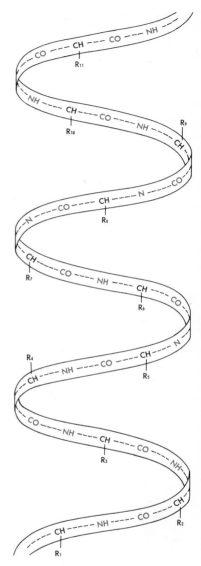

Figure 8.12 The spiral pattern of a polypeptide chain. The peptide bonds are shown in color. The R groups represent the remainder of the amino acids. Hydrogen bonds along the axis of this helical structure stabilize it and give rise to what is called the *secondary structure.* The amino acid sequence (R$_1$ → R$_{11}$) is called the primary structure.

Figure 8.13. Amphoteric properties of amino acids.

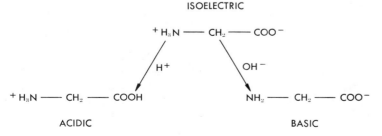

ISOELECTRIC

$^+$H$_3$N —— CH$_2$ —— COO$^-$

H$^+$ OH$^-$

$^+$H$_3$N —— CH$_2$ —— COOH   NH$_2$ —— CH$_2$ —— COO$^-$

ACIDIC   BASIC

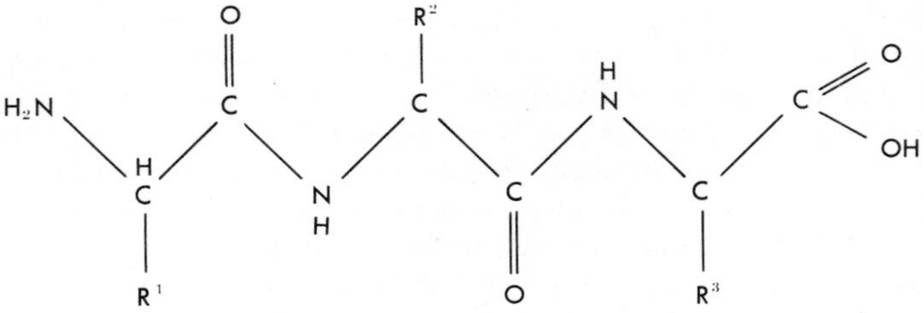

Figure 8.14 A tripeptide.

There are amino acids, however, that have more than one basic or acidic group. Aspartic acid, for example, has two carboxyl groups; consequently, even when it is in the middle of a peptide where the terminal, or *1*, carboxyl is part of the peptide bond, another acidic group (the *2* carboxyl) is still free. Some amino acids, however, carry extra basic groups. Arginine, for example, has the following structure:

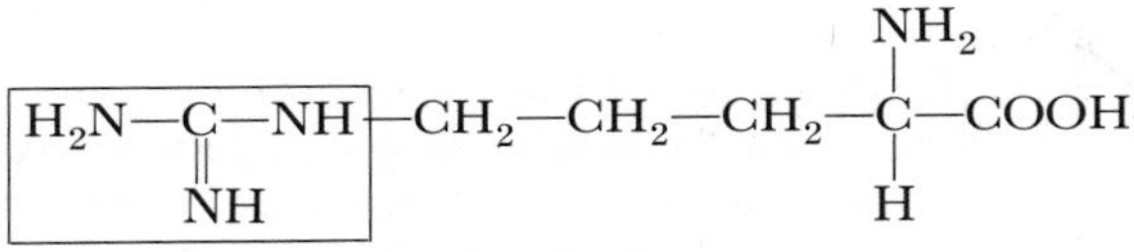

$$\boxed{H_2N-\underset{\underset{NH}{\|}}{C}-NH}-CH_2-CH_2-CH_2-\underset{\underset{H}{|}}{\overset{\overset{NH_2}{|}}{C}}-COOH$$

Arginine

The group shown in the square on the left, called a **guanido** group, is strongly basic. Another basic amino is lysine.

The behavior of proteins thus stems largely from the amino acid composition and the *p*H of the environment. What has been said about the charge properties of amino acids can be repeated for proteins, since they, too, have an isoelectric point and will migrate in an electric field. The *p*H relative to the isoelectric point determines whether the protein moves to the positive or negative pole. The isoelectric point of proteins depends on the relative numbers of free carboxyl and free amino groups, which in turn depend on the amino acid composition.

Secondary and Tertiary Structure. We now know that the long polypeptide chains of proteins are not perfectly straight chains. Instead, they exist as spirals held together by hydrogen bonding. The spiral nature of the polypeptide chains of proteins is spoken of as its secondary structure.

Many proteins also have a tertiary structure. The tertiary structure is due to the spiral polypeptide chain folding forward and backward on itself, forming a globular, rather than a long fibrous, molecule. The folding of most proteins is not a chance arrangement. It is a definite spatial configuration specific for each particular protein, and it is maintained, in part, by hydrogen bonding. Thus, some of the very important properties of proteins are related to their secondary and tertiary structure.

Figure 8.15 A schematic representation of a protein molecule showing the spiral polypeptide chain folding in a number of ways to give a *tertiary* structure to the proteins. Hydrogen bonding and the formation of disulfide (S—S) cross links stabilize the structure.

Among other important properties of a protein is its electrical charge. Electrical charge depends on the acidic and basic properties of the individual amino acids. Thus, the pH of a solution in which a protein occurs determines many of the protein's chemical properties. At neutral pH many of the free carboxyl groups are dissociated and give a negative charge (COO⁻) to the protein. As the hydrogen ion (H⁺) concentration is increased this negative charge is neutralized. The ability of various small molecules to combine with the protein often depends on the electric charge.

growth requirements of organisms

Although we have mentioned some of the organic chemicals that are either found in cells or are produced by them, we have not said anything about the growth requirements of organisms. Many different types of molecules are necessary for the formation and maintenance of protoplasm, the living material of cells. Organisms vary widely in their ability to manufacture certain molecules with their own metabolic machines. Some cells, for example, cannot make certain of the amino acids; consequently, if the organism is to survive, these substances must be supplied in its diet. They are, therefore, **essential growth factors** for the organism. A study of growth requirements is called **nutrition,** and it encompasses all the ways in which food is used.

Vitamins are essential for the normal functioning and growth of all organisms. While some simple microorganisms and most green plants can manufacture all the vitamins they need, other organisms cannot; those that cannot must have another source of vitamins—their diet. If a required vitamin is omitted from

Courtesy USDA Photo

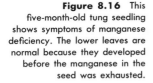

Figure 8.16 This five-month-old tung seedling shows symptoms of manganese deficiency. The lower leaves are normal because they developed before the manganese in the seed was exhausted.

the diet of a higher organism, a vitamin deficiency disease will occur. If the organism is a single-cell organism, it will not grow or multiply.

When vitamin deficiency diseases were first studied, the chemical structure of vitamins was not known; therefore, the vitamins were referred to simply as vitamins A, B, C, and so forth. In the early study of vitamin nutrition it was found that certain vitamins were soluble in fat solvents (such as ether or alcohol), while other vitamins were soluble in water. It turns out that many of the water-soluble vitamins occur in what is now known as the **B vitamin complex.** Some of the B vitamins include pyridoxine, nicotinic acid (niacin), pantothenic acid, biotin, folic acid, vitamin B_{12}, and others.

The fat-soluble vitamins include vitamins A, D, E, and K. Vitamin A exists only in animal products, but there is a yellowish substance in plants, called **carotene** (because it is found in carrots), which can easily be changed into vitamin A by animal cells. Vitamin A is essential for the growth of higher organisms: for the maintenance of nerve tissue and the growth of bone and tooth enamel. A deficiency of vitamin A produces "night blindness," the inability to see in dim light.

Vitamin D is another fat-soluble vitamin. Since it is made in the body under the influence of sunlight, it is sometimes called the "sunshine vitamin." It is essential for the absorption of calcium from the intestinal tract. Vitamin E appears to be necessary to prevent sterility in male animals. In cellular metabolism it is involved in the process of oxidation of complex molecules such as carbohydrates and fats. Vitamin K also seems to play a role in a general oxidative process, and in adult mammals it is essential for the normal coagulation of blood.

AUTOTROPHIC ORGANISMS Autotrophic organisms are able to grow and multiply in a purely inorganic medium. In other words, they do not depend on an outside source for vitamins, amino acids, or other complex organic molecules. From a nutritional viewpoint, autotrophs are the least exacting group of organisms, for they produce their own sugars, fats, amino acids, and so forth from carbon dioxide and ammonia (NH_3). They obtain their energy in one of two ways and are subclassified on the basis of this energy source.

Chemosynthetic Autotrophs. Organisms of this group make their own protoplasm from carbon dioxide, ammonia, or nitrate (NO_3^-), and obtain the energy for the synthesis by the oxidation of inorganic substances. For example, one chemosynthetic autotroph present in the soil is the bacterium *Nitrosomonas*. It is capable of oxidizing ammonia to nitrate, thus generating useful energy:

$$2NH_3 + 3O_2 \rightarrow 2HNO_2 + 2H_2O + \text{energy (79,000 calories)}$$

This organism is a typical chemosynthetic autotroph. Since its cells contain all the complicated carbohydrates, fats, proteins, nucleic acids, and vitamins, it represents a magnificent synthetic factory for making protoplasm.

Photosynthetic Autotrophs. These organisms obtain the energy for their synthetic activities by converting light energy into

Figure 8.17 *Bacilli Nitrosomonas is a chemosynthetic bacterium present in soil.*

Walter Dawn

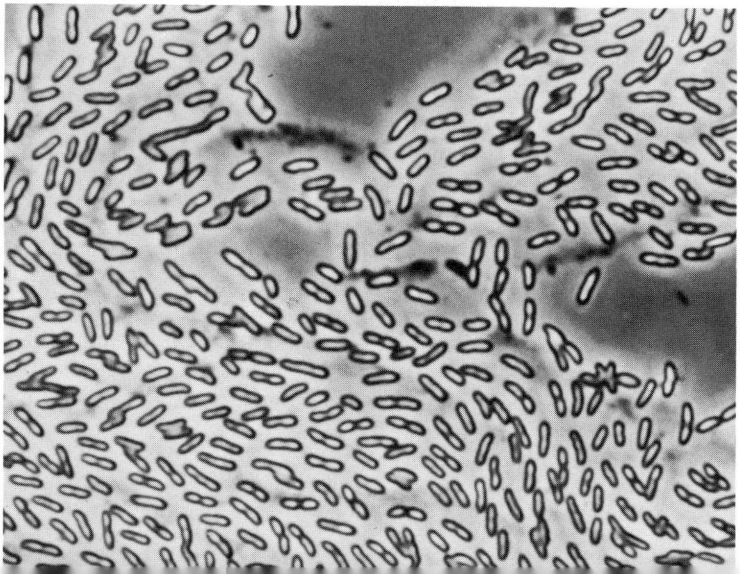

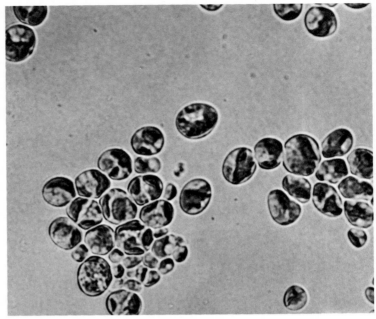

Figure 8.18 *Chlorella is a photosynthetic unicellular organism.*

chemical energy by photosynthesis. They obtain their nitrogen from ammonia or nitrate, and their carbon from carbon dioxide in the air. Some of the organisms in this group are colored sulfur bacteria, diatoms, the blue-green, red, brown, and green algae, and complex green plants. The colors in the plant result from mixtures of pigments, including the crucial one, chlorophyll, which is capable of trapping light energy. Of all the biochemical processes in nature, photosynthesis is of paramount importance. In the green plant, the end product of photosynthesis is a reserve of chemical energy (carbohydrates) that serves as a sole source of energy for most living things.

HETEROTROPHIC ORGANISMS   The other major category of organisms is the heterotrophic group. It consists of organisms that get their energy mainly from organic sources, such as carbohydrates. Heterotrophic organisms, therefore, are related to animals in their general metabolism; autotrophic organisms are related primarily to plants.

Man is a very complex heterotroph. He must eat other organisms to obtain the growth factors he himself cannot make. He requires not only certain kinds of amino acids in his diet, but other required growth factors, such as a number of vitamins

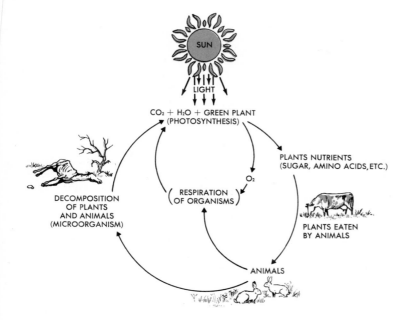

Figure 8.19 The carbon cycle: green plants use the energy of sunlight, carbon dioxide, and water to make all the organic chemicals necessary for life. During this photosynthetic reaction oxygen is also liberated. Thus, animals depend on plants for all of their food as well as oxygen. The end products of animal respiration are water and carbon dioxide. When an animal dies it rapidly decays to carbon dioxide and water—except for the skeleton. The overall effect of these various activities leads to a carbon cycle.

as well as certain fatty acids. Compared with many other organisms, the synthetic machinery of man is less versatile, hence less complex. The photosynthetic green plants, on the other hand, are remarkable in their ability to use the energy of sunlight to make essentially all the known vitamins and amino acids. It is for this reason that many higher vertebrates, such as man, depend on plants as their primary supply of food—if not directly, then indirectly by eating other animals that eat the plants (Fig. 8.19).

AEROBIC AND ANAEROBIC ORGANISMS— OXYGEN AS A NUTRIENT Some organisms cannot live and grow without molecular oxygen (O_2); such organisms are said to be **aerobic.** Those organisms which can grow in the complete absence of oxygen are said to be **anaerobic.** For anaerobic organisms, oxygen is often toxic; it inhibits growth.

There are other organisms, such as the yeast cell, that can live either anaerobically or aerobically. These organisms offer excellent material for studying the transitional changes that occur when the organism is changed from one environment to another. It appears that during embryonic development of vertebrates, for example, cells and tissues may shift from an anaerobic to an aerobic way of life. The mechanisms involved in this shift are not understood, although a great deal of interesting experimental work is now being done on this important problem.

In addition to gases such as oxygen, carbon dioxide, and, in some cases, nitrogen, there are a large number of inorganic minerals that are essential for growth. The need for these minerals varies from cell to cell, particularly from plant cells to animal cells. The mineral nutrients can be classified into two broad categories: **macro**nutrients, which are required in large quantities; and **micro**nutrients, which are required only in trace amounts.

For both plants and animals, the major macronutrients are sodium, chlorine, potassium, calcium, phosphorus, and magnesium. The principal micronutrients are iron, copper, manganese, and zinc. In addition to these, animals need cobalt, iodine, fluorine, vanadium, and selenium, and possibly silicon and chromium. Plants require, in addition, boron and molybdenum. Vandium is essential for only certain forms of plants. The essential elements for plant and animal life are listed in Table 8.1. The discoveries that vanadium, tin, chromium, and selenium are essential for various forms of life indicate the high possibility that there are unknown enzymes that require them for function. It is clear that much remains to be discovered about the function and importance of metals in biological systems.

Most of the 103 elements in the periodic table have been found in living cells, but this does not mean that all that have been found are *essential* to life. If an organism is burned, about 95 per cent of the ash is made up of potassium, phosphorus, calcium, magnesium, silicon, aluminum, sulfur, chlorine, and sodium. The remaining 5 per cent or less is accounted for by the micronutrients and other elements.

There is one important class of organic compounds that we have not mentioned in this chapter—the nucleic acids. Their role in the manufacture of protein is the subject of Chapter 12, for we need first to discuss the chemical reactions that take place within cells, and how energy is handled by the cell.

summary

In this chapter we have briefly reviewed most of the major chemical substances that are associated with living organisms, such as carbohydrates, fats, and proteins. We also saw that the source of chemical energy preserved in the carbohydrates is photosynthesis, which is carried out by green plants. Simple sugars such as glucose can combine with one another to form complex polysaccharides, such as glycogen (in animals) or starch (in plants). One chief ingredient of the cell walls of plants is cellulose, a complex polysaccharide. It can be broken down by microorganisms and made available to other organisms as a useful nutrient.

TABLE 8.1. Elements Essential for Animal and Plant Life

Element	Symbol	Atomic Number	Function
Hydrogen	H	1	In water and organic compounds
Boron	B	5	Essential in plant metabolism
Carbon	C	6	In all organic compounds
Nitrogen	N	7	In amino acids and proteins
Oxygen	O	8	In organic molecules, H_2O, respiration
Fluorine	F	9	In teeth and bone, growth in rats
Sodium	Na	11	Extracellular cation
Magnesium	Mg	12	Chlorophyll, function of many enzymes
Silicon	Si	14	Chick growth, diatom structure
Phosphorus	P	15	Energy transfer, many organic molecules
Sulfur	S	16	Some amino acids and other organic compounds
Chlorine	Cl	17	Major cellular anion
Potassium	K	19	Major cellular cation
Calcium	Ca	20	Bone structure, some enzyme function
Vanadium	V	23	Growth factor for rats and certain marine organisms
Chromium	Cr	24	Essential for insulin function
Manganese	Mn	25	Essential for some enzyme function
Iron	Fe	26	Hemoglobin structure and enzyme function
Cobalt	Co	27	Vitamin B_{12} structure, some enzyme activity
Copper	Cu	29	Hemocyanin structure, oxidative enzyme
Zinc	Zn	30	Numerous enzyme activities
Selenium	Se	34	Necessary for liver function
Molybdenum	Mo	42	Necessary for numerous enzyme activities
Tin	Sn	50	Essential in rat nutrition
Iodine	I	53	Thyroid hormone structure

Proteins are also complex polymers made up of varying combinations of simple amino acids. The primary structure of proteins is determined by the kind and number of amino acids linked together by the peptide bond. This polypeptide can form spirals and fold back and forth on itself, giving a secondary and tertiary structure to the complex macromolecule. The secondary and tertiary structures are important for certain catalytic and functional roles for the protein, and the structures are maintained, in part, by hydrogen bonds.

For organisms to make these complex structures internally— and thus grow and reproduce—certain nutrients are required. An energy source is essential (carbohydrates) as is a nitrogen source (for the synthesis of amino acids and other nitrogen-containing compounds). If an organism cannot make the vitamins it requires, then it must consume other organisms that can make the vitamins, or obtain the vitamins from other sources in the diet.

Certain autotrophic organisms can live and multiply on a diet that contains only CO_2, ammonia, minerals, and water. They obtain energy from the oxidation of ammonia (NH_3). These are called **chemosynthetic autotrophic** organisms. The green plants obtain their energy from sunlight and are called **photosynthetic autotrophs.** Man and most mammals are complex **heterotrophic** organisms. They must obtain essential amino acids, fatty acids, and vitamins from other organisms that are capable of making them.

Oxygen is essential for a large number of organisms, but there are some that can live and reproduce without this gas (**anaerobic** organisms). All organisms require certain major elements, such as sodium, chlorine, potassium, calcium, phosphorus, and magnesium. In addition, trace amounts of iron, copper, manganese, and zinc are also essential. Cobalt, iodine, fluorine, vanadium, and selenium, and possibly silicon and chromium, are required by animals. In addition, boron, molybdenum, and sometimes vanadium are required by plants. Because these latter elements are needed in trace amounts only does not mean that the trace elements are less important than the major elements (which are required in larger amounts). The trace elements function, as we shall see later, as catalysts; thus only small amounts are required.

for thought and discussion

This and the previous chapter attempt to give you the necessary chemical background for understanding some of the metabolic processes that we shall be discussing in the next few chapters.

1   Make sure that you understand what a carbohydrate is.

2   Can you write the straight-chain general formula for glucose?

It is also important to remember that there are six-carbon sugars, five-carbon sugars, and so on. Proteins and their structure are also very important. We shall discuss them in greater detail later.

3   Make sure you understand the composition of an amino acid.

4   How is a peptide bond formed?

5   What is a **polypeptide?**

You should have a general understanding of the "important" chemicals that are associated with living organisms. If an organism cannot make essential chemicals, it must obtain them in its diet.

6   What is a vitamin?

7   A trace metal?

8   An autotrophic organism?

9   A heterotrophic organism?

10   Is oxygen essential to all organisms?

selected readings

BALDWIN, E. *Dynamic Aspects of Biochemistry* (3rd ed.). New York: Cambridge University Press, 1957. An outstanding introduction to the concepts of cellular metabolism.

BENNETT, T. P. and E. FRIEDEN. *Modern Topics in Biochemistry.* New York: Macmillan Publishing Co., Inc., 1966. An excellent discussion in greater detail of the various subjects discussed in our last three chapters. Written for the beginning college student.

LOEWY, A. C., and P. SIEKEVITZ. *Cell Structure and Funtion.* New York: Holt, Rinehart and Winston, Inc., 1963. A detailed elementary discussion of various aspects of cell structure and biochemistry.

MCELROY, W. D. *Cell Physiology and Biochemistry* (3rd ed.). Englewood Cliffs, N.J.: Prentice-Hall, Inc., 1964. A small paperback intended for beginning college students.

NEILANDS, J. B., and P. K. STUMPF. *Outlines of Enzyme Chemistry* (2nd ed.). New York: John Wiley & Sons, Inc., 1958. An excellent elementary introduction to enzyme chemistry.

Readings from Scientific American

ALLEN, R. D., "Amoeboid Movement," February 1962.

ALLFREY, V. G., and A. E. MIRSKY, "How Cells Make Molecules," September 1961.

BRACHET, J., "The Living Cell," September 1961.

DELWICHE, C. C., "The Nitrogen Cycle," September 1970.

DOTY, PAUL, "Proteins," September 1957.

FOX, F. C., "The Structure of Cell Membranes," February 1972.

FRIEDEN, EARL, "The Chemical Elements of Life," July 1972.

FRIEDEN, E., "The Enzyme–Substrate Complex," August 1959.

FRUTON, J. S., "Proteins," June 1950.

HAYASHI, T., and G. A. W. BOEHM, "Artificial Muscle," December 1952.

HOLTER, HEINZ, "How Things Get into Cells," September 1961.

HUXLEY, H. E. "The Contraction of Muscle," November 1958.

KENDREW, J. C., "Three-Dimensional Structure of a Protein," December 1961.

LEHNINGER, A., "Energy Transformation in the Cell," May 1960.

LEHNINGER, A., "How Cells Transform Energy," September 1961.

MOORE, S., and W, H. STEIN, "The Chemical Structure of Proteins," February 1961.

ROBERTSON, D. J., "The Membrane of the Living Cell," April 1962.

SOLOMON, A. K., "Pumps in the Living Membrane," August 1962.

STEIN, W. H., and S. MOORE, "The Chemical Structure of Proteins," February 1961.

WOODWELL, G. M., "The Energy Cycle of the Biosphere," September 1970.

metabolic properties of cells

chapter 9

To observe some of the properties of a single cell, we have only to peer through a small hand lens or a microscope. The shifting of the protoplasm inside the cell, the changing size and shape of the cell, and the movements of the cell toward or away from light are only some of the cell's dramatic properties. However, there is a realm of smaller things—atoms and molecules—that we cannot see through an ordinary microscope. Our task in this chapter is to find out how a cell maintains itself as a living chemical unit.

the physical and chemical nature of protoplasm

The cells of *all* animals are almost identical to the ones in your own body and are all basically like the cells of plants. In an earlier chapter, you found that the cell is made up of three main parts: (1) the **cell membrane** (also called the **plasma membrane**), which surrounds the total contents of the cell; (2) **cytoplasm,** which contains a large number of important structures such as the mitochondria, the endoplasmic reticulum, lysosomes, and ribosomes; and (3) the **nucleus,** which contains the hereditary material.

The cytoplasm, nucleus, and cell membrane are all living

123

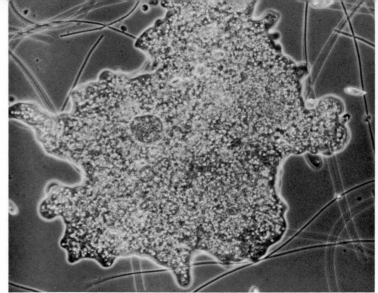

Walter Dawn

Figure 9.1 The granular material visible in this cell is part of the protoplasm. The protoplasm is the "living" material in all cells.

Figure 9.2 Fog is a colloidal suspension. Liquid water droplets are suspended in a gas (the air).

Aerofilms, Ltd. from Ewing Galloway, N.Y.

parts of a cell and collectively are called **protoplasm.** It is the chemical and physical organization of protoplasm with which we are immediately concerned at this point. We have talked about the chemical elements and how they join to make compounds. Protoplasm is a mixture of a large number of both simple and complex chemical compounds. Compounds made up of carbon, hydrogen, oxygen, nitrogen, and many other elements are found in protoplasm in many different combinations. Not only are these elements united in a variety of compounds, but the compounds themselves are often combined in very definite proportions and with a very definite organizational pattern. A cell, it should be emphasized, is not an indiscriminate mixture of compounds, but an intricate and organized arrangement.

Earlier, we found that cells contain a watery fluid. When we have such a fluid system containing molecules of various kinds, we can classify them in a number of ways. If the particles are small, such as sugar molecules, they can dissolve in the fluid and become a **true solution.** If, on the other hand, the particles are very large, such as grains of sand which can sink to the bottom, the system is called a **coarse suspension.** If particles are of intermediate size, and neither settle out nor form a true solution, we call the fluid system a **colloid.**

PROTOPLASM AND There are many different types of
COLLOIDAL SYSTEMS colloidal systems. Probably the one best known to you is mayonnaise—a mixture of fat and protein in water. Fog also is a colloidal system, one made up of small water particles suspended

in air. Most biological systems, however, are liquid–liquid combinations. Protoplasm is in part a true solution and in part a colloidal system. There are several important properties of protoplasm that make us classify it as a colloid. For example, the dispersed particles in the cytoplasm of the cells—large protein molecules and large oily fat molecules—are of such size and charge that they do not settle out under ordinary gravitational force. Instead, they keep moving about constantly. This is called **Brownian motion,** and its effect is easily observed under the microscope. Since the speed at which molecules move about depends on the amount of thermal or kinetic energy they have, Brownian motion becomes more vigorous as the temperature is raised. You can easily see this for yourself in a pot of water as it is heated to a boiling temperature.

One of the more important properties of a colloid is the electric charge of the dispersed particles. Protein molecules, fat droplets, and other large molecules of biological interest usually carry a net positive or negative charge. Because of these differences in charge, many of the molecules are prevented from aggregating. If two molecules carry a net positive charge, they repel one another; thus, the particles are kept apart. If the charges are neutralized by a negative charge, they settle out (Fig. 9.4).

Another interesting property of colloids is that they can form **gel**-like substances. For example, if we greatly increase the concentration of protein in a water solution, the protein molecules line up parallel to one another, making a compact rod-shaped particle. When many such particles interlock and form a network with water droplets throughout, the gel state of the colloid is formed. Your skin, jellyfishes, jello, and gelatin are colloids in this gel condition (Fig. 9.5). High temperatures tend to dissolve the gel state and convert it into what appears to be a viscous solution. This change from a gel to a **sol** is very important in a number of biological processes. The movement of an ameba or the ability of a jellyfish to maintain its structure depends on the gel-like property of protoplasm.

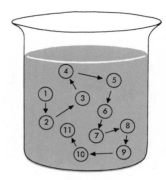

Figure 9.3 Small particles suspended in water tend to move about with a haphazard motion. This is due to the random bombardment of the suspended particles by the water molecules and is called *Brownian motion.* Molecules of all kinds are in constant motion, and when they collide with another object some of their kinetic energy imparts motion to that object.

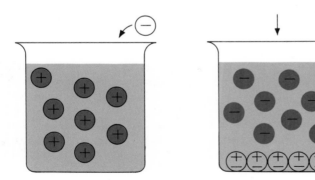

Figure 9.4 Large particles tend to stay dispersed if they have a positive or negative charge (colored circles). Neutralized particles tend to settle out.

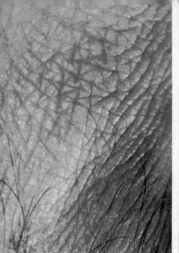

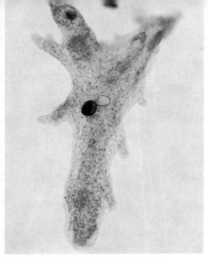

Figure 9.5 Examples of living and nonliving gel-like structures: human skin (left), jello (center), and an ameba.

Protoplasm, then, is a colloid with gel-like properties, and it is composed largely of protein, lipids, and water droplets. Such is the internal environment of a cell. Outside the cell membrane, there is usually a water environment. Let us now find out what exchanges take place between these two environments, and the role of the cell membrane in permitting or preventing an exchange of material. (For a review of the structure and composition of the cell membrane, see Chapter 2.)

the cell membrane

PERMEABILITY   When a cell is placed in a suitable nutrient medium, it takes up the nutrients and thrives. Just because a nutrient is close at hand outside the cell, however, does not mean that the cell is capable of using it. The substance must first pass through the cell membrane.

All the food products a cell takes in must be soluble, to a certain degree, in the water outside the cell in order to pass through the cell membrane. Likewise, waste products within the cell must be soluble in the protoplasm in order to pass out of the cell. Since not all dissolved substances can penetrate the membrane with equal ease, the membrane is said to be **selectively permeable.** This selectivity is vital in maintaining the life of a cell. Although the cell membrane is the major structure safeguarding the cell's internal environment, other parts of the cell, such as the nucleus and mitochondria, are also bounded by selective membranes that control their internal environments.

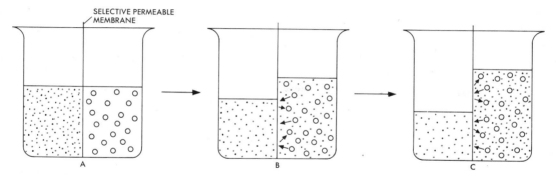

SELECTIVE PERMEABLE MEMBRANE

A   B   C

Figure 9.6 A schematic representation of osmosis. Initially, we have a vessel divided into two chambers by a selectively permeable membrane (A). On the left side we have a solution containing small molecules that are capable of passing through the membrane. On the right side we have a solution of larger molecules that cannot pass through the membrane. In time—(B) and (C)—the small molecules will be distributed throughout both chambers. However, since the larger molecules cannot pass through the membrane, there is a tendency for the water molecules to move from left to right because of the differential bombardment of the two sides of the membrane by the water molecules. The pressure that must be applied on the right-hand column to prevent the increase in volume is known as the *osmotic pressure*.

OSMOSIS   Whenever two different solutions are separated by a selectively permeable membrane, an **osmotic system** is established. Each cell, therefore, represents an osmotic unit. **Osmosis** may be defined as the movement of water molecules through a selectively permeable membrane. In cells, water is exchanged between the protoplasm and the solution surrounding each cell. Take a simple osmotic system in which water is separated by a selectively permeable membrane from a solution of sugar (Fig. 9.6). In this case, both the **solute** (sugar) and **solvent** (water) tend to diffuse, because of Brownian movement, from an area of higher concentration to one of lower concentration (Fig. 9.7). In a perfect system, where the membrane keeps the sugar isolated, only the water is able to penetrate. The two solutions can reach equilibrium, that is, a state in which further change does not occur, only by the transfer of water into the sugar solution.

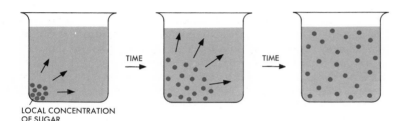

TIME   TIME

LOCAL CONCENTRATION OF SUGAR

Figure 9.7 A high concentration of sugar in water diffuses. In time the concentration of sugar will be the same throughout the water.

127

Since the water concentration on one side of the membrane is higher than on the other side, water molecules will pass into the sugar solution and increase the volume of water. Eventually, the pressure of the water on the sugar solution side of the membrane will stop the entrance of more water. This pressure is called **osmotic pressure.**

The cell is not a perfect osmotic system. Not only water but many dissolved substances commonly present in and around the protoplasm are able to penetrate the cell membrane. This enables the exchange of water between the cell and its surroundings to be accompanied by the exchange of other substances as well. Oxygen and carbon dioxide, for instance, pass readily into and out of the cell. The permeability of the complex cell membrane depends not only on the nature of the surrounding particles, but also on the changing conditions inside and outside the cell.

Although permeability varies in different cells, and sometimes on different sides of the same cell membrane, we can make certain generalizations about osmosis. For example, we know that water rapidly penetrates most cells. Gases such as carbon dioxide, oxygen, and nitrogen, and fat-solvent compounds such as alcohol, ether, and chloroform easily penetrate all cell membranes. Somewhat slower to penetrate are such organic substances as glucose, amino acids, and fatty acids. Inorganic salts, acids, and large disaccharide molecules such as sucrose, maltose, and lactose penetrate even more slowly by the process of simple diffusion.

The bulk of all solutes present in protoplasm (including proteins and most sugars and inorganic salts) penetrate the cell membrane very slowly, if at all. The solvent water, on the other hand, enters and leaves the cell very quickly. This, coupled with the fact that water is more abundant than all other components combined, means that the water must bear the main burden of establishing osmotic equilibrium between the cell and surrounding solutions. If the cell is placed in a solution with a water concentration drastically different from that in the protoplasm, so much water enters or leaves the cell that the cell may be destroyed. The cell wall in plants and most microorganisms is usually rigid enough to prevent the swelling of the cell when water rushes in, but an animal cell, bounded only by the cell membrane, can be easily ruptured in this way.

In an **iso-osmotic** solution, the concentration of the water outside the cell is the same as that in the protoplasm. This concentration occurs only when the total concentration of solute particles in the external solution equals that of solute particles in the protoplasm (Fig. 9.8). The water balance is achieved because the water molecules that are continuously escaping from the cells are matched by an equal number of water molecules entering the cells. A true iso-osmotic solution, therefore, contains a concentration of nonpenetrating or very slowly penetrating solute molecules that approximates the total concentration of

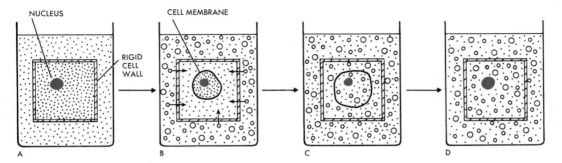

Figure 9.8 The effect of the osmotic environment on cell structure. When a plant cell (A) is placed in a hypertonic solution of a slowly penetrating solute (B), water is rapidly lost from the cell and the cell membrane shrinks away from the cell wall. As the solute slowly penetrates—(C) and (D)—water re-enters and the cell swells, resuming its original size.

nonpenetrating solutes in the protoplasm. An equal water concentration inside and outside the cell could not otherwise exist.

A **hypotonic** solution has a relatively low concentration of nonpenetrating solutes compared to the protoplasm of the cell it surrounds. Thus, the water concentration in it is relatively high. Animal cells placed in a hypotonic solution tend to take in water and swell. If the solute on the outside is of very low concentration, the swelling will continue until the cell membrane ruptures. When human red blood cells are placed in a solution containing only 0.2 per cent sodium chloride, instead of the usual 0.9 per cent, the corpuscles swell and burst, even before they can be observed under a microscope.

Cells placed in a **hypertonic** solution, on the other hand, tend to shrink. This happens because the solution contains a higher concentration of solute molecules than the protoplasm does. When animal cells are placed in a hypertonic solution, they shrivel up so much that we cannot detect their original shape. When a red blood cell, for example, is placed in a concentrated sucrose solution, it shrinks from loss of water. If this process is not carried too far, it is reversible (Fig. 9.9).

If we place a plant cell in a hypertonic solution whose solute particles slowly penetrate the cell, the plant membrane contracts away from the cell wall (Fig. 9.8). This process is called **plasmolysis.** As the solute particles slowly enter, water returns and the cell returns to its original size. This reverse process is called **deplasmolysis.** By observing deplasmolysis we can determine the rate of penetration of solute particles into the cells.

ACTIVE TRANSPORT The selective cell membrane is not a *passive* organ. While simple diffusion and osmosis account for much of the exchange of material between cells and surrounding fluids, the cell membrane also

129

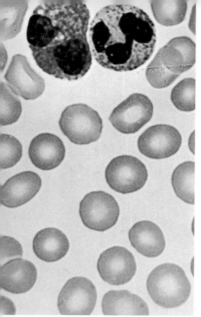

The Bergman Associates

Figure 9.9 Human red blood cells in a hypertonic solution (below) are shown here compared with normal red cells.

The Bergman Associates

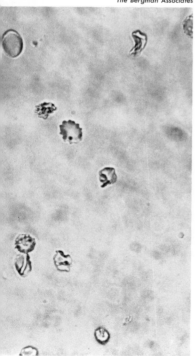

plays an active part in the exchange process. Many marine algae, for instance, accumulate iodine to a concentration more than 1 million times greater than that of the sea. Such situations cannot be accounted for by the simple laws of diffusion or osmosis alone. Cells, therefore, have a means of forcing the molecules of a particular substance to move in a direction opposite to that dictated by the laws of diffusion. This is the process of **active transport.**

There seem to be specific proteins and other molecules in the cell membrane that act as catalysts for moving substances across the membrane and into the cell. We call this transport machinery "metabolic pumps," and speak of the "sodium pump," the "potassium pump," and so on.

One of the best ways to show how the metabolic pumps work is to use **protoplasts** of bacteria. Protoplasts are bacteria with their cell walls removed. Certain bacteria accumulate so much sugar lactose, for example, that the protoplasts swell and burst owing to the internal increase of osmotically active particles. The point is that the concentration of lactose within the cell is higher than it is outside the cell, and it continues to get higher still because of the active transport process. Thus, the cell does not depend entirely upon the external concentration of a substance.

In normal conditions, active transport makes it possible to regulate the entry of substances at a rate that satisfies the requirements of the cell. Let us assume, for example, that an organism is in an environment where the concentration of an essential nutrient is extremely low. Because of active transport carried out by the cell membrane, the organism does not have to depend upon the process of free diffusion to obtain an adequate internal concentration of the essential nutrient. It is obvious that the process of active transport has tremendous survival value for an organism.

Figure 9.10 Normally when bacteria are placed in a solution that contains fewer osmotically active molecules (hypotonic) than the interior of the cell, the bacterium tends to swell because of the inward movement of water. The strong cell wall, however, prevents the cell membrane from expanding. If we add penicillin to the medium the synthesis of the cell wall is inhibited and the wall begins to break down. When this happens the cell is enlarged by the inward movement of water.
If the molar concentration of molecules outside the cell is too low,
the bacterium bursts and dies.

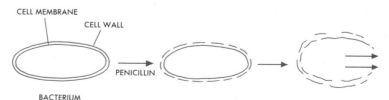

CELL MEMBRANE
CELL WALL
PENICILLIN
BACTERIUM

PINOCYTOSIS
AND PHAGOCYTOSIS
Cells such as white blood cells, epithelial cells of the intestines, and others can bring substances into themselves by forming pockets, or **invaginations,** in the cell membrane. The invagination is pinched off, and the captured material floats free in the cytoplasm. Molecules that are too large to pass through the cell membrane can be brought into the cell in this way. The small free-floating vacuoles are called **pinosomes,** and the general process is called **pinocytosis. Phagocytosis** is a similar process by which amebas and white cells are able to engulf large particles or bacteria by extending "arms," called **pseudopods,** of the cell's surface out and around the particle to be taken up (Fig. 9.11).

Pinocytosis seems to take place only under certain conditions. For example, if proteins or salts are added to water, a cell begins the process of pinocytosis, but continues for a limited time only. Then, after a lull, the process is begun again. There appear to be specific sites on the cell surface capable of forming the pinosomes. We do not yet understand how the metabolic machinery of the cell begins and stops pinocytosis according to its needs.

The material in the pinosomes is not inside the cell in a metabolic sense. The membrane around the engulfed material persists and behaves very much like the cell membrane itself. Small molecules can diffuse into the cytoplasm, but larger ones appear to be left behind. Gradually, the pinosome decreases in size and eventually becomes part of the cytoplasmic granules.

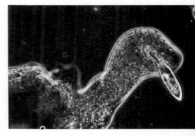

Eric V. Grave

Figure 9.11 An ameba engulfing a unicellular microorganism.

enzymes

Cells must be able to take up nutrients from the surrounding fluid and must contain the machinery for creating new parts of themselves from the food material. As pointed out earlier, while some of the nutrients become part of the cell itself, others are broken down and provide the energy needed to synthesize new molecules. In general, all these nutrients are called "food" and include the carbohydrates, fat, proteins, minerals, vitamins, and water. Taken as a whole, these cellular processes are rather complicated; but if we isolate and examine selected parts of the elaborate machinery of the cell, we can study the individual steps that lead to the synthesis of complex molecules. This is best done by following certain reactions as they take place in a test tube.

One group of molecules particularly important to all organisms is the large complex protein molecules called **enzymes.** For every essential chemical reaction that occurs in a living cell, there is a specific enzyme capable of speeding up that step. Without enzymes, these chemical reactions would not take place fast enough at normal temperatures to sustain life. Enzymes, then, are very efficient catalysts. They are also effective in very small

131

amounts, and are usually unchanged by the chemical reaction they promote. Cells can duplicate themselves in a few minutes or hours because enzymes are capable of catalyzing the chemical changes associated with life processes. Enzymes are intimately associated with all life processes, such as muscle contraction, nerve conduction, excretion, and absorption.

In some instances, for an enzyme to function it must be associated with a small molecule called a **coenzyme.** Parts of these coenzymes are often vitamins such as **riboflavin** (vitamin B_2) and **thiamin** (vitamin B_1). In addition to the organic **cofactors** (coenzymes) like the vitamin-containing compounds, several inorganic cofactors are required. For example, iron is necessary for the transfer of electrons; copper, too, apparently plays a role in electron transport. Magnesium is essential for the transfer of phosphate, and so on. In some cases, then, the enzyme protein alone is not enough to speed a chemical reaction. One or more cofactors may be needed.

Naming enzymes is very easy once we know the type of reaction, or the substance acted on by an enzyme. Enzymes are denoted by the suffix *"ase."* For example, an enzyme that catalyzes the breakdown of proteins is called a protein*ase.* One that catalyzes an oxidation is called an oxid*ase,* and so on.

THE EFFECT
OF TEMPERATURE
ON AN ENZYME
REACTION

The individual molecules of a cell are in ceaseless motion. Occasionally, they react with one another when they collide. If we deprived a cell of its enzymes, the chance that a molecular collision would result in chemical reaction is very small. However, add the right enzyme and the chance of a collision resulting in a chemical reaction is greatly increased. The major question is how enzymes perform this catalytic activity.

All molecules in a given "population" do not have the same kinetic energy, or speed. Some, through collision, acquire more energy than others. These energy-rich molecules are more likely to react with energy-poor ones than are other energy-poor ones. In other words, there is an "energy barrier" to reaction. The energy required to hurdle molecules over this barrier is called the **energy of activation.**

Figure 9.12 shows a hypothetical reaction in which compound A is converted into the product compound B. This diagram in general holds true for all chemical reactions, although the height of the energy barrier varies from one reaction to another. Note that for A to be converted into B it must first acquire the necessary energy to form the activated molecule A*. The rate of a chemical reaction must depend, therefore, on the number of A* molecules existing at any one moment and the speed with

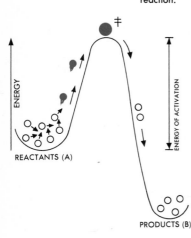

Figure 9.12 Energy of activation for a chemical reaction.

ENERGY

ENERGY OF ACTIVATION

REACTANTS (A)

PRODUCTS (B)

which they break down and form product B. Thus, the rate of a chemical reaction can be given by the following simple equation: **rate $= [A^*] \cdot k$,** where k is a constant (which is essentially the same for all chemical reactions) and $[A^*]$ means concentration of A^*.

The number of molecules activated depends on the temperature. As the temperature is raised, the number of A^* also increases; thus the rate of the reaction increases. One significant feature of enzymes is their ability to lower the energy of activation, thus *increasing the rate of chemical reaction.* We do not know exactly how this is done, but we are certain that reactions that take place only at boiling temperatures in a test tube can take place with relative ease at body temperature in the cell when enzymes are present (Fig. 9.13).

High temperatures tend to destroy enzymes so that no further reaction can take place. Thus, biological processes that depend on enzyme reactions have an optimal temperature at which they function best. Above certain critical temperatures, enzymes do not perform, as shown in Fig. 9.14. In the example shown, although the reaction rate increases rather rapidly from 0°C up to about 25°C, above this temperature the rate begins to slow; and at approximately 35°C it starts to decrease. If the damaging temperature is maintained for too many minutes, the enzyme will be completely inactivated. Other enzymes have different temperature optima.

This thermal behavior of enzymes imposes serious limitations on organisms. Most cells lose their capacity to carry on metabolism at temperatures above 40°C. Only the few organisms that have heat-resistant enzymes are able to survive in exceptionally hot places, such as in hot springs. In most cells, the rate of metab-

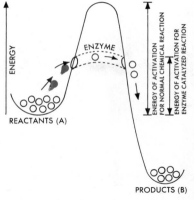

Figure 9.13 The effect of enzymes on a chemical reaction is shown in this diagram. Enzymes, in some unknown way, lower the energy of activation for reactions so that the rate is greatly accelerated. It is as if the enzyme makes a tunnel in the energy barrier.

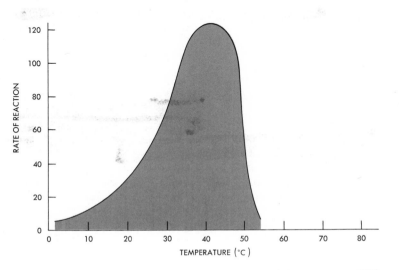

Figure 9.14 The effect of temperature on an enzyme-catalyzed reaction.

Figure 9.15 A chipmunk coiled up in its nest underground in winter hibernation. In the state of hibernation the heart beat slows down extensively; in general, the overall metabolic rate is greatly depressed. In contrast, the high metabolic rate of the Alaskan brown bear, which regularly fishes for salmon in cold water, maintains the body temperature at constant value.

olism, and hence the intensity of the life processes, changes as the temperature varies from day to day and from season to season. The winter metabolism of most organisms declines considerably. In general, plants in the colder winter climates cease to grow. We refer to this state of low metabolism as **dormancy**. Warm-blooded organisms, mammals and birds, are the only creatures that have evolved a method of controlling their body temperature.

HOW AN *ENZYME WORKS* The single molecule that an enzyme breaks down into separate new molecules, or the molecules that an enzyme combines into a single new molecule, are called **substrate** molecules, or simply the substrate. An enzyme and a substrate fit together like pieces of a jigsaw puzzle. By combining with the enzyme, some of the bonds of the substrate molecule lose their form, thus producing a condition that favors chemical reaction.

Figure 9.16 shows something of the geometry of enzyme-substrate activity; Fig. 9.17 shows on a chemical level how a single maltose molecule is split into two glucose molecules (in

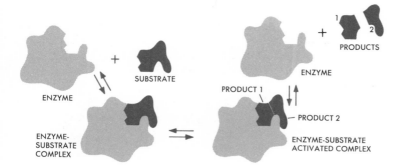

Figure 9.16 The formation of an enzyme-substrate complex, followed by catalysis.

the presence of water) by the enzyme maltase. Since the enzyme molecule is so large compared with the substrate maltose molecules, only an outlined section of the maltase molecule is shown. The first step in the reaction is the bonding of the maltose molecule to a proper position (**active site**) on part of the maltase molecule. Next, the oxygen bond is broken by hydrolysis (the addition of a molecule of water). One hydrogen and one oxygen atom take up a position on the end carbon atom of the glucose subunit, while the second hydrogen atom from the water molecule joins the old oxygen of the other glucose subunit. The final step is for the two new glucose molecules to break away from the enzyme maltase, which has in no way been permanently altered during the reaction. The enzyme is now ready to be used again.

If we isolate an enzyme and dissolve it in a water solution, we can study the rate of the chemical reactions it catalyzes. In general, if we add only a few substrate molecules so that our solution is of low concentration, the reaction rate will be relatively low; but if we add many substrate molecules so that the solution is of high concentration, the reaction rate will be relatively high. This variation of rate is over a limited range of concentration. As illustrated in Fig. 9.18, the rate of enzymatic reaction at first rises as we increase the substrate concentration. Then there is a

Figure 9.17 Maltose combines at a specific site (the catalytic site) on the enzyme molecule. When bound to the site, water can readily split the linkage between the two sugar molecules to give free glucose.

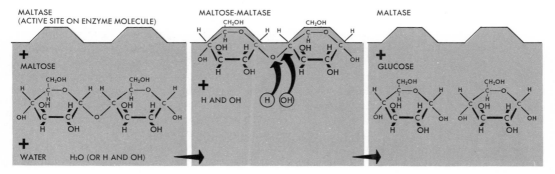

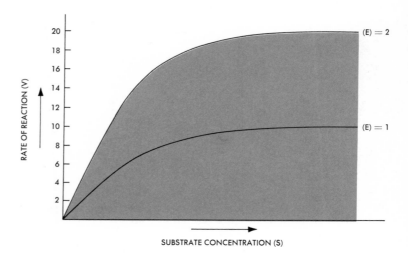

Figure 9.18 The effect of substrate concentration on the rate of enzyme-catalyzed reactions.

slowing of the rate, and eventually the rate becomes constant—no matter how much more substrate we add. When the rate of the reaction becomes constant, we assume that the surfaces of all the enzyme molecules are completely covered, or saturated, with the substrate molecules. Two different enzyme concentrations are shown in Fig. 9.18.

The combination of a substrate with an enzyme is a very specific process, and each enzyme has a unique chemical arrangement containing precise sites where the substrate molecules join it. Because of the unusual folded surface of proteins, only very specifically shaped substrate molecules can gain access to the active region of the enzyme.

The specificity and the geometry of an enzyme reaction are beautifully illustrated by the use of enzyme inhibitors, "intruder" molecules that compete with the substrate molecules for the active site on the enzyme. Let us consider **succinic dehydrogenase,** an enzyme that catalyzes the removal of hydrogen from succinic acid. If malonic acid, whose molecule is very similar in shape to a succinic acid molecule, is added to the solution, the enzyme's effectiveness is considerably reduced. Malonic acid apparently attaches itself to the enzyme at a position that would normally be filled by succinic acid. Therefore, malonic acid competes with succinic acid for the active sites of the enzyme molecules (Fig. 9.19). By so doing, it prevents the enzyme from acting as a catalyst. Since substrate and inhibitor compete for the same active sites as the enzyme molecules, the degree of inhibition depends on the relative amounts of substrate and inhibitor. If there is more substrate, the inhibitor might not be able to compete.

There are many different types of **competitive inhibitors.**

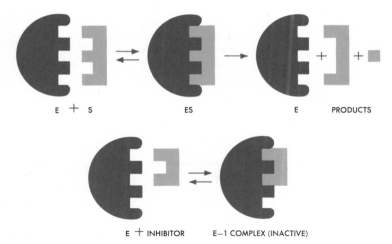

E + S   ES   E   PRODUCTS

E + INHIBITOR   E—1 COMPLEX (INACTIVE)

Figure 9.19 When a substance has the same shape or geometry as the normal substrate, it can combine at the active site of the enzyme, thus inhibiting enzyme activity. This is called *competitive inhibition* because the inhibitor competes with normal substrate for the active site.

Some are very effective in killing bacteria. One of these is **sulfanilamide** (Fig. 9.20). The success of sulfanilamide and other bacterial killers opens the exciting possibility that all disease-producing organisms, as well as the abnormal growth in cancer, might be susceptible to the antimetabolite approach, that is, to inhibitors that make an enzyme inactive. Unfortunately, the host and disease-producing organisms usually have very similar enzyme systems. This means that the inhibitor might be just as toxic to the host organism as it is to the disease-producing organism. However, by careful screening we may be able to find an appropriate chemical that will kill the disease-producing organisms without harming the host organism.

The current, exciting studies of the mechanism of enzyme action are concentrating on the specific reactive groups on the protein molecule and their relationship to the neighboring amino acids, making it imperative that we know the amino acid sequence in the protein as well as the chemical nature of the active site. In some cases, it has been possible to remove some of the amino acids from the polypeptide without affecting its catalytic activity, providing further evidence that there is an active site and that the entire protein is not essential for catalytic activity.

Catalytic activity is sometimes lost when a small peptide is removed from the protein, but it can be restored to the protein residue by adding equal amounts of the small peptide to the reaction mixture. Since no peptide bonds are formed under these conditions, hydrogen bonding and other noncovalent linkages must be essential in restoring the catalytic site, a fact that rein-

Figure 9.20 Sulfanilamide is a competitive inhibitor. p-Aminobenzoic acid (PABA) is one of the B vitamins and is essential for normal metabolic reactions. Sulfanilamide is very similar to PABA in structure and can interfere with the metabolic processes that depend upon PABA.

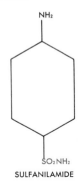

SULFANILAMIDE

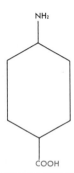

P-AMINOBENZOIC ACID

137

forces the theory that the secondary and tertiary structures of proteins play a key role in the functioning of enzymes. It will be up to future studies to determine whether any loss in specificity results when part of the polypeptide chain is removed or altered.

We shall be considering many different types of enzymes when we discuss the metabolism of various foods. There are, however, certain points that should be made at this time. In some instances, for an enzyme to function it must be associated with a small molecule called a **coenzyme.** In recent years, biochemists have made the significant discovery that parts of the coenzymes are often specific vitamins, such as riboflavin (B_2) and thiamin (B_1); we shall discuss the function of coenzymes when we consider their need in metabolism.

In addition to the organic cofactors (vitamin-containing compounds), a number of inorganic cofactors are required for enzyme catalyses. For example, iron is necessary for electron transport. Copper, too, apparently functions in electron transport; magnesium is essential for the transfer of phosphate groups, and so on. The mechanisms of action of these various cofactors constitute one of the most active fields in enzyme research today, but much is yet to be learned. It is clear, however, that in some cases the enzyme protein alone is not enough to speed the chemical reactions.

Specific enzymes are extremely useful to the biochemist as tools for determining the structure of biological compounds. But they also have many practical applications. The enzymes best known to the general public are the proteolytic enzymes added to detergents for removing protein stains from textiles. A promising anticancer agent is the enzyme asparaginase, which is isolated from the bacterium *Escherichia coli.*

summary

Cells of both plants and animals are all basically alike. The cell membrane surrounds a **colloid** with gel-like properties. The colloid is composed largely of proteins, lipids, and water droplets. Within this gel-like structure are subcellular particles such as the nucleus, mitochondria, and ribosomes. The cell membrane is **selectively permeable.** Water moves back and forth across the membrane with great ease while other molecules have difficulty penetrating this barrier. The separation of two solutions (one inside the cell, the other outside) by a selectively permeable membrane creates an **osmotic system.** Molecules can often move through the cell by free diffusion. However, the cell membrane is not a passive barrier. In some cases metabolic energy is used to **actively transport** nutrients into the cell. Thus, cells may accumulate a high concentration of a particular substance. In such a system, water also moves in and maintains osmotic equilibrium.

Consequently, the cell may increase in volume owing to the increased osmotic pressure inside. Cells may take in larger particles by the processes known as **pinocytosis** and **phagocytosis.**

Cells are capable of carrying out chemical reactions at high speeds and at room temperature because of the presence of specific catalysts called **enzymes.** Enzymes are complex proteins that often require **coenzymes** in order to function. Parts of these coenzymes are often vitamins such as riboflavin (vitamin B_2). Inorganic metals such as copper and iron are essential cofactors for some specific enzymes. As in ordinary chemical reactions, enzyme-catalyzed reactions increase as the temperature is raised. Because very high temperatures destroy enzymes, biological processes have an optimal temperature at which they function best. The enzyme molecule must combine with the substrate molecule in a specific way for catalysis to occur. Each enzyme has a specific shape at the site where the substrate combines. Only substrate molecules with this specific geometry can combine at the **active site.** Certain antibiotics inhibit enzyme reactions because they have shapes similar but not identical to the substrate. They compete with the substrate for the active site on the enzyme, thus inhibiting the catalytic process. These are called **competitive inhibitors.**

1	What happens to the colloidal properties of protoplasm when cells are heated at high temperature?	*for thought and discussion*
2	If a cell that contains the equivalent of 0.5 molar osmotically active particles is placed in a 1 molar solution of sucrose or glucose, what will ultimately happen to the water content of the cell?	
3	Draw a sequence of rough pictures showing the process of pinocytosis.	
4	Draw a scheme depicting the mechanism of enzyme reaction.	

See the books listed at the end of the previous chapter.   **selected readings**

metabolism
and energy
chapter 10

When nutrients pass through the cell membrane, they enter a new environment which contains the metabolic machinery of the cell. One of the machine's chief purposes is to convert these nutrients into *useful* energy. When an organism breaks or splits certain atomic links during metabolism, we want to know whether the energy released is free to do useful biological work. If it is, we call it **free energy.** In biochemical systems, free energy is not literally liberated into the environment. It is conserved as a special **bond energy** (special bonds between elements) that is passed on to other molecules and is used to form new chemical bonds.

The liberation of energy in a biological system is really energy distribution and the formation of new chemical bonds of different **potential** energy. Potential energy is stored energy that can be released to do work. During oxidation and reduction reactions, large amounts of free energy are transferred and used to form energy-rich bonds (potential energy). In oxidation, you will recall, electrons are removed; and in reduction electrons are gained. (Oxidation is not merely the use of molecular oxygen, a special case of oxidation.) Oxidation and reduction, then, involve the movement of electrons from one place to another, with a resulting transfer of energy that can be used to create new bonds of biological interest. Some compounds act as strong reducing

Courtesy Charles J. Ott from National Audobon Society

Figure 10.1 The metabolic machinery of this short-tail weasel (*Mustela erminea*) enables the animal to convert nutrients into useful energy. The animal's energy needs in winter are greater than in summer, because in winter the weasel loses more energy (in this case heat) to the environment.

agents (giving up electrons); others act as strong oxidizing agents (taking up electrons). Let's take a hypothetical situation in which a reducing substance (AH) combines with an enzyme (E) and is subsequently oxidized by another substance (X), according to the following equations:

$$E + AH \rightarrow E{-}AH$$

$$E{-}AH + X \rightarrow E{\sim}A + XH$$

Normally, when AH reacts with X and forms A plus XH, the energy in the A—H bond is lost as heat. However, in the illustration here, when hydrogen is removed [removal of an electron (e) plus a proton (H$^+$)], the energy of the oxidation is conserved in the bond between A and the enzyme (E$\sim$A), indicated by $\sim$. The energy of this bond can be used to combine A with other molecules to make new compounds, as illustrated here:

$$E{\sim}A + B \rightarrow E + AB$$

This is what we mean by a *special bond energy* that is conserved in oxidation reactions. **It is one of the most important concepts**

141

in chemical energy transformations in organisms. It is essential in the synthesis of all known substances in the cell.

OXIDATION
AND ENERGY

The oxidation of carbohydrates or related compounds is the main source of energy for most organisms. During oxidation, *energy-rich* bonds are formed. The organism may then use these bonds to make other bonds, as discussed above. Energy-yielding reactions, then, can be coupled with energy-consuming reactions, with the result that energy is not actually liberated, but is redistributed in the reacting molecules.

The simplest type of oxidation is called **dehydrogenation,** in which hydrogen (an electron plus a proton) is removed from a compound. Every substance undergoing oxidation must be accompanied by a substance undergoing reduction, and vice versa. Hydrogen is not released as a gas in the process. Thus, in the reaction below, AH_2 is the reductant and is being oxidized by B.

$$AH_2 + B \rightleftharpoons A + BH_2$$

If the reaction is reversible, we can start with the chemicals on the right; then BH_2 is the reductant and A is the oxidant. Usually, when electrons and protons flow from a reducing system to an oxidizing system, energy is liberated. We now want to find out how the energy is liberated, trapped, and used for a typical biological oxidation.

First let us consider a typical reaction, the oxidation of an aldehyde. You will recall from Chapter 7 that an aldehyde is a compound having a

$$-C\overset{\textstyle O}{\underset{\textstyle H}{\diagdown}}$$

group. You will also recall from the previous chapter that enzymes are essential for most reactions and combine with a substrate. Enzymes often contain an SH group **(sulfhydryl),** as indicated in Fig. 10.2. In this case, the aldehyde first combines with the enzyme. During the combining process, the double-bond oxygen in the aldehyde part of the molecule reacts with the SH group. This produces a compound that contains an OH and an H on the end carbon atom. One of the bonds that was formerly connected to the oxygen reacts with the sulfur on the enzyme molecule. The hydrogen normally associated with the sulfur moves to the oxygen. This can be described, then, as the enzyme–substrate complex, which is the important intermediate that must be formed before catalytic oxidation can occur. In the next step, the oxidation process occurs, and energy in the mole-

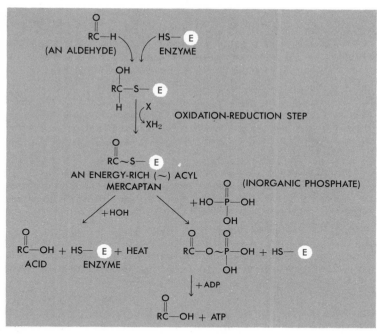

Figure 10.2 The oxidation of an aldehyde to an acid and the generation of an energy-rich group. Note that in the oxidation of the aldehyde, an energy-rich ($\sim$) intermediate is formed on the enzyme. If water splits this intermediate, the energy is lost as heat. If phosphate splits the intermediate, the energy is conserved in the phosphate bond and can be stored in the form of ATP (see text for details).

cule is redistributed to form the energy-rich bond, as indicated in Fig. 10.2.

If this energy-rich bond between the carbon and sulfur is broken by water (hydrolysis), a large amount of energy in the form of heat will be liberated. If the molecule is hydrolyzed, it forms an acid (carboxyl group), and the SH group is restored on the enzyme. In short, the aldehyde is oxidized in the presence of water, forming an acid and heat. During oxidation, therefore, an energy-rich group is formed on the enzyme. Ultimately, this bond energy is lost as heat if it is split by water. On the other hand, if the organism is to use this bond energy, it must transfer the energy-rich group to some other molecule, or use the energy to make a new and useful compound. When this occurs, a new compound is synthesized by drawing on the energy of the carbon–sulfur link. If the organism does not immediately need the reacting group for synthesis, it can store the energy in some other chemical form as a chemical energy reservoir. The enzyme is then freed for further work.

Coupling of bond energy in the reservoir to other systems is the key reaction in all biosynthetic processes. In the illustration used above, we can conserve the oxidation energy by splitting the high-energy bond with phosphoric acid (H_3PO_4) instead of water (HOH). Perhaps now you can begin to see why phosphate (PO_4^{-3}) is extremely important in biological reactions, particularly in the conservation of energy. For some reason that we do not completely understand at the present time, the phosphate group can split the carbon–sulfur link and prevent the loss of energy of the link. This results in the formation of an energy-rich link between the carbon and the phosphorus compound. The energy of the link is conserved, and at the same time the enzyme is freed to catalyze the oxidation of another substrate molecule (an aldehyde) and thus form more energy-rich groups.

As more and more such energy-rich groups are formed, they are transferred to another molecule, called **adenosine diphosphate (ADP)**, whose structure is shown in Fig. 10.3; and from ADP a compound called **adenosine triphosphate (ATP)** is formed (Fig. 10.3). Notice that on the right-hand side of the ATP molecule there are three phosphate atoms. The last two are energy-rich bonds. When one is broken by water, large amounts of heat are liberated, showing its energy-rich character.

ATP is found in all plants, microorganisms, and animals. It serves as the initial storehouse for the energy generated by oxidation reactions taking place in the cell. This bond energy of the phosphate groups directly or indirectly drives all the energy-requiring processes of life—talking, walking, and building cellular material (Fig. 10.4).

The concepts of cellular metabolism can generally be described, therefore, in basic, simple chemical reactions. General oxidation (removal of electrons) is essential for the generation of energy and its subsequent use for the synthesis of the cellular components required for cell growth and division. Although the overall picture is complicated, when it has been possible to isolate

Figure 10.3 Adenosine triphosphate.

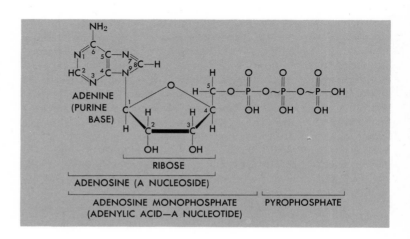

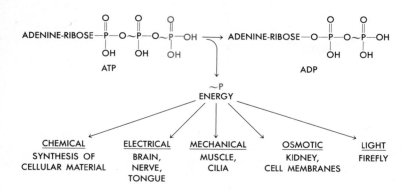

Figure 10.4 All biological reactions depend on the energy derived from converting ATP into ADP.

a given reaction from other complicated cellular processes, the analysis usually results in a relatively simple chemical explanation.

When one studies specific organisms it soon becomes clear that much more complicated interrelationships have developed. For example, organisms that use molecular oxygen as electron acceptors (aerobic) generate much more energy than those that use other, metabolic-generated electron acceptors (anaerobic organisms). For this reason aerobic metabolism is much more efficient than anaerobic metabolism in so far as energy generation is concerned. We shall analyze these two processes in detail, because they have much to tell us about some of the major principles that have been established in biochemistry during the past 75 years.

ALCOHOLIC FERMENTATION (ANAEROBIC METABOLISM) Although man has known about alcoholic fermentation since prehistoric times, its cause was not discovered until about 1860. A short time before this, however, the French chemist Gay-Lussac had described the production of alcohol from sugar with the following equation:

Glucose $\rightarrow$ 2 Carbon dioxide + 2 Ethyl alcohol

$(C_6H_{12}O_6)$ $(2CO_2)$ $(2CH_3CH_2OH)$

In the middle of the 19th century, there were two conflicting ideas about alcoholic fermentation. One group thought the process to be strictly chemical; the other claimed that the process was intimately associated with living organisms. The great French chemist, biochemist, and physiologist Louis Pasteur concluded that fermentation takes place in the absence of oxygen **(anaerobic)** and only in the presence of certain microorganisms. Without these organisms, fermentation would not occur. Pasteur's experiments indicated that fermentation is a physiological process

Figure 10.5 Yeast being grown in a large commercial vat.

closely bound up with the life of cells. This contradicted the idea that all living things needed oxygen. Pasteur believed that there were substitutes for molecular oxygen. In short, he believed that fermentation is the result of life activities being carried on in the absence of air.

More than 20 years passed before the next major breakthrough in the investigation of fermentation occurred. In 1897, the German physiologist Eduard Buchner accidentally stumbled onto the discovery that opened the door to the secrets of fermentation, and the whole field of modern enzyme chemistry as well. He was primarily interested in making what he called "protoplasmic extracts" from yeast which were to be injected into animals. Hopefully, the injections might make the animals live longer. He ground yeast with sand, mixed it with certain other compounds, and finally squeezed out the juice with a hydraulic press. Since it was difficult to prepare this material daily, he made various attempts to preserve the cell-free extract. Because it was to be injected into animals, ordinary antiseptics, such as chloroform, could not be used, so he tried the usual kitchen chemistry method of preserving fruit by adding large amounts of sugar.

To Buchner's surprise, the sugar was rapidly decomposed

into carbon dioxide and alcohol, or fermented, by the yeast juice! So far as we know, he was the first person to observe fermentation in the complete absence of living cells. At last it was possible to study the process of alcoholic fermentation outside the living cell. Buchner's work was soon followed by intensive studies of the properties of yeast juice. Yeast juice was found to ferment many sugars, such as glucose, fructose, mannose, sucrose, and maltose. Glucose was converted by the juice into ethyl alcohol and carbon dioxide, according to the Gay-Lussac equation. The next question to be answered was *how* yeast juice brought about the fermentation of sugar.

THE PHOSPHORYLATION OF GLUCOSE The first important analysis of the activity of yeast juice was made by two English scientists, Harden and Young, when in 1905 they added fresh yeast juice to a solution of glucose. At *p*H 5, fermentation began almost at once. Although the rate of carbon dioxide production soon fell off, they could restore fermentation by adding inorganic phosphate. The recovery was only temporary, however, because the phosphate was soon used up; the rate of fermentation dropped as the phosphate concentration declined. But by adding more phosphate they sparked another burst of fermentation.

The work of Harden, Young, and others has shown that phosphate is essential in the metabolism of carbohydrates. Phosphate aids in breaking down carbohydrates to smaller molecules, and is also essential for the conservation of energy released during the breakdown process.

Subsequent investigations have led to the conclusion that when yeast juice metabolizes glucose, ATP energy is required for the initial reaction. As Fig. 10.6 shows, a phosphate from ATP is transferred to glucose, making a glucose phosphate compound plus ADP. Through a series of reactions, glucose phosphate is converted to the hexose sugar **fructose phosphate.** Fructose phosphate accepts a second phosphate from another ATP molecule, forming a hexose sugar (a C_6 sugar) with two phosphate groups—fructose **di**phosphate. The first product of the phosphorylation reactions that prepare glucose for metabolic breakdown, then, leads to the formation of a hexose diphosphate, the Harden and Young ester.

Next, the fructose diphosphate is split into two compounds, each containing three carbons **(triose)** and one phosphate. These triose phosphate compounds are interchangeable and are, in effect, identical. The important one for fermentative metabolism is an aldehyde compound called **glyceraldehyde 3-phosphate** (Fig. 10.7). This compound is oxidized in exactly the same way as we described aldehyde oxidation on page 143. During the process, energy is liberated in the form of phosphate bond energy,

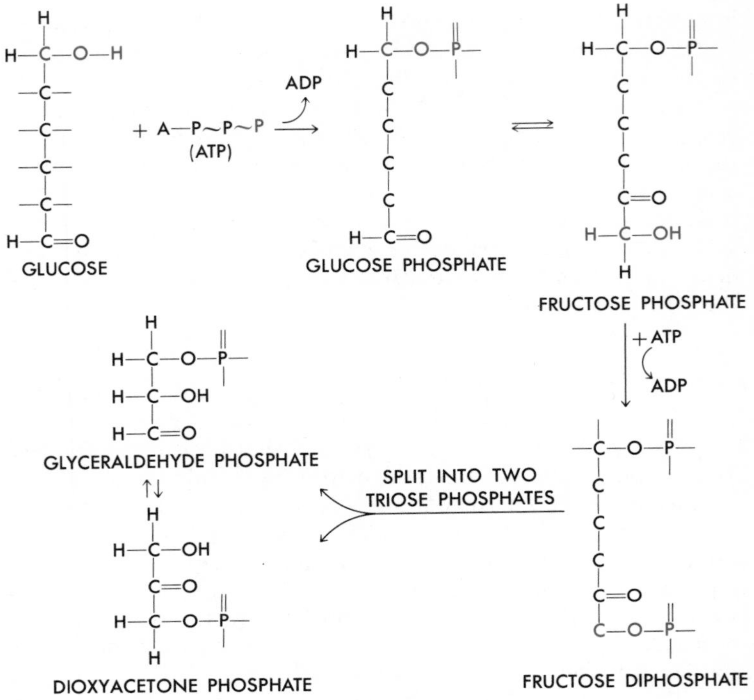

Figure 10.6 Phosphorylation of glucose and the formation of aldehyde phosphates. The hydrogen and oxygen atoms are not shown on all the phosphorus and carbon atoms.

which leads to the synthesis of a molecule of ATP. The other product of the reaction is **phosphoglyceric acid** (Fig. 10.7). It is this reaction that requires the inorganic phosphate, which makes the ATP that is essential to the introduction of glucose into the metabolic pathway.

In additional reactions, water is removed from phosphoglyceric acid in such a way that the phosphate in the phosphoglyceric acid also becomes energy-rich. This group is transferred to ADP and forms ATP. We end up also with a molecule of **pyruvic acid** (Fig. 10.8). Thus, if we start with a glucose molecule, we put in two ATP molecules to make two molecules of glyceraldehyde 3-phosphate. During the oxidation and dehydration reactions, we recover two energy-rich groups for each triose. Consequently, from the two trioses we recover four ATP molecules. The net gain in terms of bond energy in the fermentative metabolism of glucose is two ATP molecules.

During the oxidation of the glyceraldehyde 3-phosphate, an enzyme is essential for catalysis. In addition, the enzyme requires a coenzyme capable of accepting the hydrogens during the oxidation of the aldehyde. The coenzyme required for the oxidation of an aldehyde is a derivative of the B vitamin, **niacin.** This finding provided the first revealing clue to the function of vitamins in cellular metabolism. The particular coenzyme that is

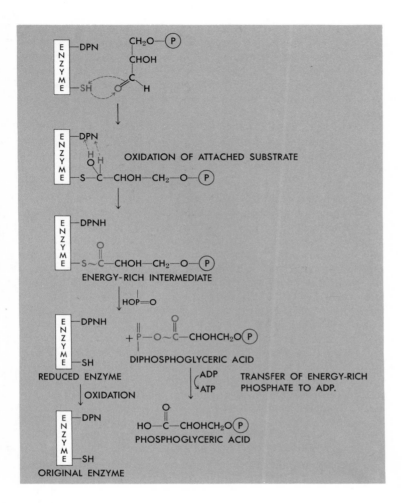

Figure 10.7 The formation of ATP from the oxidation of glyceraldehyde-phosphate.

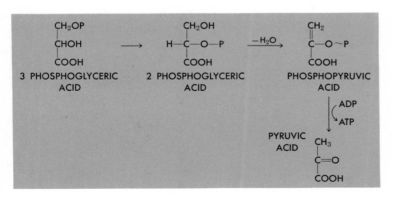

Figure 10.8 The formation of ATP and pyruvic acid from phosphoglyceric acid.

involved in the dehydrogenation of the aldehyde is known by its initials as **DPN (diphosphopyridine nucleotide).** Recently it has been suggested that the name be changed to nicotinamide adenine dinucleotide (NAD). The name of the enzyme requiring DPN as a coenzyme is **glyceraldehyde 3-phosphate dehydrogenase.** Sometimes it is referred to as **triose phosphate dehydrogenase.** Thus, in the oxidation of the aldehyde, DPN is reduced. It is during this oxidation process that the energy-rich intermediate is formed.

For each molecule of glucose fermented, two molecules of alcohol and two of CO_2 are formed. The final key step in the formation of alcohol is the removal of CO_2 from pyruvic acid (Fig. 10.9). The B vitamin, thiamine, is required for this reaction. Thus, when pyruvic acid loses the carboxyl group by **decarboxylation** it forms a new compound, acetaldehyde. Certain organisms can oxidize acetaldehyde and form acetic acid. Yeast cells, however, do not have the enzyme capable of oxidizing this short-chain aldehyde. But they do have an enzyme capable of catalyzing the transfer of hydrogen from reduced DPN to aldehyde (Fig. 10.10). The reduction of acetaldehyde leads to the formation of **ethyl alcohol;** the enzyme that catalyzes the reaction is called **alcohol dehydrogenase** (that is, the enzyme catalyzes the reverse reaction and has been named as indicated).

In the final steps of fermentation, we have decarboxylation and reduction—forming one molecule of alcohol and one molecule of CO_2 from each pyruvic acid molecule. Since two pyruvic acid molecules are formed during the breakdown of one of glucose, we end up with two molecules of alcohol and two of CO_2, thus satisfying the Gay-Lussac equation for this process.

MUSCLE METABOLISM—
GLYCOLYSIS

For a long time after Harden and Young's discovery of the phosphorylation of sugars in alcoholic fermentation, the process was not considered significant, except as a means of shaping the hexose molecule for fermentative breakdown. However, similar studies of other cells, particularly muscle cells, revealed that phosphate plays a dominant role in energy transformation. This is especially true in muscle contraction.

Among the questions asked by the earlier investigators of muscle contraction were these: What is the chemical source of

Figure 10.9 The formation of alcohol from pyruvic acid.

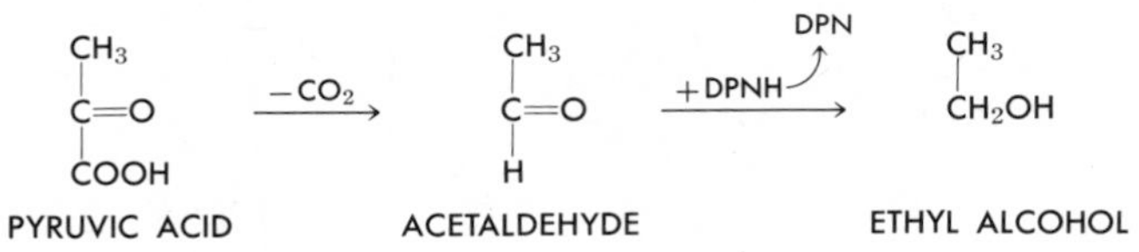

PYRUVIC ACID   ACETALDEHYDE   ETHYL ALCOHOL

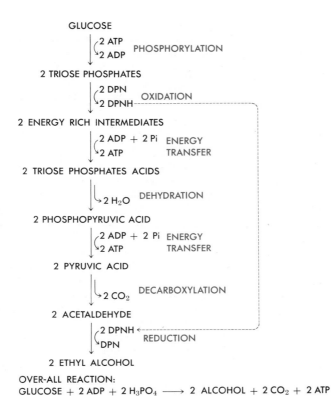

OVER-ALL REACTION:
GLUCOSE + 2 ADP + 2 H$_3$PO$_4$ $\longrightarrow$ 2 ALCOHOL + 2 CO$_2$ + 2 ATP

Figure 10.10 Summary of key reactions in alcoholic fermentation.

the energy that powers the muscle machine? And how is the chemical energy of the cells transformed into the mechanical energy of muscle contraction? The investigators soon discovered that muscle can contract in a normal manner in the complete absence of oxygen. They also discovered that **lactic** acid is produced during the contraction process and is associated with muscle fatigue. If a fatigued muscle, however, is supplied with oxygen, it soon recovers its ability to contract, and the lactic acid disappears.

By the late 1920s, it became evident that most of the energy spent by muscle contraction came from the metabolism of reserve **glycogen,** which is stored in muscle. Investigators also found that glycogen was converted to lactic acid in a process very similar to that of alcoholic fermentation. We call this process **glycolysis,** that is the breakdown of glycogen. Eventually, other investigators pointed to ATP as the immediate energy source for muscle contraction. The breakdown of glycogen is essential for the formation of ATP, and the reactions are the same as in the breakdown of glucose by the yeast cell.

An additional compound, however, was isolated from mus-

$$O$$

$$H-N \quad P-OH$$
$$OH$$
$$C=NH$$
$$N-CH_3$$
$$CH_2$$
$$COOH$$

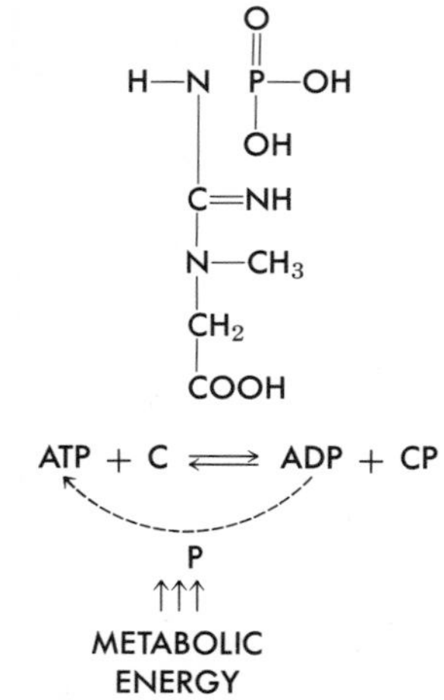

$$ATP + C \rightleftharpoons ADP + CP$$

$$P$$
$$\uparrow\uparrow\uparrow$$

**METABOLIC
ENERGY**

Figure 10.11 Creatine phosphate and the relationship to ATP. Creatine phosphate acts as an important storehouse of energy-rich phosphate groups. When ATP reacts with creatine, ADP and creatine phosphate (CP) are formed. The ADP is now free to "pick up" another energy-rich phosphate. The process continues until essentially all the creatine is converted into creatine phosphate. When ATP is used rapidly, the ADP formed can be rapidly converted back into ATP by using the CP reserve.

cle, and it proved to be a very important storehouse of energy-rich phosphate. This compound is **creatine phosphate** (Fig. 10.11). Creatine, it turned out, could react with ATP and form creatine phosphate, which accumulates in muscle cells in very high concentrations. When phosphate from ATP is transferred to creatine, ADP is produced; and ADP is essential for the continued metabolism of glucose.

When a muscle contracts, it uses ATP vigorously, but the mechanism that converts the chemical energy of ATP into mechanical work remains obscure. When we study muscle for its chemical composition, we find two important proteins. One is called **actin** and the other **myosin.** When actin and myosin are mixed they form **actomyosin,** which can be made into threads that rapidly decompose ATP. At the same time, the threads contract. Actin and myosin are the major proteins of contractile muscle fiber (Fig. 10.12).

Gradually, the mechanical arrangements of these fibers in the muscle, and the function of ATP in causing contraction, are being cleared up. We now think that muscle contracts when its actin and myosin fibers slide over one another. ATP can dissociate, or separate, actin and myosin. It is possible that when a nerve impulse stimulates a muscle to contract, the first thing that happens may be the activation of the enzyme **ATPase,** which splits ATP. The destruction of the ATP would allow the actin and myosin to combine. The actomyosin threads may then contract (Fig. 10.13).

One important difference exists between the breakdown of glycogen and carbohydrate by muscle and yeast. When a muscle breaks down carbohydrate, alcohol and CO_2 are not produced! Muscle does not have the enzyme that is capable of catalyzing

Figure 10.12 Skeletal muscle is made up of a number of parallel bundles of very large multinucleated cells. These muscle cells contain a number of parallel myofibrils, which are the functional units of the contractile machinery. The cross-striated appearance is due to the differences in the optical properties of the myofibril substances.

Eric V. Grave

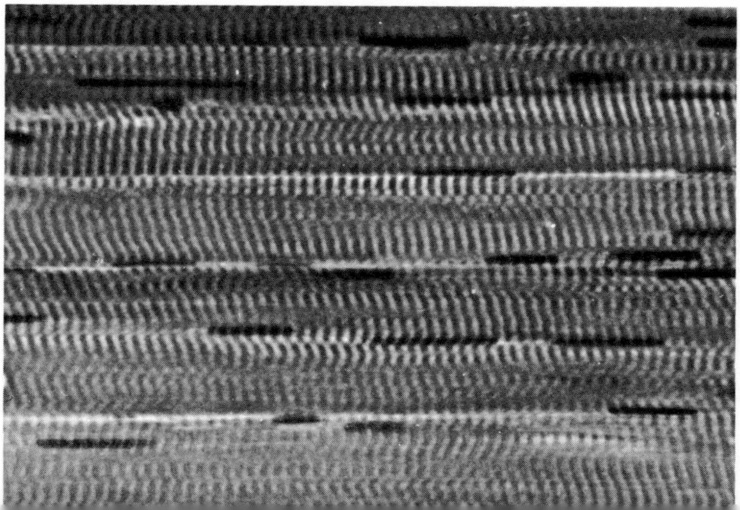

the removal of CO_2 from pyruvic acid; thus, the latter substance must accept the hydrogens from reduced DPN. For this reaction to occur, the muscle has an enzyme called **lactic acid dehydrogenase.** It is this enzyme that catalyzes the transfer of hydrogens from reduced DPN to pyruvic acid, making lactic acid. As the name indicates, this enzyme will catalyze the reverse reaction. The reaction is shown in Fig. 10.14. Thus, pyruvic acid in muscle acts very much like acetaldehyde in alcoholic fermentation. The net effect of the anaerobic reaction sequence in muscle is as follows: For every glucose unit derived from glycogen, two molecules of lactic acid are produced. The coenzyme DPN is alternately reduced and oxidized. In muscle extract, as in yeast juice, the sequence generates four new energy-rich phosphate bonds for each glucose molecule metabolized.

CONTROL OF GLYCOGEN SYNTHESIS AND BREAKDOWN Glycogen is a complex polymer of glucose. As shown in Fig. 10.16, the sugar molecules are linked together by a bond **(glycosidic)** between carbon 1 of one glucose and carbon 4 of the next glucose. Branching from this straight chain, which may have as many as 18 glucose units, are a number of 1–6 glycosidic linkages, which then continue as 1–4 linkages. Glycogen, therefore, is a branched, complex polysaccharide with a molecular weight as high as 4 million. Enzymes attack it at either the 1–4 or the 1–6 linkages. Many hydrolytic enzymes (glycosidases and amylases) can break down glycogen to smaller polysaccharides or free-glucose units. In muscle, when glycogen is broken down it reacts first with inorganic phosphate instead of water at the 1–4 glycosidic linkage; the process is thus a phosphorolysis, not a hydrolysis.

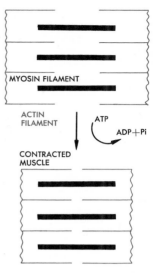

Figure 10.13 This schematic representation shows the relationship between actin and myosin filaments from muscle and what is thought to occur when ATP energy induces contraction.

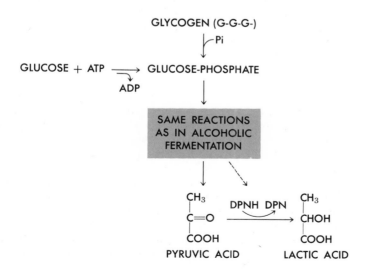

Figure 10.14 Glycolysis: the formation of lactic acid from glycogen. The energy in the glucose-glucose links in glycogen is such that inorganic phosphate (Pi) can be used to form the glucose phosphate. When free glucose is used, ATP energy is necessary to add phosphate to the sugar molecule. In the synthesis of glycogen from free glucose, phosphate bond energy must be used.

153

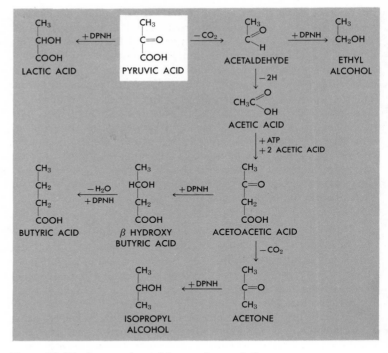

Figure 10.15 Some products of fermentative metabolism.

The enzyme that catalyzes this reaction is **phosphorylase,** and the product of the reaction is glucose-1-phosphate and the remainder of the glycogen molecule. Once the outer tier of 1–4 glycosidic linkages is broken, the reaction stops. There are hydrolytic enzymes, however, that split the 1–6 linkage and thereby expose additional 1–4 glycosidic linkages, which phosphorylase will attack. The new phosphate ester formed in the breakdown of glycogen is readily converted into glucose-6-phosphate by the enzyme phosphoglucomutase; this phosphate ester, you will recall, is the compound formed when ATP reacts with glucose. All these reactions are reversible, and if there is an excess of glucose and ATP, glycogen synthesis will occur.

The control of the synthesis and degradation of glycogen is an intriguing and informative chapter in the history of biochemistry. Muscle phosphorylase exists in two forms, a and b. Phosphorylase a has a molecular weight of 500,000 and is made of four identical polypeptide chains. Each polypeptide contains a serine residue whose hydroxyl group is esterified to phosphate. Muscle also contains a specific phosphatase that will remove these phosphates. When this happens, phosphorylase a dissociates into two inactive dimeric phosphorylase b molecules. Phosphorylase b can be converted into phosphorylase a by using ATP to phos-

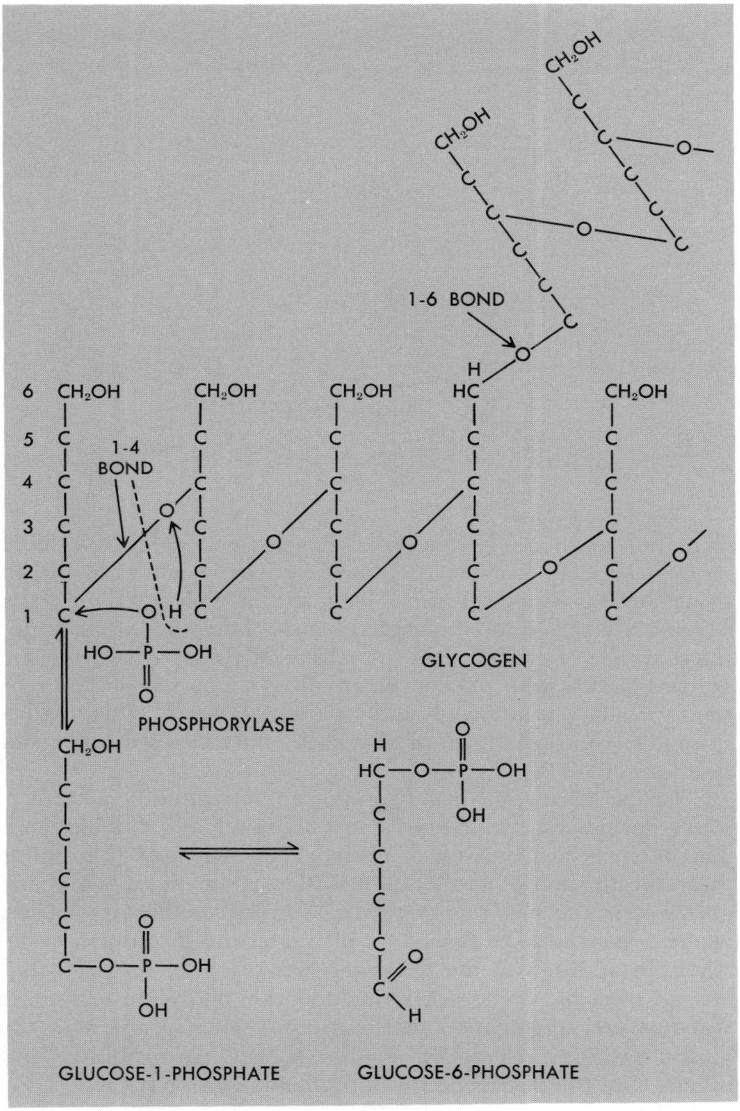

Figure 10.16 The initial step in glycogen metabolism.

phorylate the serine residues, as shown in the following equation:

$$2 \text{ Phosphorylase b} + 4\text{ATP} \rightarrow \text{Phosphorylase a} + \text{ADP}$$

The interconversion of phosphorylase a and b (presented schematically in Fig. 10.17) is the basic mechanism for the control of glycogen breakdown. The conversion of phosphorylase b into a requires the presence of an active phosphorylase b kinase. The

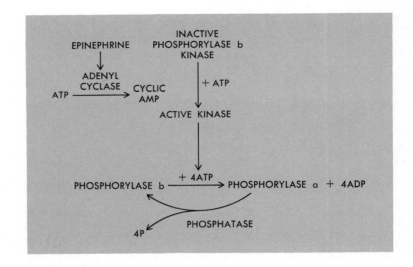

Figure 10.17 The conversion of phosphorylase a and b: control of glycogen breakdown.

Figure 10.18 Structure of cyclic adenylic acid.

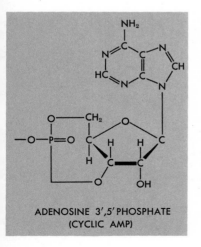

ADENOSINE 3′,5′ PHOSPHATE
(CYCLIC AMP)

formation of this active kinase depends upon a second kinase that requires a specific cofactor, 3′,5′-cyclic adenylic acid (Fig. 10.18). **Cyclic AMP** is formed from ATP by the enzyme adenyl cyclase, whose activity is under hormonal control by epinephrine. Thus, the sequence of events upon the release of epinephrine from the adrenal glands, which occurs when one is frightened, leads eventually to the formation of active phosphorylase a. This in turn stimulates the breakdown of glycogen, which leads to the rapid synthesis of ATP.

Cyclic AMP appears to be important in regulating a number of cellular processes in various stages of growth and development. This often occurs by way of hormones, which appear to effect the concentration of cyclic AMP in the target organ. Thus, the hormone is the first messenger that travels to the target cells, where it increases the formation of the second messenger, cyclic AMP. In addition to the phosphorylase reaction, activities that are affected by cyclic AMP include the phosphofructokinase reaction, cell aggregation in slime mold, HCl secretion in the gastric mucosa, release of insulin, kidney permeability, and others. (See the discussion on hormone action.)

The synthesis of glycogen follows a pathway different from the reversal of the phosphorylase reaction; however, glucose-1-phosphate is the initial substrate. The second substrate is a compound very similar to ATP, **uridine triphosphate** (UTP). As shown in Fig. 10.19, UTP reacts with glucose-1-phosphate to form **uridine diphosphate glucose** (UDPG) and inorganic pyrophosphate. UDPG is referred to as "active" glucose, since in the presence of the enzyme glycogen synthetase the glucose of UDPG is transferred to glycogen to increase the chain length by way of 1–4 glycosidic linkages. Glycogen synthetase occurs in a phosphory-

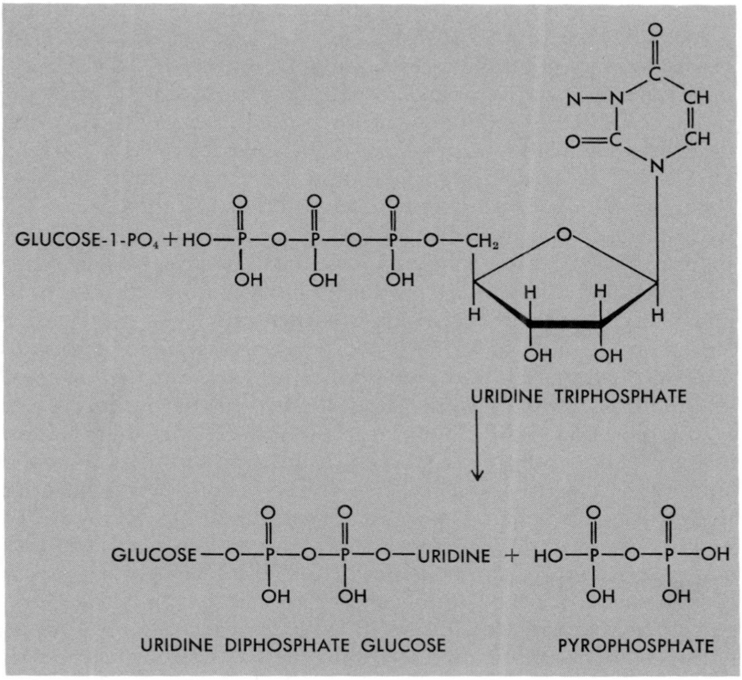

Figure 10.19 Formation of uridine diphosphate glucose from UTP and glucose-1-phosphate.

lated and dephosphorylated form. The dephosphorylated enzyme is enzymatically active, and can be converted into the inactive form by a specific kinase and ATP. This inactive enzyme can be activated by glucose-6-phosphate, thus assuring glycogen synthesis only if excess sugar and phosphate bond energy are available (UTP). The glycogen synthetase kinase, like the phosphorylase b kinase, is active only in the presence of cyclic AMP; thus, epinephrine also affects glycogen synthesis. The relationship between glycogen synthesis and breakdown is shown in Fig. 10.20.

After the formation of glucose-6-phosphate, glycolysis and

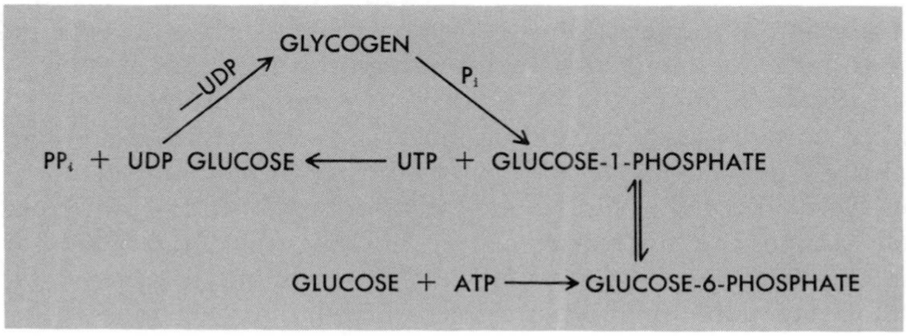

Figure 10.20 Synthesis and breakdown of glycogen.

alcoholic fermentation follow a common path until pyruvic acid is formed. Here the pathways again diverge, for muscle, unlike yeast, does not contain carboxylase, the enzyme that decarboxylates pyruvic acid. Pyruvic acid will, nevertheless, react with the reduced DPN and oxidize it in much the same way as the acetaldehyde did in the yeast extract. When pyruvic acid is reduced by DPNH, lactic acid is formed, and the enzyme that catalyzes this reversible oxidation–reduction reaction is called lactic dehydrogenase. The net effect of the anaerobic reaction sequence in muscle is that for every glucose unit (from glycogen) two molecules of lactic acid are produced. DPN is alternately reduced and oxidized, as it is in alcoholic fermentation.

In yeast juice, as in muscle extract, the sequence generates four new energy-rich phosphate bonds for each glucose molecule metabolized. In the fermentation of free glucose, however, two of the new bonds are used in the preliminary phosphorylation of the glucose molecule, so there is a net gain of only two molecules of ATP. In the first stage in glycolysis, on the other hand, the glycogen is split by inorganic phosphate and not by ATP; so when the product, glucose-1-phosphate, is converted into glucose-6-phosphate, only one molecule of ATP is needed to produce each molecule of hexose diphosphate. Muscle glycolysis, therefore, gains a net of three molecules of ATP for each glucose unit of glycogen metabolized, thus enabling us to classify glycogen as an energy reservoir. Although the ester link of glucose phosphate that is formed in the uptake of free glucose uses ATP, the energy is not lost when the glycosidic linkage in glycogen is created, for this energy is such that inorganic phosphate can react with it to form again a hexosephosphate ester, which can be metabolized.

The control of glycogen or glucose metabolism in muscle is very similar to the process described for yeast. The reoxidation of DPNH, the action of inorganic phosphate, and the utilization of ATP are the essential processes. Can you describe now the chemical events that take place in muscle during a 440-yard dash? What happens to ATP? If the inorganic phosphate increases in concentration, what happens to glycogen breakdown?

In muscle glycolysis, then, the same fundamental mechanisms, oxidation–reduction, dehydration, and phosphorylation, are in operation as are present in alcoholic fermentation. These reactions are the basic ones involved in energy and carbon transformation in fermentations. In the past 50 years, we have found that in the metabolism of various organisms these relatively simple steps account for almost all the fermentation products formed from carbohydrates by the organisms. For example, lactic acid bacteria produce lactic acid by a sequence of reactions identical to those observed in muscle. Outlined in Fig. 10.15 are a number of reactions that are carried out by different organisms. Note that in the formation of most of these compounds, only simple oxidation–reduction, hydrolytic, or decarboxylation reactions occur. Their net effect is to produce a number of electron acceptors, which then oxidize the reduced DPN that is formed in the triose phosphate dehydrogenase reaction. Pyruvic acid is formed in the same way by these organisms.

OXIDATION OF REDUCED DPN BY OTHER SUBSTANCES (AEROBIC METABOLISM) So far, we have discussed the oxidation reactions in which DPN acts as the initial hydrogen or electron acceptor. This step is the important initial dehydrogenation in the metabolism of several substrates. In a series of brilliant studies beginning in the early 1920s, the German biochemist Otto Warburg found traces of two other important coenzymes in cells grown under aerobic conditions.

He was intrigued by the ability of iron-containing compounds to catalyze the oxidation of many different organic substances by using molecular oxygen. Warburg suspected that the iron contained in these compounds was responsible for their catalytic activity. He reasoned that an iron-containing substance is needed in the cell if oxygen is to be activated and used. A search led to the discovery of several iron-containing compounds, the function of which is to carry electrons to molecular oxygen. The compounds are called **cytochromes.**

It turned out that a second conenzyme was needed to transport electrons (plus the protons H^+) from reduced DPN to molecular oxygen—**flavin,** a derivative of the vitamin **riboflavin.** It soon became clear that when hydrogen atoms (electrons plus protons H^+) were removed from various substances they were transported by DPN flavin and iron-containing cytochromes to molecular oxygen, with a resulting production of H_2O. In addition to the compounds mentioned so far, there are other electron acceptors that are important in the metabolism of carbohydrates.

[It should be noted here that we often use the terms *electron transport* and *hydrogen transport* interchangeably. We do so because, although both processes occur, the net effect of each one is the same. Thus, a compound such as AH may dissociate into A^- and H^+ before oxidation (electron removal). Thus,

$$AH \rightarrow A^- + H^+$$

$$A^- + B \rightarrow A + B^-$$

B^- may now take up a proton from the environment as follows:

$$B^- + H^+ \rightarrow BH$$

The net effect is to transfer a hydrogen atom:

$$AH + B \rightarrow A + BH$$

In some cases, this actually happens without the proton dissociation, that is, hydrogen transport instead of electron transport.]

Figure 10.21 The chain of electron transport from DPNH to oxygen.

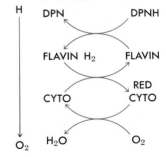

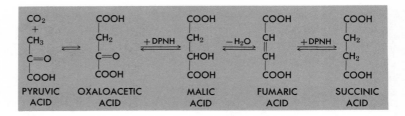

Figure 10.22 During the fixation of carbon dioxide by pyruvic acid adenosine triphosphate is needed. The oxaloacetic acid that is formed can be reduced and in several steps, as indicated, can form succinic acid.

CO_2
+
CH_3
|
$C=O$
|
COOH
PYRUVIC
ACID

$\rightleftharpoons$

COOH
|
CH_2
|
$C=O$
|
COOH
OXALOACETIC
ACID

+DPNH

COOH
|
CH_2
|
CHOH
|
COOH
MALIC
ACID

$-H_2O$

COOH
|
CH
||
CH
|
COOH
FUMARIC
ACID

+DPNH

COOH
|
CH_2
|
CH_2
|
COOH
SUCCINIC
ACID

CARBON DIOXIDE FIXATION

Green plants take up CO_2 in the presence of water; then they make use of light energy and form carbon–carbon links, which, eventually, are reduced to a carbohydrate. The reaction is called CO_2 fixation and reduction. This is the ultimate source of all complex organic molecules on the surface of the Earth. However, it is not the only way in which CO_2 is **fixed.** In the presence of ATP, CO_2 can be added to several compounds. Pyruvic acid is one such compound. When pyruvic acid takes up CO_2, it forms a new four-carbon compound capable of accepting hydrogens, which leads to the formation of other types of acids, in particular, succinic acid (Fig. 10.22). These acids are very important in stimulating cell respiration, and, as we shall find in a moment, are very important in converting carbohydrates completely into CO_2 and H_2O. Large amounts of energy are released in this process.

Figure 10.23 Oxidative decarboxylation of pyruvic acid forms the energy-rich acetic acid molecule.

CH_3
|
$C=O$ + THIAMINE
|
COOH
PYRUVIC ACID

$\downarrow \searrow CO_2$

LIPOIC ACID
COENZYME A
FLAVIN
DPA

$\downarrow \searrow$ DPNH

O
||
$CH_3—C$ S—CoA
ACETYL COENZYME A
(ACTIVE ACETIC ACID)

OXIDATIVE METABOLISM

One additional key reaction in the metabolism of pyruvic acid is the removal of the carboxyl (—COOH) group, forming CO_2. The two-carbon aldehyde molecules remain attached to a coenzyme and are oxidized; the result is the formation of an energy-rich acetic acid (Fig. 10.23). The activated acetic acid combines with the four-carbon compound **(oxaloacetic acid),** which is formed from another pyruvic acid by the CO_2 fixation reaction discussed above. This leads to the formation of the complex molecule called **citric acid** (Fig. 10.24). Eventually, citric acid is broken down step by step until two of its carbons are converted into CO_2, and oxaloacetic acid is reformed. The electrons removed are transported to oxygen, and water is formed. The oxaloacetic acid regenerated accepts another two-carbon unit from the decarboxylation of another pyruvic acid molecule. The cyclic process of adding two-carbon units to oxaloacetic acid to form citric acid, and then breaking it down to reform oxaloacetic acid, is called the **citric acid cycle** (Fig. 10.25).

This integrated group of enzymes—the dehydrogenases and the decarboxylases—which engineer the complex series of reactions in the citric acid cycle are located in the mitochondria of

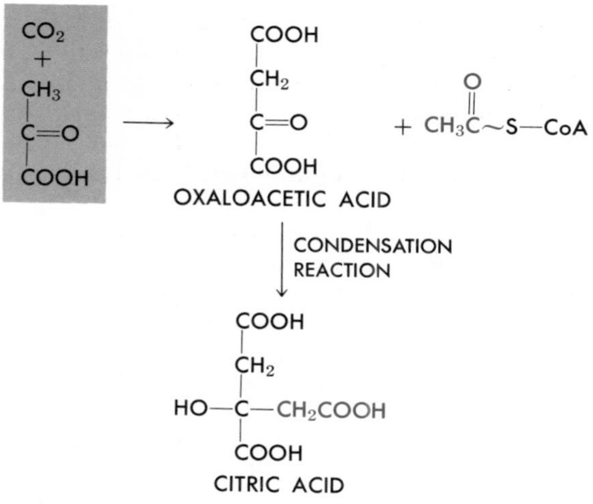

$$CO_2 + CH_3\text{-}C\text{=}O\text{-}COOH \longrightarrow$$

OXALOACETIC ACID

$$+ \; CH_3\overset{O}{\overset{\|}{C}}\sim S\text{—CoA}$$

CONDENSATION
REACTION

CITRIC ACID

Figure 10.24 The synthesis of citric acid from active acetic acid and oxaloacetic acid.

the cell. Bound to this complicated structure, in some unknown way, are the various enzymes and coenzymes. The mitochondria from several different kinds of cells are known to be capable of carrying on this final oxidative stage of cell metabolism by themselves.

The significance of these oxidative reactions is that energy-rich phosphate groups are generated during the complete combustion of pyruvic acid. These oxidative processes generate ATP much in the same way that the oxidation of the aldehyde does. The process is called **oxidative phosphorylation** and is one of the crucial events in oxidative metabolism. The energy liberated in the oxidation of these coenzymes is thus used to synthesize ATP. A mitochondrion seems to have three sites for the formation of

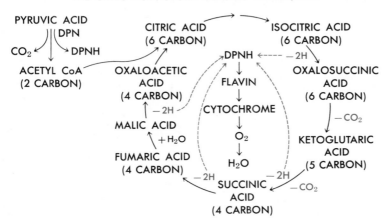

THE CITRIC ACID CYCLE—AEROBIC METABOLISM

Figure 10.25 When active acetic acid condenses with oxaloacetic acid, citric acid is formed. The citric acid then goes through a series of oxidation and decarboxylation reactions leading ultimately to the reformation of oxaloacetic acid. The net effect is the complete combustion of pyruvic acid (active acetic acid) to carbon dioxide and water. The hydrogens removed are transported to oxygen by means of the electron transport chain. All these reactions occur in the mitochondria.

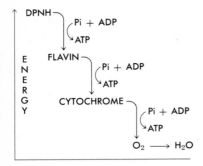

Figure 10.26 Oxidative phosphorylation, the formation of ATP during the transfer of electrons over the electron-transport chain.

ATP during the oxidation of reduced DPN by molecular oxygen (Fig. 10.26). The oxidation of one molecule of reduced DPN by the mitochondrial enzyme complex produces three molecules of ATP. The total combustion of pyruvic acid to CO_2 and water leads to the net synthesis of 15 ATP molecules. It is no wonder that the mitochondria have been called the "powerhouses" of the cell.

Since 15 ATP molecules are formed from the metabolism of pyruvic acid, it is clear that the majority of the energy in a compound becomes available as ATP. But this occurs only when a compound is oxidized all the way to CO_2 and water. Thus, aerobic organisms are capable of obtaining large amounts of energy from very small amounts of food. This is in contrast to those organisms that must live anaerobically. For example, a yeast cell that must grow and multiply under anaerobic conditions uses more than 30 times as much sugar as it would use if it grew and metabolized under aerobic conditions. In other words, there is still a large amount of energy in the alcohol that is formed by the yeast cell.

Thus, in the aerobic metabolism of glucose the initial pathways of breakdown are the same as those we discussed for the fermentation or anaerobic system. Under aerobic conditions the DPNH that is formed in the triose phosphate dehydrogenase reaction is not used to form alcohol (in yeast) or lactic acid (in muscle); instead it is oxidized by molecular oxygen by way of the flavin and cytochrome systems. When the oxygen concentration is varied, it is possible to get some fermentation products as well as aerobic products (Fig. 10.27).

A muscle under vigorous activity will consume oxygen so fast that some of the hydrogens of DPNH may be diverted to pyruvic acid and form lactic acid. Thus, there may be competition between two systems for the hydrogen derived in the dehydrogenase reactions. When the muscle accumulates lactic acid because of the exhaustion of oxygen, it acquires what is called

Figure 10.27 Comparison of the anaerobic and the aerobic phase of metabolism. Note the large amount of ATP that is synthesized under aerobic conditions.

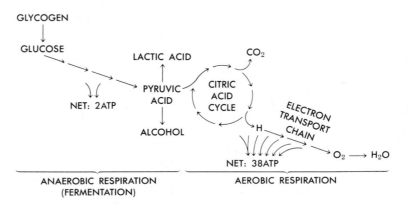

an **oxygen debt.** When resting, the muscle will then use more oxygen than it normally does in order to metabolize the accumulated lactic acid.

Although we have talked about the citric acid cycle as one that liberates large amounts of energy, it is also valuable in the synthesis of carbon skeletons, which can be used for making many other compounds. For many years the breakdown of carbohydrates was thought to be primarily a "tearing-down" process **(catabolism),** whose sole purpose was the liberation of energy. This is true so far as energy utilization is concerned, but we know that the formation of key intermediates in the breakdown of carbohydrates is also significant in the "building" **(anabolism)** of compounds of biological interest. During the rapid growth of cells, the principal function of the citric acid cycle may be to supply the cellular carbon skeletons for biosynthesis.

FATTY ACID OXIDATION The mitochondrion also functions as the cellular furnace for the combustion of fatty acids, amino acids, and other fuels. A typical fatty acid, **butyric acid,** is shown in Fig. 10.28. The metabolism of this acid is very much like the metabolism of pyruvic acid. It is first oxidized and eventually split into two carbon fragments, which can add to oxaloacetic acid in the citric acid cycle (Fig. 10.25).

Because there are more hydrogens on fatty acids than there are on compounds such as glucose or pyruvic acid, the metabolism of fatty acids liberates much more energy than the consumption of an equivalent length of carbohydrate. For example, from butyric acid we obtain two active C_2 units; on entering the citric acid cycle they lead to the generation of 24 ATP molecules. Five additional oxidative steps lead to the formation of five more ATP molecules, giving a total of 29, which come from a single C_4 unit. Fatty acids, then, are a much better source of energy than carbohydrates are.

Synthesis of Fatty Acids. The *de novo* synthesis of fatty acids involves enzymes and coenzymes that are different from those used in degradation, although acetyl coenzyme A (CoA) again functions as a major carbon source for lipogenesis. Furthermore,

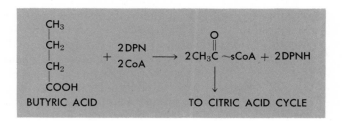

Figure 10.28 Metabolism of a fatty acid (butyric acid).

the synthesis occurs in the cytoplasm in a complex that contains all the lipogenic enzymes in a unit with a particle weight of well over 2 million. The two key carbon sources for fatty acid synthesis are acetyl CoA and malonyl CoA. The latter is formed from acetyl CoA by a CO_2 fixation reaction. The enzyme that catalyzes this reaction contains the B vitamin biotin. Details of these reactions can be found in advanced biochemistry texts.

The enzymes used in fatty acid synthesis are associated as a complex in the cell. Attempts to separate the complex into distinct enzymes have been unsuccessful. It would appear that the separated enzymes are inactive, perhaps because they do not have the proper shape; but when they are aggregated into a **multienzyme complex,** they can carry out all the necessary reactions in sequence. There are obvious advantages to such a system. If we are dealing with a chemical series, and the product of one reaction is the substrate for the next enzyme, the enzyme complex acts as a cellular assembly line. The complex would be considerably more efficient than a system in which each enzyme and substrate is separate in solution, and they come together haphazardly to react. Also, control is more effectively exercised, because blocking one reaction of the series blocks the entire system. The mitochondrion and the chloroplasts in the cell are other examples of multienzyme complexes.

AMINO ACID
METABOLISM

Proteins occupy a central place in both the structural and dynamic aspects of living matter. Along with the nucleic acids, they compose the most important macromolecular structures of cells. Proteins are composed of amino acids, and it is partly the difference in amino acid composition that gives the proteins some of their unique properties. A supply of amino acids, either from the diet or from the biosynthetic machine, is obviously essential to cells and tissues for the synthesis of proteins. When these essential amino acids are acquired in the diet, the remainder of the nitrogen needed for protein synthesis can be supplied in the form of ammonium salts. Ammonia is an intermediate in nitrogen metabolism, and most organisms, when given an adequate amount of utilizable carbon compounds and other essential growth elements, can readily employ ammonia as the principal source of protein nitrogen. A crucial reaction in the uptake and incorporation of ammonia into proteins involves one of the intermediates in the citric acid cycle. As outlined in Fig. 10.29, α-ketoglutaric acid can be converted into the amino acid glutamic acid by a process called **reductive amination,** in which TPNH is the reducing power and ammonia is the nitrogen source (TPN structure is very similar to DPN; the former contains one additional phosphate group). This reaction is reversible, and, in fact, the enzyme that catalyzes the reaction is called aspartic

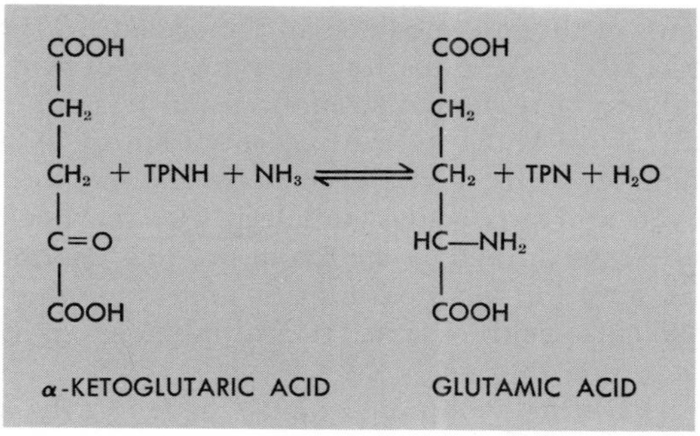

Figure 10.29 Synthesis of glutamic acid.

acid dehydrogenase. It is clear that this reaction serves as a main link between amino acid and carbohydrate metabolism.

Once ammonia is converted into amino nitrogen, it can be transferred to other carbon skeletons to form a different amino acid. This process of **transamination,** as shown in Fig. 10.30, involves an amino acid and a keto acid. Since the reaction is reversible, the process represents one of the principal metabolic pathways for the formation and deamination of amino acids. Enzymes called **transaminases,** which catalyze these various reactions, are known to occur in a variety of animal tissues, plants, and microorganisms. The B vitamin pyridoxine, in the form of pyridoxal phosphate, is an essential cofactor for all transaminases.

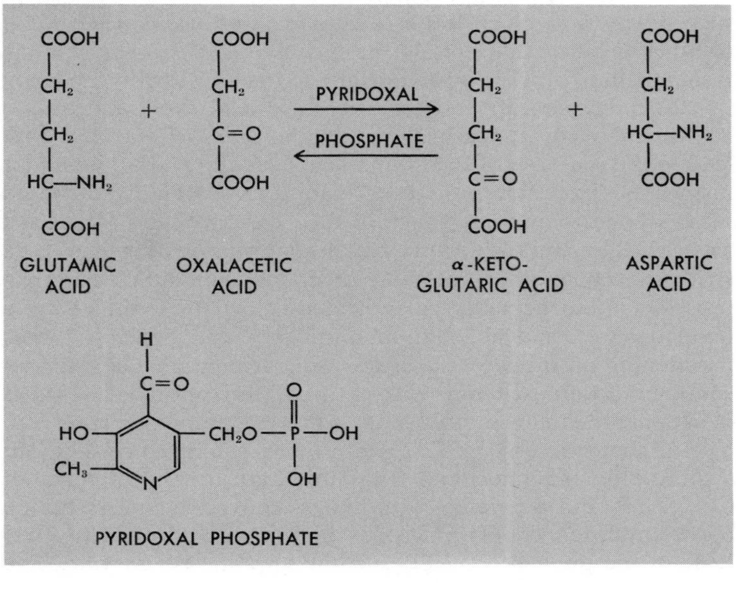

Figure 10.30 Transamination.

Two of the intermediates in the citric acid cycle, therefore, serve as carbon skeletons for the synthesis of two amino acids. The process of reductive amination and transamination can lead, as far as we know, to the incorporation of ammonia into a large number of the carbon skeletons that are formed from carbohydrate and fatty acid metabolism. How particular carbon skeletons are formed, then, is the major problem in amino acid biosynthesis, and, for the most part, it appears that the intermediates of the citric acid cycle are the initial compounds for the formation of amino nitrogen.

During evolution, animals have lost the ability to make the carbon chain for certain amino acids. These amino acids must be supplied in the diet and are therefore called **essential amino acids.** The essential amino acids for rats are arginine, histidine, isoleucine, leucine, lysine, methionine, phenylalanine, threonine, tryptophan, and valine. Although equally important, the following amino acids can be synthesized by mammals and therefore are called nonessential: alanine, aspartic acid, cystine, glutamic acid, glycine, hydroxyproline, proline, serine, and tyrosine.

Looking at the list of the essential amino acids, it becomes clear why the recently developed strain of corn that produces a high lysine content is such an important development. In most strains, about 50 per cent of the protein made in corn is **zein,** which is very low in lysine. Thus, if man's only source of amino acids were corn, he would develop a lysine deficiency and consequently could not make proteins (this condition is known as **kwashiorkor** and occurs in economically undeveloped areas, primarily in the tropical regions of the world). The new strain of corn (opaque 2, a simple single gene difference) is changed in such a way that much less zein is made, and the concentration of other proteins that contain more lysine is increased. Opaque 2 should thus provide an important source of nutrition to man.

The relationship between the citric acid cycle and amino acid metabolism is clearly shown in the series of reactions that lead to the synthesis of the amino acid arginine and to the major nitrogenous excretion product of man, urea. When excess amino acids are fed to animals, much of the excess nitrogen is excreted as urea. The unraveling of the mechanisms involved in urea synthesis was another milestone in our understanding of cyclic processes in biochemistry. As indicated below, the **ornithine cycle** regulates the removal of ammonium ions and depends for its functioning on the presence of the citric acid cycle. The essential compounds that are required to maintain the continued synthesis of urea are the following: ATP, aspartic acid, ammonia, and CO_2.

As shown in Fig. 10.31, glutamic acid can be converted into ornithine by reduction and transamination. In the presence of CO_2, ATP, and ammonia, ornithine is converted into citrulline by the addition of NH_2COOH, which is made from CO_2 and NH_3 in the presence of ATP and the appropriate enzyme. The

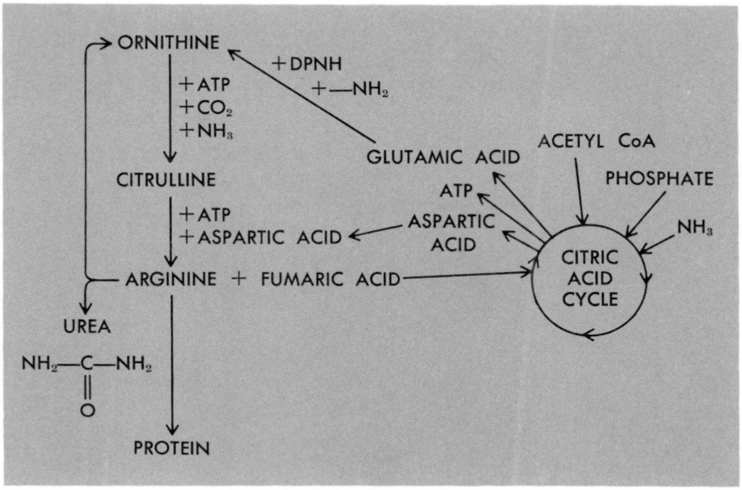

Figure 10.31 Urea and arginine synthesis.

true intermediate is carbamyl phosphate:

$$CO_2 + NH_3 + ATP \rightarrow NH_2\overset{\overset{O}{\|}}{C}O{\sim}\overset{\overset{O}{\|}}{\underset{\underset{OH}{|}}{P}}{-}OH + ATP$$

It has been shown recently that the vitamin biotin is an essential cofactor for the enzyme that catalyzes this reaction. As noted previously, biotin is essential in other reactions involving CO_2 fixation.

Aspartic acid adds a $-NH_2$ group to citrulline to form the amino acid arginine. If arginine is not used for protein synthesis, it is broken down in the liver by the enzyme arginase to form urea and ornithine. The ornithine or urea cycle thus leads to the excretion of two nitrogen atoms in the form of urea. As is evident from the scheme, the process depends on the immediate participation of the citric acid cycle for a supply of ATP and of the appropriate carbon-skeleton intermediates for the synthesis of aspartic and glutamic acids.

The metabolism of the amino acid tryptophan in animals and plants is of considerable interest. As indicated in Fig. 10.32, tryptophan is the primary source of the plant growth hormone, auxin. In man, tryptophan can be converted into 5-hydroxy-tryptamine (serotonin). Serotonin is present in the blood platelets and is released when there is damage to the blood vessels. It acts as a very potent vasoconstrictor, and therefore may be of some importance in preventing blood loss before clotting occurs. Since serotonin is concentrated rather highly in the brain, its role as a possible neurohumoral agent has been suggested. Certain drugs

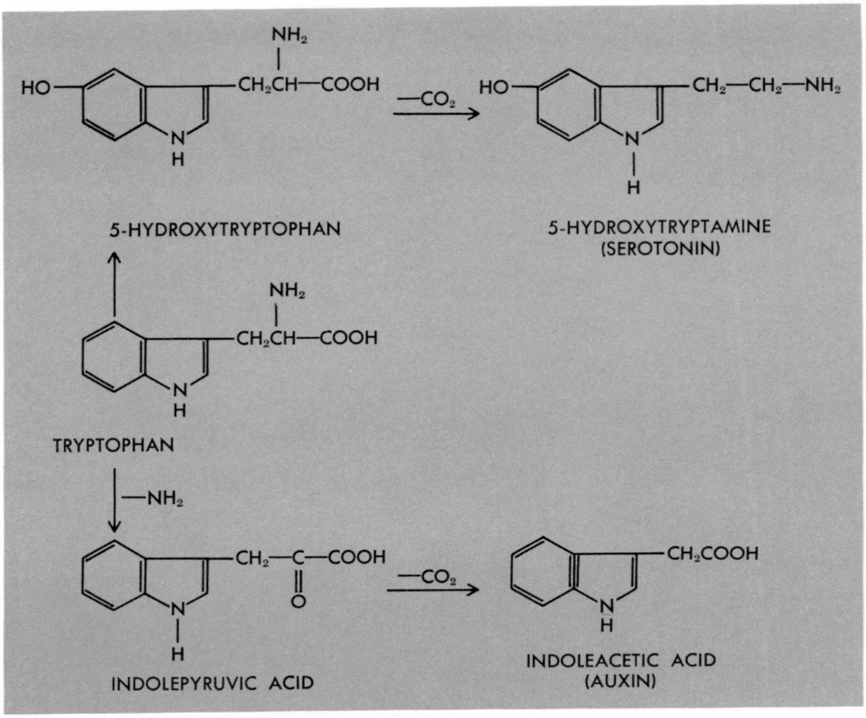

Figure 10.32 Formation of auxin and serotonin from tryptophan.

that have a profound effect on the activities of the central nervous system are potent antagonists of serotonin. Reserpine, found in the Indian plant *Rauwolfia serpentina,* when administered to man considerably decreases the activity of the central nervous system and at the same time stimulates the excretion of serotonin in the urine.

The metabolism of phenylalanine is of particular interest because it is known to be implicated in an inherited metabolic disease. In normal individuals the phenylalanine that is not used in protein synthesis is hydroxylated to form the amino acid tyrosine. This reaction is shown in Fig. 10.33. The occurrence of a genetic defect (homozygous recessive genes) in about 1 in every 200 individuals leads to a loss of the enzyme (phenylalanine hydroxylase) that converts phenylalanine to tyrosine. In the absence of this enzyme, excess phenylalanine is metabolized by way of phenylpyruvic acid, which is excreted in large amounts in the urine. This condition, known as **phenylketonuria** (PKU), causes mental retardation. It can be detected in infancy, and by feeding

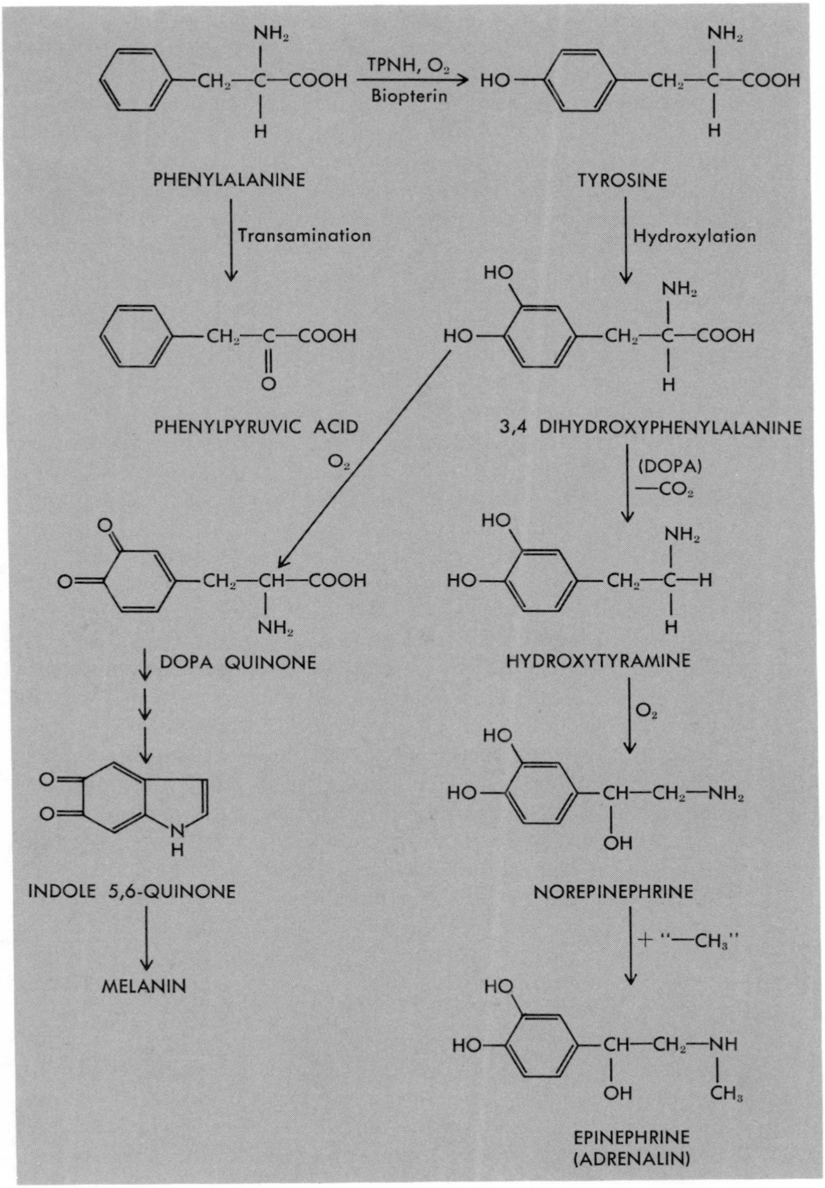

Figure 10.33 Metabolism of phenylalanine and tyrosine.

tyrosine and keeping the phenylalanine low in the diet one can prevent, in part, the mental retardation. Little is known about how to prevent this inherited metabolic defect; clearly, this is an important area for future research.

As indicated in Fig. 10.33, tyrosine is the precursor to the adrenal hormone, epinephrine, and to the dark pigment, melanin.

The mechanism of the synthesis and degradation of a number of the amino acids is a fascinating subject, but it is not necessary for us to go into details of all these reactions at this time. The principles outlined for the synthesis and breakdown of the amino acids discussed hold for the metabolism of other amino acids. To pursue the many other interesting metabolic products of amino acid metabolism, some of which are extremely important to cellular function, you should consult a more advanced book on biochemistry.

NITROGEN FIXATION— INORGANIC NITROGEN METABOLISM

As indicated in the previous section, inorganic ammonia is essential for the synthesis of amino acids, the building blocks of proteins. The original source of inorganic ammonia is atmospheric nitrogen (N_2), which makes up close to 80 per cent of the air we breathe. Most organisms cannot use gaseous nitrogen directly. Fortunately, there exist certain bacteria in the soil that are capable of taking up nitrogen gas and converting it into ammonia and other inorganic nitrogen compounds, such as nitrate (NO_3) and nitrite (NO_2). One important group of **nitrogen-fixing** bacteria is called *Rhizobium;* these bacteria are associated with small nodules that grow on the roots of specific plants called legumes (clover, alfalfa, soybeans). By growing soybeans or other such plants it is possible to enrich the soil with fixed nitrogen, thus reducing the need for extensive fertilization. Unfortunately, we do not understand the detailed mechanism of nitrogen fixation and reduction. Ammonia is the first stable product that can be detected; thus, gaseous nitrogen must be split and finally reduced by an appropriate electron donor to two molecules of ammonia. We do know that molybdenum, a plant hemoglobin-type molecule (leg hemoglobin), and an iron-containing coenzyme (ferredoxin) are essential for the overall process.

Many plants and microorganisms are capable of using inorganic nitrate as the sole nitrogen source. These plants contain several enzymes that catalyze the reduction of nitrate to ammonia. The first step involves the utilization of reduced flavin and a molybdenum-containing enzyme known as nitrate reductase. The nitrate is reduced to nitrite, and the latter is ultimately reduced to ammonia. Thus, the two major sources of ammonia are from (1) nitrogen fixation and reduction and (2) the formation of ammonia from nitrate or nitrite.

In nature the nitrogen continuously cycles. Gaseous nitrogen is fixed in the form of ammonia or nitrate. The nitrate and ammonia are used by plants to build their protein and other nitrogen-containing compounds. The plants are ultimately eaten by animals, which upon death return the fixed nitrogen to the soil. Some of this nitrogen can be converted into the gaseous form by special dinitrifying bacteria, thus returning some nitrogen gas to the atmosphere.

Cycles such as these are very important for a chemical balance in nature. The carbon dioxide and oxygen cycles along with the nitrogen cycle are three very important cyclic processes.

summary

When organisms break down **(metabolize)** sugars and other food substances, energy is trapped in special energy-rich bonds. The energy in these "rich" bonds can be used to make new compounds that are important to the organism. Useful energy that can be used in synthetic reactions is called **free energy.** This special form of chemical energy can be stored in the terminal phosphate bonds of a compound called **adenosine triphosphate (ATP).**

Large amounts of ATP are formed when compounds are oxidized aerobically by the mitochondria of the cell (citric acid cycle). During the cycle, molecular oxygen is used as the final hydrogen acceptor, thus leading to the complete combustion of foodstuff to CO_2 and water. The formation of ATP from inorganic phosphate and ADP in this process is called **oxidative phosphorylation.** This does not occur under anaerobic conditions. Yeast without oxygen forms alcohol and CO_2, while muscle cells make lactic acid. Even under these conditions, the limited amount of ATP formed is due to oxidative reactions.

Coenzymes that contain the B vitamins niacin and riboflavin are essential for oxidation (electron transport), and iron-containing cytochromes are necessary for activating molecular oxygen so that it can accept electrons (e) and protons (H^+) to form water.

Fatty acids, amino acids, and other simple organic compounds can also be metabolized by the mitochondria to form large amounts of ATP. In addition, the intermediates formed from carbohydrate breakdown can be used as carbon skeletons for the synthesis of specific amino acids or fatty acids. By the addition of ammonia to one of the intermediates in the citric acid cycle, the cell can make the amino acid, aspartic acid; or by addition to another intermediate, it can make glutamic acid.

Thus, in the breakdown of carbohydrates energy liberation is important. In addition, key intermediates are formed. The intermediates are essential for building new compounds of biological interest.

---

1   Vitamins are important in the functioning of enzymes. Describe the function of niacin in alcoholic fermentation by yeast and lactic acid formation in muscle.

for thought and discussion

2   What do we mean by the **phosphorylation** of glucose? Describe the role of ATP.

3   What is **fermentation?** How does it differ from aerobic metabolism?

4   Define **oxidation–reduction,** and show the process in a typical reaction.

5   Describe a hypothetical oxidation reaction and show how an **energy-rich** bond can be formed.

6   Do you get more ATP energy from glucose or ethyl alcohol when both are metabolized completely to CO_2 and water? Explain.

selected readings

BALDWIN, E. *Dynamic Aspects of Biochemistry* (3rd ed.). New York: Cambridge University Press, 1957. An outstanding introduction to the concepts of cellular metabolism.

BENNETT, T. P., and E. FRIEDEN. *Modern Topics in Biochemistry.* New York: Macmillan Publishing Co., Inc., 1966. An excellent discussion in greater detail of the various subjects discussed in our last three chapters. Written for the beginning college student.

LEHNINGER, A. L. *Biochemistry.* New York, Worth Publishers, Inc., 1970. An advanced textbook of biochemistry.

LOEWY, A. G., and P. SIEKEVITZ. *Cell Structure and Function.* New York: Holt, Rinehart and Winston, Inc., 1963. A detailed elementary discussion of various aspects of cell structure and biochemistry.

MCELROY, W. D. *Cell Physiology and Biochemistry* (3rd ed.). Englewood Cliffs, N.J.: Prentice-Hall, Inc., 1971. A small paperback intended for beginning college students.

NEILANDS, J. B., and P. K. STUMPF. *Outlines of Enzyme Chemistry* (2nd ed.). New York: John Wiley & Sons, Inc., 1958. An excellent elementary introduction to enzyme chemistry.

WHITE, A., P. HANDLER, and E. SMITH. *Principles of Biochemistry* (4th ed.). New York: McGraw-Hill Book Company, 1968. An advanced textbook of biochemistry.

Readings from Scientific American

ALLFREY, V. G., and A. E. MIRSKY, "How Cells Make Molecules," September 1961.

DICKERSON, R. E., "The Structure and History of an Ancient Protein (Cytochrome c)," April 1972.

GREEN, D., "The Synthesis of Fat," February 1960.

HAYASHI, T., and G. A. W. BOEHM, "Artificial Muscle," December 1952.

HOLTER, HEINZ, "How Things Get into Cells," September 1961.

HUXLEY, H. E., "The Contraction of Muscle," November 1958.

KENDREW, J. C., "Three-Dimensional Structure of a Protein," December 1961.

LEHNINGER, A., "Energy Transformation in the Cell," May 1960.

LEHNINGER, A., "How Cells Transform Energy," September 1961.

MARGARIA, R., "The Sources of Muscular Energy," March 1972.

PASTAN, IRA, "Cyclic AMP," August 1972.

SIEKEVITZ, P., "Powerhouse of the Cell," July 1957.

SOLOMON, A. K., "Pumps in the Living Membrane," August 1962.

STEIN, W. H., and S. MOORE, "The Chemical Structure of Proteins," February 1961.

STUMPF, P. K., "ATP," April 1953.

light and
life
chapter 11

The oxygen we breathe and most of the food we eat are formed by plants through **photosynthesis.** The power to drive the photosynthetic machinery of plant cells comes from sunlight, which is absorbed by **chlorophyll,** the green pigment in the plant (Fig. 11.2). All life that we know on Earth depends directly or indirectly on photosynthesis. We know a great deal about the nature of the machinery that traps the sunlight and converts carbon dioxide into carbohydrate; however, our knowledge is still incomplete. So far, man has not been able to devise a chemical system that can serve as a substitute for photosynthesis. Only the green plant is able to use light energy to convert carbon dioxide into organic matter, and at the same time produce free oxygen from water.

One of the earliest experiments in photosynthesis was performed by the Belgian scientist Jean Baptiste van Helmont in the latter part of the 16th century. He planted a small willow shoot in a tub of soil, and for a few years added only water to the growing plant. After five years he removed the plant and weighed it. He found that it had gained well over 150 pounds; meanwhile, the soil had lost only a few ounces. He concluded that the extra weight of the tree must have come from the water, the only thing he had added. Although his conclusion was not

Walter Dawn

Figure 11.1 Sunlight provides the power needed to drive the photosynthetic machinery of plant cells. Light and life are intimately related.

Figure 11.2 Chloroplasts, the cell components that give green plants their green color, are clearly visible in the cells of this moss (*Minium*) leaf.

Photo by Hugh Spencer

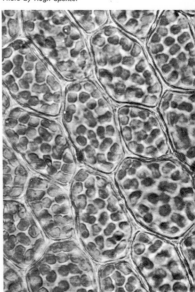

entirely right, it opened the way to a more careful study of photosynthesis.

In 1771, the British scientist Joseph Priestly showed that green plants could regenerate "good air" that had been converted to "bad air." Priestley placed a mouse in a closed container of air. After a short time the mouse died. He then placed a lighted candle in the same container of air and found that the flame quickly went out. He said that this air was "bad," that some life-giving part of it had been taken away. When he put a sprig of mint in a similar closed container in which a candle had been extinguished, he found that the bad air became good air, and that a candle could be burned in it. We now know that Priestley's experiment demonstrated that the mouse required oxygen and produced CO_2; also, that the plant used the CO_2 and some water, and released oxygen.

The real quantitative aspects of the use of light by plants were not considered seriously until about 1845. At that time the German physician Robert Mayer formulated the law of conservation of energy. The idea that energy could be changed from one form to another—that it does not just disappear—was important to an understanding of photosynthesis in plants. Mayer realized that photosynthesis was a conversion of light energy to a form of chemical energy, which was stored by the plant cell.

We can now summarize photosynthesis as being the process in which CO_2 is taken up by green plants in the presence of water. Sunlight is then used to form carbohydrate and molecular oxygen, as shown in the following equation:

$$6CO_2 + 6H_2O + 672,000 \text{ calories (light)} \rightarrow C_6H_{12}O_6 + 6O_2$$

What this equation says is that sunlight energy is absorbed by the plant and converts six units of CO_2 and water into one unit of carbohydrate and six units of oxygen. But this general equation tells us only what goes in and what comes out. The problem is a more difficult one when we ask what takes place inside the cell, and *how* CO_2 is taken up and *how* light energy is used.

photosynthesis and the atmosphere

The oxidation of organic compounds during respiration, and during fermentation, gives off CO_2 to the air. In addition to this source, the burning of oil and coal also releases large amounts of CO_2 into the air. It is remarkable that the concentration of CO_2 within the air remains very nearly constant. Evidently the total rate of CO_2 production is exactly balanced by CO_2 photosynthetic consumption. The exact mechanism of the regulation of the atmospheric CO_2 by photosynthesis is not clearly understood. However, we do know that it is of great importance in maintaining the Earth's temperature.

THE ROLE OF LIGHT Before describing photosynthesis and the general effect of light on biological processes, we should consider some of the properties of light. When a fine beam of sunlight is passed through a prism, it is separated into its compo-

Figure 11.3 A spectrum of colors is formed when light passes through a glass prism. Light of all wavelengths is slowed down when it passes from the air into the glass of the prism. Violet light is refracted the greatest amount and travels slower than red light, which is refracted least.

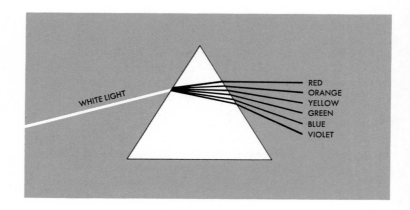

nent colors: violet, blue, green, yellow, orange, and red. If the different colors are then passed through a second (reversed) prism, the colors recombine as white light. But if only a single color is selected from the spectrum, no treatment can change it in any way. The individual colors can be regarded as resulting from the behavior of discrete units, or particles, called **photons.** The energy of the photons determines the color of the light. The photons of violet light have more energy than the photons of blue light, blue has more energy than green, green more than yellow, and so on to red, whose photons are least energetic of all.

The word "light" is usually applied to that range of radiation that can be detected by the eye. Scientists, however, speak of ultraviolet light and are able to "see" other types of radiation by means of photocells. Then there would seem to be more to light than meets the eye. While light behaves as particles (photons), it also behaves as waves. When the wavelength of violet light is measured, it is found to be shorter than the wavelength of red (Fig. 11.4). From violet to red along the spectrum, the wavelengths become progressively longer; and the longer the wavelength of a particular color of light, the less energy its photons have. Or, the shorter the wavelengths, the greater the energy. Visible light represents only a small part of the total range of radiation of the **electromagnetic spectrum.**

Ultraviolet light and x-rays, beyond the violet end of the spectrum, have very short wavelengths and extremely high energies. Infrared and radio waves, beyond the red end of the spectrum, have very long wavelengths and low energies.

During photosynthesis, only the light in the visible part of the spectrum is used as an energy source. Let us now examine in some detail what happens when light energy falls on a green plant. In the process of photosynthesis, light energy of various wavelengths excites a chlorophyll molecule. As Fig. 11.5 shows, a photon strikes an atom of the chlorophyll molecule, causing an electron to jump to a higher energy level, called the **excited state.**

RED λ = 7,000 ANGSTROMS

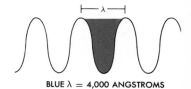

BLUE λ = 4,000 ANGSTROMS

Figure 11.4 Comparison of wavelengths of red and blue light.

Figure 11.5 Formation of reduced TPN by chlorophyll and light energy.

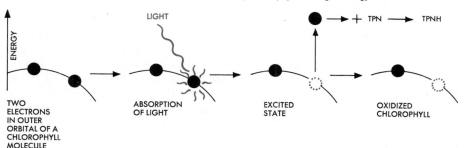

177

The primary process in photosynthesis, therefore, is the absorption of light quanta by the chlorophyll molecule. Since chlorophyll absorbs in the red region of the spectrum (656 nm)—for this reason the leaves look green—we should suspect that the red quanta are the effective ones in photosynthesis. In general, this has been found to be true. Although we are not sure of the exact mechanism, it is clear that light energy must cause some of the electrons of the chlorophyll molecules to jump to an excited state. This excitation energy somehow splits water to form a highly reducing H and an oxidant OH. Thus, in the light phase of photosynthesis, the photochemical decomposition of water produces a tremendous amount of potential energy.

Recent evidence indicates that there are probably two types of chlorophyll in green plants and that both must be excited by light in order for photosynthesis to proceed. One of these chlorophyll molecules absorbs light quanta in the red region of the spectrum, and the other absorbs light quanta in the far-red region.

As shown in Fig. 11.6, when chlorophyll absorbs red light quanta it gives up an electron to cytochrome 550. The iron in the cytochrome accepts the electron. The oxidized chlorophyll reacts with water and becomes reduced, and in the process oxygen is liberated. Unfortunately, we do not understand the mechanism of this process; we do know, however, that at least four light quanta are required for each oxygen molecule that is liberated. Thus, the overall reaction for the red-light process can be described as follows:

$$4H_2O \rightarrow 2H_2O + O_2 + 4H$$

The four hydrogens or electrons are used to reduce four chlorophyll molecules. The reduced cytochrome formed in the initial photochemical event is oxidized by another cytochrome at a lower energy level. When the electron is transported from cytochrome 550 to cytochrome b, the potential energy release is conserved as phosphate bond energy, and consequently ATP is formed.

The reduced cytochrome b is apparently complexed closely with a chlorophyll molecule, because when this chlorophyll absorbs a red light quantum, the electrons in the reduced cytochrome b are raised to a high energy level and are now capable of reducing another iron-containing molecule, called **ferredoxin** (Fig. 11.6). The photo-oxidized cytochrome b is now available to accept additional electrons from reduced cytochrome 550.

The reduced ferredoxin passes its electrons on to TPN to form TPNH, the reducing potential needed to synthesize carbohydrate. Since two light quanta are needed to activate the phototransfer of one electron to ferredoxin, it is obvious that four quanta of light are needed to form one molecule of TPNH (two electrons). Since we need four red light quanta to form one

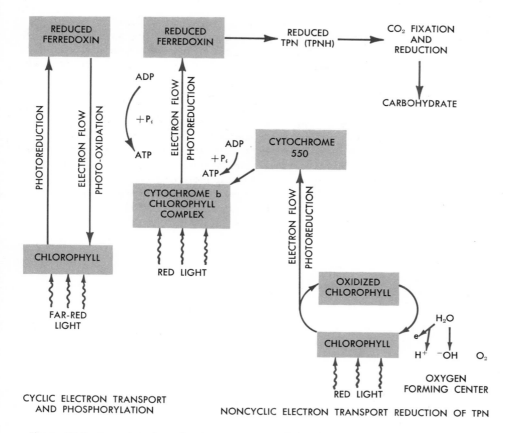

Figure 11.6 Suggested scheme for electron transport and phosphorylation during photosynthesis in green plants.

molecule of oxygen in the primary event, we can write a balanced equation for the formation of reduced TPN which requires eight quanta, as follows:

$$2TPN + 4HOH \xrightarrow{\text{8 light quanta}} 2TPNH + 2H_2O + O_2 + 2H^+$$

A second independent photochemical reaction occurs in the chloroplasts of green plants and makes use of **far-red** light. In this system the reduced ferredoxin formed is oxidized by the reduced chlorophyll, and in the process ATP is formed (Fig. 11.6). Since the electron transfer in this system is cyclic, the phosphorylation that occurs has been called **cyclic phosphorylation.** The phosphorylation occurring in both systems is known as **photophosphorylation.**

Reactions similar to the ones described above for green

179

plants also occur in certain photosynthetic bacteria, except that the latter do not evolve oxygen. Instead of water they use other reducing agents, either inorganic or organic. The overall equation for TPN reduction in photosynthetic bacteria can be summarized as follows:

$$2TPN + 2H_2A \rightarrow 2TPNH + 2A + 2H^+$$

One primary role of TPNH is to reduce CO_2. Before this can occur, CO_2 must be taken up by some reaction in the plant. Recent isotopic evidence indicates that the CO_2 adds to the five-carbon sugar **ribulose diphosphate,** which immediately splits into two molecules of **3-phosphoglyceric acid.** By reversing the glycolysis scheme and making use of reduced pyridine nucleotides and ATP, sugar can be synthesized (Fig. 11.7).

Figure 11.7 CO_2 fixation by the pentose pathway.

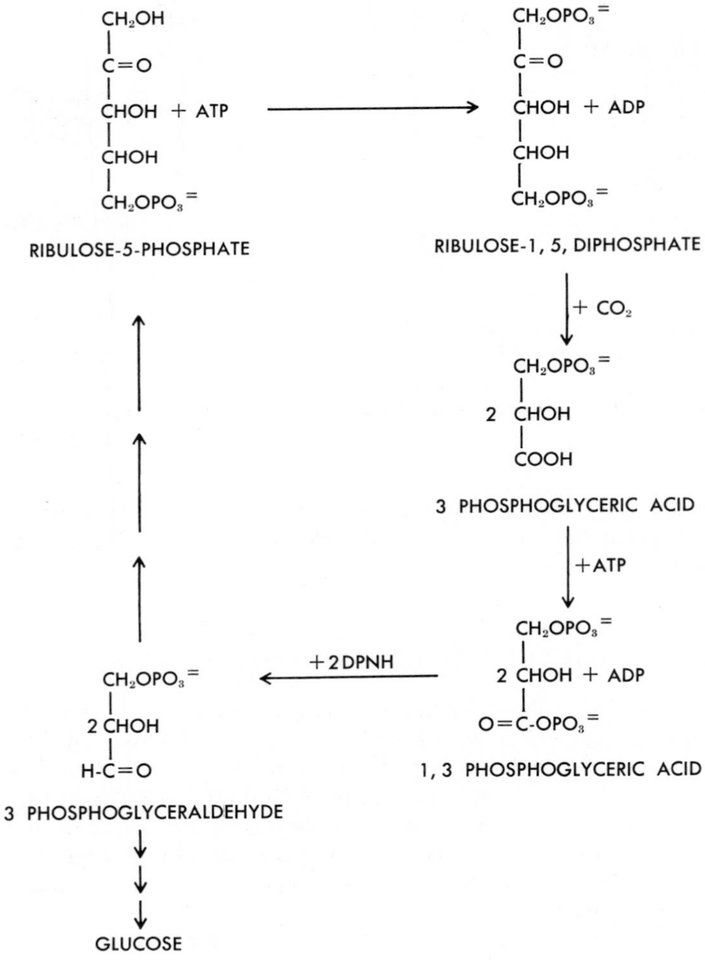

One particularly fascinating thing about photosynthesis is that only about 2 per cent of the total energy of sunlight falling on a plant is used. A large amount of the energy is never absorbed. About 30 per cent of the light goes right through the leaf, while about 20 per cent is radiated away as heat (long-wave radiation). Close to 48 per cent is used up as heat in the evaporation of water.

The 2 per cent of sunlight used in photosynthesis provides enough energy to maintain the whole plant, which, in turn, directly or indirectly is the energy source for all other living things. Considering the plant itself, only some parts of it are useful for human consumption; a large fraction of a typical plant consists of inedible fibers. If a crop is eaten, only a very small amount of it is converted into animal tissues: possibly 90 per cent of the energy in a crop fed to animals is lost as heat or waste. This means that when we eat meat, we are obtaining only about 10 per cent of the energy originally fed into the animal. Using plants to feed farm animals to produce food for humans is a very wasteful and inefficient process. It would be much more efficient—if not quite so pleasant—to omit the animal stage and eat only plants.

THE RATE OF
PHOTOSYNTHESIS

Several things affect the rate of photosynthesis. An adequate supply of CO_2 and water is essential. In bright sunlight, the relative rate of photosynthesis increases as the CO_2 concentration increases up to a given level (Fig. 11.8). The same can be said for water. In addition to the CO_2 and water concentration, temperature influences the rate of photosynthesis. When the light intensity is high and the CO_2 concentration is great, the effect of temperature becomes very apparent.

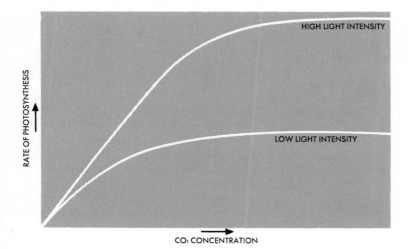

Figure 11.8 Effect of light intensity and carbon dioxide concentration on the rate of photosynthesis.

The rate of photosynthesis at different temperatures increases up to about 30 or 35°C and then starts to decline because of the destruction of the plant enzymes by heat.

THE PRODUCTION OF CHLOROPHYLL Light is essential not only for photosynthesis, but also for the production of chlorophyll. Chlorophyll begins as a substance called **protochlorophyll.** Protochlorophyll is different from chlorophyll in that it needs two additional hydrogens to make it into chlorophyll. Only in the presence of light can protochlorophyll change into chlorophyll. You can see that this is so by growing bean seedlings in the dark and then exposing them to light until various amounts of chlorophyll are formed. If you use light bulbs of different colors, you can study the effectiveness of different wavelengths of light in making the leaves become green.

There are other pigments in green plants capable of absorbing light and transferring its energy to chlorophyll. Called **accessory pigments,** they are very important in that they allow the plant to use light of different wavelengths for photosynthesis. Were it not for accessory pigments, certain wavelengths of light would not be absorbed and used by the chlorophyll system. The blue pigment, phycocyanin, in blue-green algae is an accessory pigment.

PHOTOSYNTHESIS AS AN ENERGY SOURCE Not only is photosynthesis a supplier of food for living organisms, it is also a supplier of energy for other processes. Our industrial civilization has depended on a reservoir of coal and oil, which we have been removing from the ground at a rapid rate during the past century. Both coal and oil are derived from plants that grew in past geologic ages. Although we cannot say exactly how long these reservoirs will last, at the present we need not worry about what seems to be an inexhaustible supply of fossil fuel. However, the day will come when this stored supply of coal and oil will be used up, and a continuing source of energy will have to be found.

We could, of course, make use of the abundant energy coming from the Sun. It has been calculated, for example, that $1\frac{1}{2}$ square miles of the Earth's surface receives from the Sun during one day approximately the same amount of energy released by the explosion of one small atomic bomb. The difficulty at the present time is that man does not understand how to make really good use of the sunlight directly. About all we can do now is convert sunlight into electricity by using solar batteries. Possibly in the future we may learn how to store this energy in some other more stable form; it could then be released as we needed it.

Figure 11.9 Spacecrafts are equipped with solar panels that collect solar energy and convert it into electrical power.

The need for power sources is one of the reasons for our great interest in photosynthesis. A process which, on a large scale, captures light energy and stores it in a chemical form is of considerable economic interest. Photosynthesis takes place over the surface of our planet and fixes CO_2 at the rate of about 1 million tons of carbon per minute. Unfortunately, we do not yet know if we can duplicate these feats in the laboratory in an economical way. If we do manage to build a machine capable of converting solar energy to power on a large scale, the machine will probably be quite different from the green plant. Considering that the Earth's energy sources are limited, a photosynthesizing machine, or something akin to it, must one day become a necessity. As one investigator has pointed out, unless more progress is made in solar energy conversion during the next hundred years, we may again go back to the horse-and-buggy method of transportation.

vision

Photosynthesis is but one aspect of light and life. Although there are many, many others, we shall consider only two—vision and bioluminescence. The visual process is another striking biological example where electrons in excited states must be involved. In spite of a great deal of outstanding work on the eye, we know very little about the basic mechanisms underlying vision. By attaching electrodes to the optic nerve, which connects the eye to the brain, we can tell that a nerve impulse is produced when photons strike the retina. A man's eye contains about 4 million units called **cones** and an additional 125 million units called **rods.** The rods lead to about 1 million optic nerve fibers. They control

183

our dim light vision, which is mostly black and white vision. The cones control bright light, or color, vision.

We assume that the triggering mechanisms of the nerve impulse must lie in the rods and cones because most of the visual pigments are found in the outer segment of them. One of the visual pigments, **rhodopsin,** has been extracted and has been shown to consist of a protein, called **opsin,** attached to a vitamin A derivative called **retinene.** Vitamin A is converted into retinene by the reduction of the aldehyde group to the alcohol by reduced DPN. Alcohol dehydrogenase catalyzes this reaction. When light strikes the visual pigment rhodopsin, the pigment immediately dissociates into opsin and retinene (Fig. 11.10). It is during this time that the nerve impulse is triggered.

The relationship between vitamin A and vision was first noticed when a vitamin A deficiency was associated with the eye's ability to adapt to the dark. If you do not have enough vitamin A, the reconstruction of the visual pigment is retarded, and you become night-blind. The **absorption spectrum** of rhodopsin is identical to the sensitivity of the eye to different wavelengths of light. The peak absorption and sensitivity is around 500 mm. How the photoexcitation in the pigment's molecules produces an impulse in the nerve fibers is not understood. Rhodopsin dissociates into opsin and retinene when exposed to light, and recombines in the dark. These facts led to the following conclusion.

The photocurrent must be associated with the chemical changes in the retinene that follow after dissociation. Recent studies, however, show that these changes probably are quite slow compared to the rate of initiation and conduction of the nerve impulse. We now believe that the small chemical changes that might have occurred in the unfolding of the protein during the photochemical event may be sufficient to initiate the electrical impulse which is conducted by the nerve to the brain. The im-

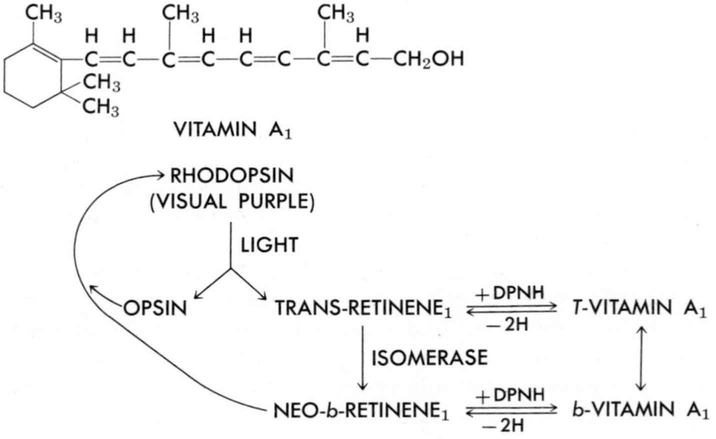

Figure 11.10 When light strikes the visual pigment the rhodopsin immediately dissociates into opsin (a protein) and retinene (the pigment). It is during this time that the nerve impulse is triggered in some unknown way and is registered in the brain as a visual impulse.

pulse is registered in the brain as a picture and gives us vision. The method of decoding this electrical message to give a visual sensation still remains a mystery.

Research into brain function is rapidly becoming one of the most interesting and challenging fields of research in biology. It seems possible, by using special electrical and computer techniques, that we can come to understand these and other coded messages transmitted by nerves to the brain.

Considerable progress has been made recently in understanding certain aspects of color vision in man, goldfish, and other animals. There seem to be three distinct types of cones, each of which is sensitive to three different colors of light: blue, yellow-green, and red. The difference in sensitivity is due to the presence of pigments that absorb photons primarily at the three different wavelengths. The combination of electrical impulses from these three basic color receptors gives us the color spectrum that we can see. But much work remains to be done on this fundamental problem before we can be certain about the basic mechanisms involved.

bioluminescence

There are many organisms that are capable of giving out light, a process called **bioluminescence.** They range from the bacteria that give off a blue light to the South American railroad worm (a larva of a beetle) that gives off a red light. The light of decaying wood is produced by luminous fungi. The luminescence of the sea is caused by a variety of forms, including various protozoans, sponges, jellyfish, brittle stars, snails, clams, squid, shrimp, small crustaceans, fish, and many other forms (Fig. 11.11). Probably the best known luminous forms on land are the fireflies and the glow worms. In addition to these forms, there are luminous spring tails, flies, centipedes, millipedes, earthworms, and snails. The mechanism of this bioluminescence is not, at present, too well understood.

In 1887, the French physiologist Raphial DuBois suggested that the light from a luminous clam, *Pholas dactylus,* is caused by the oxidization of a substance he called **luciferin.** In the presence of oxygen and an enzyme called **luciferase,** luciferin is destroyed by an oxidative reaction that gives off light. Although a great deal of progress has been made in recent years in probing the nature of the chemical substances required for light emission, the basic mechanism remains obscure. In some respects bioluminescence is very similar to photosynthesis and vision. In the latter two cases, the excited states are generated by light, while in the case of luminescence the excited state is created by chemical reactions and the energy is lost as light.

How do those organisms displaying this remarkable ability

Figure 11.11 Luminous toadstools. Upper, by ordinary light; lower, by own light.

Dr. Y. Haneda

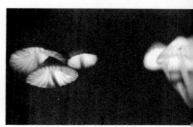

to emit light use the light? There is no clear answer yet. There are many examples in which bioluminescence has been adapted to good biological use. The reproduction cycle of many organisms in the sea is intimately tied to light emission. The flash of the firefly is used by some species as a sex signal. Light emission by deep-sea organisms living at great depths provides the only light available to them. In some cases, luminous bacteria are known to grow in special glands in fish and to provide a regular source of light. Whether light emission in the depths of the ocean triggers other photobiological processes is not known, but it is safe to guess that it does.

the effect of light on biological processes

In addition to the photosynthetic reactions in green plants that are dependent on light, there are other striking effects of light. Most green plants, for example, display striking growth responses to light, and their response may be independent of photosynthesis. The bending and growth of a green plant toward light is known as **phototropism** and is attributed to the effect of light on the metabolism of one of the plant growth hormones, **auxin.** The duration of light and dark periods also regulates the flowering and reproduction of plants, an effect known as **photoperiodism.** The phenomenon of photoperiodism is not limited to plants. Most organisms that have been studied carefully show a day–night periodicity for a number of physiological processes. Thus, both animals and plants exhibit rhythms in a manner indicating that a "biological clock" plays an important role in controlling behavior and physiological function.

Plants and animals are able to "recognize" the seasons of the year by measuring the changing length of night and day. For example, the migration of Canadian geese in the autumn is initiated by the change in the length of the nights and is one of the many important and interesting displays of photoperiodic responses in animals. The length of the day or the shortness of the night have been demonstrated to have important effects in the sexual cycle of numerous fishes and reptiles, and in the reproductive cycle and migration of birds.

The flowering of the poinsettia at the Christmas season is another example of photoperiodism. Other plants grow, flower, and bear fruit at different times of the year. Thus, we come to recognize that some clocks in plants cause flowering in the spring, while other periodic clocks cause flowering in the summer, and still others in the autumn.

Unfortunately, we know very little about the details of the biochemistry of physiology or the biological clock systems that control these rhythmic responses. There must be specific pigments in the cells that are affected by light and, in turn, regulate cellular responses. One such pigment in plants, **phytochrome,** has been studied in detail. We know from studies on flowering response at various wavelengths of light that the interruption of a long dark period by red light inhibits the flowering response in short-day plants (that is, those that require long nights). If the plants are subsequently exposed to far-red light, flowering is stimulated. The pigment that absorbs the light is phytochrome. One form absorbs in the red region and is then converted into a form that absorbs in the far-red region. For flowering to occur in short-day plants (for example, soybean), the period of darkness (or far red) must be long enough to decrease the concentration of the far-red-absorbing pigment to a low level by conversion to the red-absorbing form. This state must be maintained for several hours if flowering is to be initiated. If the dark period is interrupted with a brief flash of red light, the far-red-absorbing pigment produced suppresses flowering.

Unfortunately, we do not know the mechanisms involved in the flowering response. One important lead has been obtained by grafting experiments. By such techniques we can show that the phytochrome is in some way associated with the production of a plant hormone that initiates flowering. Basic research in this area is of considerable importance since it could lead to important agricultural advances.

Flowering in plants is not the only response to the red–far-red light system. Leaf expansion, stem elongation, seed germination, and other processes also demonstrate a red–far-red antagonistic response.

summary

Light is essential for all life. Through the process of photosynthesis in green plants, light energy is used to convert carbon dioxide into complex organic matter (food). At the same time, molecular oxygen is produced from water. The green pigment, chlorophyll, is the primary trapping agent for the light quanta. When light is absorbed, an electron in the chlorophyll molecule is raised to an excited state. The electron in the high-energy state can be used to reduce DPN and to make ATP in a process called **photophosphorylation.** The ATP energy and the reducing power of DPNH are essential for the fixation and reduction of CO_2 to carbohydrate. The oxidized chlorophyll is reduced in the light by an electron from water, and oxygen is formed.

The eyes of animals are a very special organ for trapping light quanta. The eye of man contains both rods and cones. The rods are associated with black and white vision, while the cones control color vision. When light quanta strike the rods and cones, the quanta are absorbed by a special pigment called **rhodopsin.** Rhodopsin consists of a protein called **opsin** and a vitamin A derivative called **retinene.** In some way the absorption of light triggers a nerve impulse from the eye along the optic nerve to the visual center in the brain. Unfortunately, we do not understand the mechanism of this aspect of the visual process.

There are many organisms that produce light by special chemical reactions. This process of **bioluminescence** is particularly noticeable among beetles (fireflies) and in the ocean. There are many examples in which bioluminescence has been adapted to biological use. However, we are not certain how this unique ability arose in nature.

In addition to photosynthesis and vision, light has other effects on biological processes. The bending of plants toward light **(phototropism)** is an excellent example. The length of the day or the shortness of the night has been demonstrated to have important effects in the sexual cycle of numerous marine invertebrates, fish, and reptiles, and in the reproductive cycle and migration of birds. Duration of light and dark periods also regulate flowering and reproduction in plants. This is called **photoperiodism.** Numerous studies of this type have lead to the general notion that there are "biological clocks" that regulate specific functions. Light seems to have a pronounced effect on the rhythm or period of these clocks.

for thought and discussion

1   Draw a graph indicating the approximate relationship between the energy in a light quantum and the wavelength.

2   If a green plant absorbs light in the red and far-red regions of the spectrum, does this give you an idea as to why the leaf looks green?

3   What is an **excited state?** How is it used by the green plant?

4   What would happen to the CO_2 and oxygen content of the air if photosynthesis by plants was stopped?

5   What is **photophosphorylation?**

6   In human vision, which wavelength of light is the eye most sensitive to? What determines this sensitivity?

7   Can you think of any good reasons why organisms give off light?

8   Can you describe any example where photoperiodism in either plants or animals may be of economic importance?

selected readings

Readings from Scientific American

ARNON, D. I., "The Role of Light in Photosynthesis," November 1960.
BASSHAM, J., "The Path of Carbon in Photosynthesis," June 1962.
BROWN, F. A., JR., "Biological Clocks and the Fiddler Crab," April 1954.

EVANS, R. M., "Seeing Light and Color," August 1949.

LAND, E. H., "Experiments in Color Vision," May 1959.

MCELROY, W. D., and H. H. SELIGER, "Biological Luminescence," December 1962.

RUSHTON, W. A. H., "Visual Pigments in Man," November 1962.

WALD, G., "Life and Light," October 1959.

WATERMAN, T. H., "Polarized Light and Animal Navigation," July 1955.

DNA—
the molecule
of life
chapter 12

Up to this point we have considered the nature of some molecules of crucial biological importance. In particular, we have dealt in some detail with the carbohydrates, fats, and proteins: their molecular structure, their involvement in cellular structures and reactions, and the flow of energy in and out of these molecules. We recognize that these molecules and the role they play in the life of a cell form an organized pattern, that a cell is an organized structure in which organized reactions take place. A bacterial cell does what is proper for a bacterial cell to do, and it does not do those things that are proper for an elephant, grass, or human cell. We need, therefore, to consider the control mechanisms that operate within cells.

Two major controlling systems are recognized. The first governs the general character of cells. That is, a system exists which determines whether a cell is part of a man or part of a man-eating shark. If the cell is human, its control system determines whether it is part of Mary Smith or Tom Jones. Both of these aspects are part of the problem of inheritance. If the cell is part of Mary Smith, the system also determines whether a given cell will be part of her liver, muscles, or skin—a problem of cellular differentiation. This system has already been discussed briefly in Chapter 5, and we now know that the crucial molecule is deoxyribonucleic acid, or DNA. Its structure and function will be discussed in this chapter.

The second controlling system is that which governs cellular metabolism. Consideration of it will be the subject of the next chapter.

We want now to do three things: (1) inquire into the chemical and physical nature of the DNA molecule; (2) consider how such a molecule can be a source of information which determines the character and, in part, the activities of the cell; and (3) explore how this information passes from DNA to other parts of the cell where it is put to use.

The story of the inquiry, which has taken place in the last 20 years, is one of the most exciting chapters in the history of the biological sciences. Although the existence of nucleic acids has been known for nearly 100 years, only within the last quarter century have their structure and role been fitted into general biological theory. An understanding of the role of these molecules has revolutionized biology.

DNA is unique in three respects. First, it is a very large molecule, having a certain outward uniformity of size, rigidity, and shape. Despite this uniformity, however, it has infinite internal variety. Its varied nature gives it the complexity required for information-carrying purposes. One can, indeed, think of the molecule as if it had a chemical alphabet somehow grouped into words that the cell can understand and to which it can respond.

The second characteristic of DNA is its capacity to make copies of itself almost endlessly, and with remarkable exactness. The biologist or chemist would say that such a molecule can **replicate,** or make a carbon copy of itself, time and again with a very small margin of error.

The third characteristic is its ability to transmit information to other parts of the cell. Depending upon the information transmitted, the behavior of the cell reflects this direction. As we shall see, other molecules play the role of messenger, so that DNA exercises its control of the cell in an indirect manner.

the structure of DNA

DNA can be isolated from nearly every organism—from viruses, bacteria, and fungi to man and the plants and animals with which we are all familiar. Only certain of the viruses lack DNA; in such viruses DNA is replaced by a comparable molecule, RNA. No matter where it is found, all DNA has much the same chemical and physical properties.

DNA is a **polymer**—a very large molecule made up of repeating units. In this sense it is much like rubber or many plastics, but with the single, important exception that the repeating units of DNA can vary. To give you some idea of the size of the

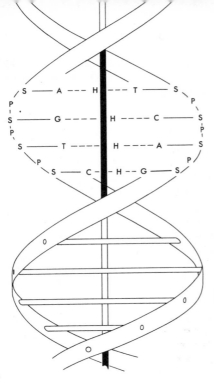

Figure 12.1 DNA molecule.
In the model above, the double
helix is shown with the base
pairs extending across the
molecule from the sugars on
either side. The sugar residues
(S) are linked by phosphates (P)
on both sides, forming the two
continuous "backbones" of the
long molecule. Only the base
pairs A–T, T–A, C–G, and G–C
are possible, but they can vary,
giving internal variety to the
molecule. The vertical rod
running through the center of the
molecule is an imaginary axis
around which the helix entwines.
At right, the same molecule is
shown, but with the atoms
indicated.

COLOR KEY

BASE PAIRS   H   O   P   C

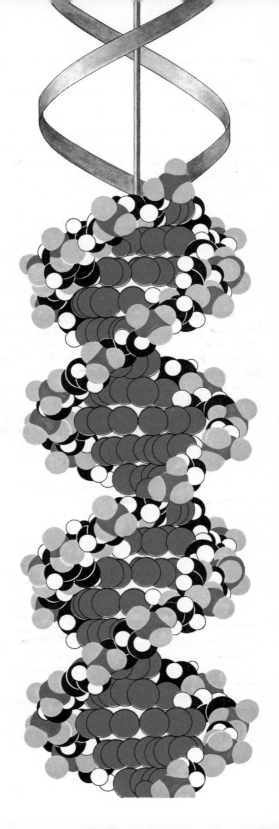

molecule, Fig. 12.2 shows the length of a bacterial virus DNA molecule in relation to the length of the virus itself. For such a large molecule to be compacted into so small a space, DNA must be capable of considerable folding.

The repeating units of DNA are called **nucleotides;** these are rather complicated molecules. All DNA nucleotides contain three units: **phosphoric acid,** a five-carbon sugar called **deoxyribose,** and a **base,** in this case a ring structure that can take up hydrogen ions. The way these components are arranged to form repeating units is shown in Fig. 12.3. The bases project from one end of each of the sugars, while the phosphoric acid serves as a link binding the successive sugars.

The bases are of four major types. Two of the four are **purines (adenine** and **guanine),** and two **pyrimidines (cytosine** and **thymine).** Four nucleotides can consequently be formed from these four bases by the addition of phosphoric acid and a sugar to each one. The two purine nucleotides are **deoxyadenylic acid** and **deoxyguanilic acid.** The pyrimidine nucleotides are **deoxycytidylic acid** and **deoxythymidilic acid.**

It is the number, type, and arrangement of the bases of DNA that determine what kind of an organism will develop. Since these bases can vary in sequence along the length of the DNA molecule, much as the amino acid sequences of protein can vary, the DNA's of different organisms are composed differently. We can, therefore, characterize the various DNA's by their base composition, that is, by the relative proportions of the four nucleotides to each other. As yet we know little about the ordered arrangement of the bases in the DNA from any given organism. The discovery that DNA from a variety of organisms can vary was an important step in our understanding of this molecule as a source of hereditary information.

The biochemist Irving Chargaff and his collaborators at Columbia University, however, made two additional important discoveries. First, when they analyzed DNA for its base composi-

Figure 12.2 A drawing of the entire DNA molecule contained within a bacterial virus. The dimensions of the virus and the length of the molecule are drawn to scale, so the molecule must be highly compacted to fit within the hollow head of the virus.

Figure 12.3 In this diagram of a DNA molecule, the base pairs are stretched in flat planes perpendicular to the sugar–phosphate backbones. Actually there are 10 base pairs for every full turn of the double helix. Only five are shown here.

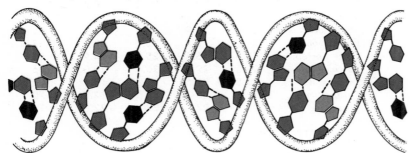

tion, they found that the number of purines was always equal to the number of pyrimidines; second, they found that the number of adenines was equal to the number of thymines, and that the number of guanines equals the number of cytosines. We can now discuss DNA in terms of its base ratios. Using capital letters to represent the bases we can say the following:

$$A + G = T + C$$

Also, A = T and C = G; but A does not have to equal C or G. The base ratios for several well-known organisms are shown in Table 12.1.

The importance of these discoveries is that we can now visualize how each organism can pass a specific kind of information to its cells and to its offspring. For example, in a given small section of DNA, the base sequence of ATCCGATT may mean something very different from ACCGTTAT, even though the number and kinds of letters are the same. In a sense, this is not very different from the words *heat* and *hate*, which have the same letters, but a different arrangement and, consequently, a different meaning.

A MOLECULAR MODEL OF DNA   With the discovery that DNA is the key molecule of heredity, scientists asked the question: "What is the structure of DNA?" The answer came in the early 1950s through the efforts of an English crystallographer, M. H. F.

TABLE 12.1.   Base Ratios of Several Well-Known Organisms*

	Adenine	Thymine	Guanine	Cytosine	Ratio of $\frac{A + T}{C + G}$
Man	29.2	29.4	21.0	20.4	1.53
Sheep	28.0	28.6	22.3	21.1	1.38
Calf	28.0	27.8	20.9	21.4	1.36
Salmon	29.7	29.1	20.8	20.4	1.43
Yeast (fungus)	31.3	32.9	18.7	18.1	1.19
Staphylococcus	31.0	33.9	17.5	17.6	1.85
Pseudomonas	16.2	16.4	33.7	33.7	0.48
Colon bacterium	25.6	25.5	25.0	24.9	1.00
Vaccinia virus	29.5	29.9	20.6	20.0	1.46
Pneumococcus	29.8	31.6	20.5	18.0	1.88
Clostridium	36.9	36.3	14.0	12.8	2.70
Wheat	27.3	27.1	22.7	22.8	1.19

* Values are arbitrary but accurate as ratios.

Wilkins, an American biologist, James Watson, and an English chemist, Francis Crick.

The findings of these men suggested that a molecule of DNA was a double-strand structure. The two polynucleotide strands are intertwined in the form of a long helix, or spiral. The outside edges of the helix have alternating molecules of phosphoric acid and sugar. These bases project to the inside of the helix and form base pairs, A pairing with T, and C pairing with G. One very important aspect of the model, therefore, is that the sequence of bases in one strand of the helix determines, or is complementary to, the sequence of bases in the other strand. Thus, if strand X has the sequence shown, strand Y must have the complementary sequence. That is, if X strand has the sequence of ATGGC, then Y strand must be TACCG. The base pairs are connected to each other by hydrogen bonds (Figs. 12.4 and 12.5.)

The usefulness of this model of DNA has been demonstrated several times. For example, biologists have long known that the hereditary materials in the cell—that is, the chromosomes in the nucleus—replicate themselves exactly at each cell division. In a gross way, we can see the result of this replication in the microscope. The doubleness of each chromosome at metaphase and the separation of the two chromatids at anaphase tell us that the chromosome has replicated itself. The Watson–Crick model demonstrates how this can be accomplished chemically. A double-stranded piece of DNA (Fig. 12.6) is gradually separated into two single polynucleotide strands. Each of these strands directs the synthesis of a new strand, producing two double-stranded structures. If we were to label the two original strands X and Y, it would become evident, because of the manner of accurate base pairing, that X would direct the formation of a new Y strand. At the same time, the original Y strand would direct the formation of a new X strand.

This synthesis of new DNA, like other cellular reactions, is under the direction of a nuclear enzyme called **polymerase.** Two double helices, apparently identical in all respects to the original double helix, can thus be formed, and the piece of DNA replicates itself. It is through this mechanism of chemical synthesis that new chromosomes, which are exact replicas of previous chromosomes, are formed. Hereditary information is passed from one cell to another in division, and from one generation to the next through reproduction.

PROOF OF The DNA story is an attractive
DNA REPLICATION and exciting one. It is attractive
because it ties together a lot of information about cell behavior, and exciting because it provides an opening wedge into some of the mysteries of the cell.

Let us now play the role of the skeptic, a role that comes

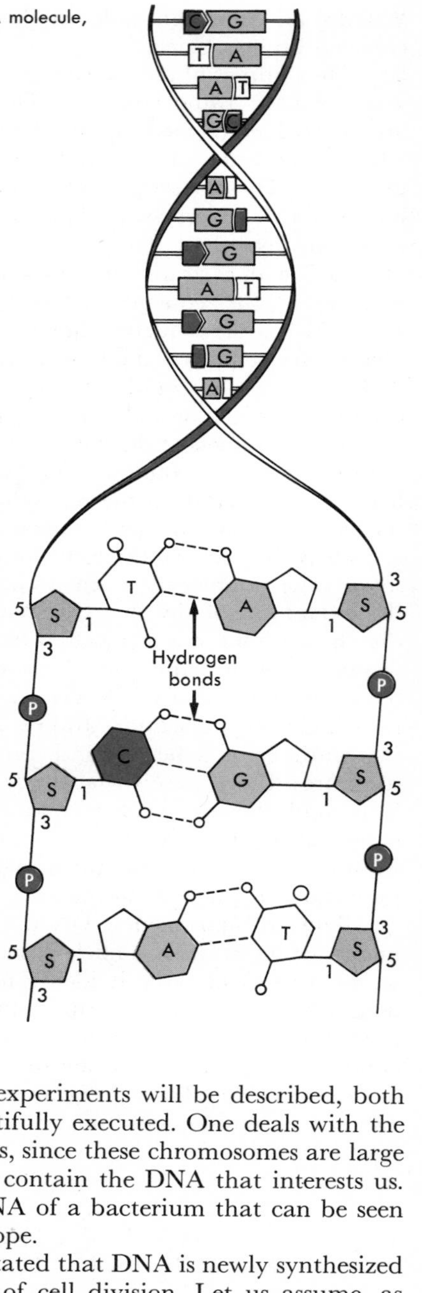

Figure 12.5 A schematic diagram of the DNA molecule, base pairing and the helical structure.

Figure 12.4 Pairing of purine and pyrimidine bases by means of hydrogen bonding.

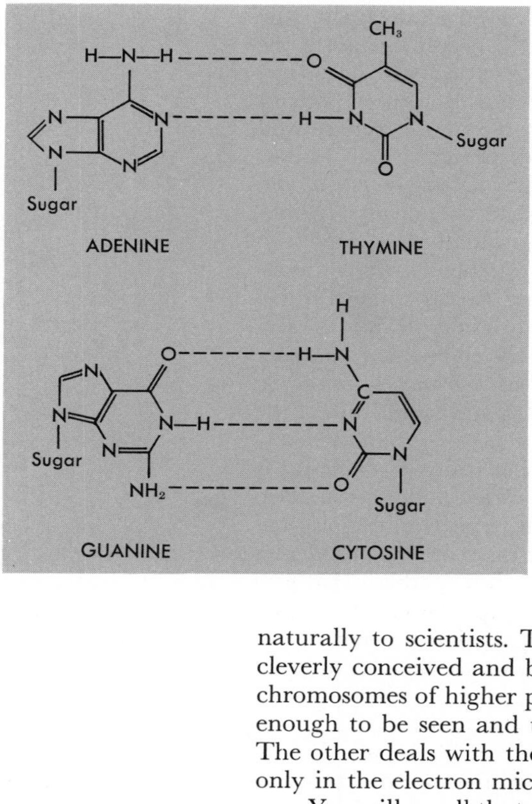

ADENINE   THYMINE

GUANINE   CYTOSINE

Hydrogen bonds

naturally to scientists. Two experiments will be described, both cleverly conceived and beautifully executed. One deals with the chromosomes of higher plants, since these chromosomes are large enough to be seen and they contain the DNA that interests us. The other deals with the DNA of a bacterium that can be seen only in the electron microscope.

You will recall that we stated that DNA is newly synthesized during the interphase stage of cell division. Let us assume, as we have indicated in Fig. 12.6, that the original (X and Y) strands of DNA will be preserved intact and that the X′ and Y′ strands will be the newly synthesized ones. If this is so, it should then

196

be possible to identify the old from the new if the new one can be "tagged" in some way. This was accomplished by J. H. Taylor of Florida State University by using the following procedure.

At the time the DNA was being synthesized, he fed the roottip cells of a broad bean plant a solution containing radioactive thymidine. This molecule is incorporated *only* into DNA, where it becomes the deoxythymidilic nucleotide. Since DNA is a very stable molecule, the old strand would not pick up any of the radioactive thymidine. During formation, however, the new strand would. When the chromosomes reach metaphase and anaphase, at which time they are large and distinct enough to see, the distribution of the radioactivity can be determined by the use of **autoradiographic** techniques.

If you understand how a photographic negative is produced, you can understand **autoradiography.** A roottip is squashed on a glass slide so that the cells are flattened out. Then, a thin photographic emulsion is placed over the flattened cells. As the radioactive atoms in the thymidine decay, atomic particles or rays are released and pass through the emulsion. These darken the emulsion much as ordinary light affects a photographic film.

Taylor showed that cells fed radioactive thymidine during interphase would show radioactivity in both chromatids during metaphase stage. If, however, he allowed these cells to pass through another division, and in a solution containing no radioactive thymidine, then the chromosomes in the next metaphase would show one chromatid radioactive and the other "cold," or nonradioactive. These events are shown in Fig. 12.7.

Let us now relate these results to the model of DNA we have discussed. Assume that the chromosome in interphase, before DNA synthesis, consists *only* of a double helix. (Actually it has other molecules—protein and RNA—but we can disregard these for the moment.) Assume, also, that when the DNA replicates itself the old strands are preserved intact, but when the new ones are synthesized they take up radioactive thymidine. The diagram shows what will be expected at the first metaphase and also after another division. The results of the Taylor experiment are, therefore, in agreement with what we might have predicted on the basis of the Watson–Crick model and its mode of replication.

Let us now take stock of what we *know* as opposed to what we have *assumed.* Taylor showed by means of autoradiography that radioactive thymidine is incorporated into DNA. He also demonstrated that only the newly synthesized strands pick up the radioactive thymidine; the old strands are unaffected. To prove the correctness of this, the old strand of DNA had to be identified with certainty. This problem was attacked by Matthew Meselson and Frank Stahl, working at the California Institute of Technology, in a beautiful experiment with bacteria (Fig. 12.8).

The most common nitrogen has an atomic weight of about

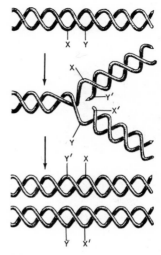

Figure 12.6 As a DNA molecule separates into its two polynucleotide strands (X and Y), each strand directs the formation of a complementary strand to pair with it; that is, X directs the formation of Y′, and Y the formation of X′. The entire strand is progressively replicated in this way, and the two new helices are identical to each other in nucleotide sequence.

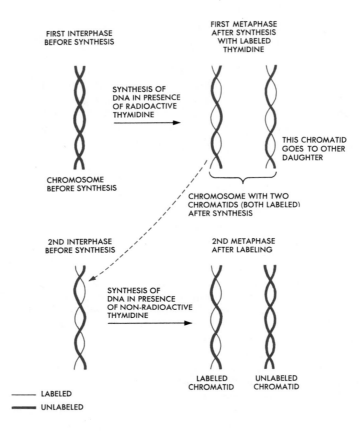

FIRST INTERPHASE
BEFORE SYNTHESIS

FIRST METAPHASE
AFTER SYNTHESIS
WITH LABELED
THYMIDINE

SYNTHESIS OF
DNA IN PRESENCE
OF RADIOACTIVE
THYMIDINE

THIS CHROMATID
GOES TO OTHER
DAUGHTER

CHROMOSOME
BEFORE SYNTHESIS

CHROMOSOME WITH TWO
CHROMATIDS (BOTH LABELED)
AFTER SYNTHESIS

2ND INTERPHASE
BEFORE SYNTHESIS

2ND METAPHASE
AFTER LABELING

SYNTHESIS OF
DNA IN PRESENCE
OF NON-RADIOACTIVE
THYMIDINE

LABELED
CHROMATID

UNLABELED
CHROMATID

———— LABELED
▬▬▬ UNLABELED

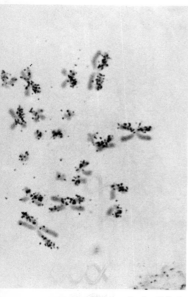

Courtesy T. C. Hsu

Figure 12.7 The experiment of H. J. Taylor demonstrated that the method of replication shown in Fig. 12.6 is correct for chromosomes as well as for the double helices of DNA. The newly synthesized strands showed that both chromatids of each chromosome were radioactive at the first metaphase following labeling. At the second metaphase, each chromatid had replicated again, but in the presence of nonradioactive thymidine, and one chromatid showed itself radioactive. The other was not. The results at the second metaphase can be seen in the photograph (above) of hamster chromosomes.

14. Some nitrogen atoms, however, have an extra neutron, which gives the atom a weight of 15. Ordinary nitrogen is written ^{14}N; heavy nitrogen, ^{15}N. If nitrate, NO_3^-, is used as a food source, the bacteria can grow equally well on $^{15}NO_3^-$ as on $^{14}NO_3^-$. Also, if the nitrate used is the only source of nitrogen available to the bacteria, the N atoms from the NO_3^- will become incorporated into the cells; in particular, for our discussion, the N atoms will go into the nucleotides of DNA.

Therefore, if one culture of bacteria is fed $^{15}NO_3^-$ and another culture is fed $^{14}NO_3^-$, the DNA extracted from the first should be heavier, or denser, per unit of volume than that from the second. If a mixture of heavy and ordinary DNA is put into a centrifuge, the heavy DNA can be separated from the light DNA, since the movement of a molecule in a centrifugal field (that is, its sedimentation rate) depends upon its weight, or density. Stated in another way, heavy DNA will move a greater distance before coming to rest, or equilibrium, in a centrifugal field than will ordinary DNA. A molecule of DNA made up of one-half ^{14}N and one-half ^{15}N will come to rest halfway between ^{14}DNA and ^{15}DNA.

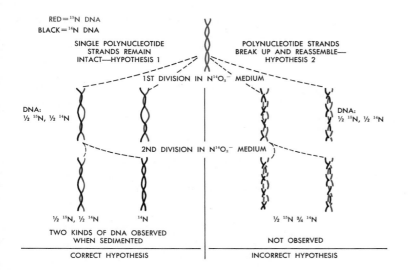

RED = ^{15}N DNA
BLACK = ^{14}N DNA

SINGLE POLYNUCLEOTIDE
STRANDS REMAIN
INTACT—HYPOTHESIS 1

POLYNUCLEOTIDE STRANDS
BREAK UP AND REASSEMBLE—
HYPOTHESIS 2

1ST DIVISION IN N^{14}O$_3^-$ MEDIUM

DNA:
½ ^{15}N, ½ ^{14}N

DNA:
½ ^{15}N, ½ ^{14}N

2ND DIVISION IN N^{14}O$_3^-$ MEDIUM

½ ^{15}N, ½ ^{14}N ^{14}N

½ ^{15}N ¾ ^{14}N

TWO KINDS OF DNA OBSERVED
WHEN SEDIMENTED

NOT OBSERVED

CORRECT HYPOTHESIS

INCORRECT HYPOTHESIS

Figure 12.8 The Meselson–Stahl experiment demonstrated that the polynucleotide strands separate but remain intact during synthesis rather than breaking down and reassembling. Compare these results with that of the Taylor experiment in Fig. 12.7. The results are the same whether whole chromosomes or only DNA molecules are used.

Meselson and Stahl grew bacteria for several generations in ^{15}NO$_3^-$ until virtually all the nitrogen of the DNA was ^{15}N. Cells were then removed from the culture medium containing ^{15}NO$_3^-$ and placed in a fresh medium containing ^{14}NO$_3^-$, *but only long enough to allow each cell to divide only once.* During this time the number of cells doubled, and each molecule of DNA had replicated itself. Of the total amount of DNA, one half should have been ^{15}DNA, the other half ^{14}DNA. When the DNA from these cells was removed, its sedimentation rate was determined to be between that of ^{14}DNA and ^{15}DNA. So far, so good. But let us now consider the results in relation to our prediction: that the old strands of ^{15}DNA are preserved intact, and that the new ones will contain only ^{14}N. Essentially, this is what Taylor found when he used radioactive thymidine.

However, let us suppose, as a contradictory hypothesis, that the old DNA breaks down completely instead of being preserved intact. The new DNA will then be assembled at random from bases containing ^{15}N (from the old DNA) and from newly synthesized bases containing ^{14}N. Both strands of the double helix will have a mixture of bases of ^{15}N and ^{14}N, and we could label this as $^{1/2N15,1/2N14}$DNA. This kind of DNA would be indistinguishable, insofar as sedimentation rate is concerned, from DNA in which one strand had only ^{15}N and the other only ^{14}N. It is the amount of ^{15}N and ^{14}N bases—not their distribution in the strands—that determines sedimentation rate. The experimental results described so far, then, cannot distinguish between the two hypotheses.

However, Meselson and Stahl carried the experiment one step further. They allowed some of the bacteria to divide once more in ^{14}NO$_3^-$, harvested the cells, and extracted the DNA. Let us now see what would be predicted on the basis of our two hypotheses. If the DNA breaks down, as our second hypothesis supposes, each strand of DNA would have $\frac{1}{4}$ ^{15}N and $\frac{3}{4}$ ^{14}N, with the ^{15}N and ^{14}N randomly distributed. *There would be only one kind of DNA.*

On the other hand, if the old ^{15}N strand had been preserved intact, two kinds of DNA would be found: $^{1/2N15, 1/2N14}DNA$ and ^{N14}DNA (Fig. 12.8). Also the heavy DNA should contain one-half of the original ^{15}N. These two kinds were found, and we can, therefore, state that our first hypothesis has been supported.

We are not certain, at this moment, that the answers are final. The DNA molecule we have been discussing is a **macromolecule,** meaning that it is a huge molecule with a molecular weight of several million or more. It has many thousands of turns in its spiral configuration, and the base sequence can be arranged in any order. The possible variations, therefore, are astronomical in number and give an infinite variety to the DNA's from various sources. It has been estimated, for example, that every cell in the human body has approximately 8 billion nucleotide pairs making up the DNA of its 46 chromosomes. If we look upon the base pairs as the letters in a genetic alphabet, which when put together in a particular sequence form a "word" having meaning to the cell, then we can readily grasp the idea of how DNA carries information.

THE DNA–RNA PROTEIN CHAIN OF RELATIONSHIPS

The question we must ultimately ask is, "How does DNA carry out its cellular role of command?" Before trying to answer this question, we need to remember that what a cell does and how it is constructed depend largely on the kinds of proteins it contains. These include all the enzymes and membranes of the cell, and are found as well in ribosomes, spindle, plastids, and chromosomes. Therefore, if we maintain our hypothesis that DNA is the controlling agent of the cell, then we must also hypothesize further that DNA in some way controls the production of the various kinds of proteins.

However, we know that DNA is mainly in the nucleus, while protein is formed largely on the ribosomes of the cytoplasm. Some substance must consequently be a messenger, carrying the information from DNA to the ribosomes so that they will "know" just what protein to make. We now know that this messenger is **RNA,** a nucleic acid molecule that differs from DNA in having a substance called **uracil** in place of thymine and in having a **ribose sugar** in place of the deoxyribose sugar. In a very much oversimplified way we can then say that DNA makes RNA, and that RNA makes protein.

If this is so, then each RNA molecule produced by DNA must contain enough information to spell out the order, kind,

Ala—Ala—Lys—Phe—Glu—Arg—Glu(NH₂)—His—Met—Asp—Ser—Ser—Thr—Ser—Ala—Ala
Ala ... Thr—Asp(NH₂)—Cyst—Tyr—Glu(NH₂)—Ser—Tyr—Ser—Thr—Met ... Ser
Thr ... Glu(NH₂) ... S ... Ser ... Ser
Glu ... Gly ... S ... Ileu ... Ser
Lys ... Asp(NH₂)—Lys—Cyst—Ala ... Thr ... Asp(NH₂)
Val ... Asp ... Tyr
Glu(NH₂)—Ala—Val—Cyst—Ser—Glu(NH₂)—Lys—Asp(NH₂) ... Cyst—S—S—Cyst
Val ... Arg ... Asp(NH₂)
Asp ... S ... Glu ... Glu(NH₂)
Ala ... Ileu—Val—Ala—Cyst—Glu—Gly—Asp(NH₂)—Pro—Tyr—Val—Pro—Val—His—Phe—Asp—Ala—Ser—Val ... Thr ... Met
Leu ... Ileu ... Gly ... Met
Ser ... His—Lys—Asp(NH₂)—Ala—Glu(NH₂)—Thr—Thr—Lys—Tyr—Ala—Cyst—Asp(NH₂)—Pro—Tyr—Lys—Ser—Ser ... Lys
Glu ... S
His—Val—Phe—Thr—Asp(NH₂)—Val—Pro—Lys—Cyst—Arg—Asp—Lys—Thr—Leu—Asp(NH₂)—Arg—Ser

Figure 12.9 The entire molecule of ribonuclease (or insulin), showing the sequence of amino acids and the manner by which cross-links cause the molecule to fold.

and number of amino acids in each protein, and the features that give each protein its uniqueness of structure and function. (Such a protein is illustrated in Fig. 12.9.) Three different RNA's participate in the formation of protein, and each plays a special role. These are **ribosomal RNA (rRNA), messenger RNA (mRNA),** and **transfer,** or **soluble, RNA (tRNA or sRNA).** All these are formed by DNA, much in the same manner that DNA forms more DNA. This being so, each piece of RNA must have a sequence of nucleotides *complementary* to the strand of DNA that formed it—but with one exception: the thymine of DNA will be replaced by uracil in the RNA. Thus, if a piece of DNA having the sequence ATAGTCTTT makes RNA, the RNA would have the sequence UAUCAGAAA (Fig. 12.10).

The solution to the problem of how the chemical information in DNA is translated into the formation of a protein is one of the most significant biological discoveries of the 20th century. The problem was basically one of unscrambling a code. What we need is a four-letter code that can form a dictionary of 20 words. The four letters are the four nucleotides of DNA: A, T, C, and G. The 20 words are the 20 amino acids found in virtually all proteins. Immediately we know that one nucleotide cannot be responsible for one amino acid (Fig. 12.11). If it were, only four of the 20 amino acids would be accounted for. Two nucleotides for each amino acid are also not enough. The four letters, arranged in groups of two, would produce only 16 words, accounting for only 16 amino acids. What about a three-letter code: ATT, ATG, ATC, AAT, AAG, AAC, and so on? This would give us $4 \times 4 \times 4 = 64$. It would now appear that we have too many words. Possibly some of the words make sense, while others do not. For example, take the letters N, O, and W. These three letters can form six possible groups: NOW, OWN, WON, ONW, WNO,

Figure 12.10 Here is an outline of the sequence of events from DNA to RNA to protein. The determination of the amino acid sequence by means of the triplet code is also shown.

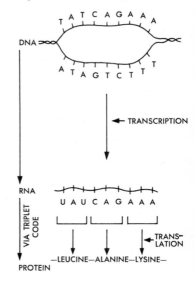

DNA

ATCAGAAA
ATAGTCTT

← TRANSCRIPTION

RNA

UAUCAGAAA

VIA TRIPLET CODE

← TRANS-LATION

—LEUCINE—ALANINE—LYSINE—

PROTEIN

201

SINGLET CODE (4 WORDS)	DOUBLET CODE (16 WORDS)				TRIPLET CODE (64 WORDS)			
					AAA	AAG	AAC	AAU
					AGA	AGG	AGC	AGU
					ACA	ACG	ACC	ACU
					AUA	AUG	AUC	AUU
					GAA	GAG	GAC	GAU
					GGA	GGG	GGC	GGU
A	AA	AG	AC	AU	GCA	GCG	GCC	GCU
G	GA	GG	,GC	GU	GUA	GUG	GUC	GUU
C	CA	CG	CC	CU	CAA	CAG	CAC	CAU
U	UA	UG	UC	UU	CGA	CGG	CGC	CGU
					CCA	CCG	CCC	CCU
					CUA	CUG	CUC	CUU
					UAA	UAG	UAC	UAU
					UGA	UGG	ʊGC	UGU
					UCA	UCG	UCC	UCU
					UUA	UUG	UUC	UUU

Figure 12.11 The possible numbers of code words based on singlet, doublet, and triplet codes are shown here. We now believe that the triplet code is correct.

and NWO. The first three are sense words while the last three are nonsense words, at least in the English language.

We now believe that the three-letter or **triplet code** is the correct one for DNA. But is nature so wasteful that it makes use of only 20 out of 64 possibilities of making amino acids into protein molecules? It now appears that several amino acids have more than one triplet code (see Table 12.2). Much still needs to be done to clarify the coding system. For example, the first two letters of the code may stand for the particular amino acid, while the third letter may have something to do with the efficiency of the process under a given set of cellular circumstances.

Amino Acid Activation. The first step in protein synthesis is the selection of specific amino acids out of the heterogeneous mixture of metabolites in the cytoplasm of the cell. This selection process is carried out by amino-acid-activating enzymes that result in the formation of amino acid adenylates (Fig. 12.12). The activation of the amino acids requires ATP energy and the reaction is very similar to the activation of fatty acids, which we discussed previously. There is at least one unique activating enzyme for each different amino acid incorporated into protein, and the specificity of this enzyme for the particular amino acids is very high, ensuring critical control of protein synthesis.

tRNA—the Adaptor. The activated amino acids remain tightly bound on the enzyme following the activation step. The next step in protein synthesis involves the attachment of the amino acid to a specific adaptor RNA molecule. This RNA, a small, soluble molecule, has been variously termed soluble (s), transfer (t), or adaptor RNA. As we shall discuss shortly, the existence of several classes of tRNA to which a single amino acid is attached is important in the coding process.

tRNA is a single chain of about 70 nucleotides that is folded back upon itself and held in a helical configuration by hydrogen bonding. Consequently the molecular weight is about 25,000. The terminal sequence of nucleotides to which the amino acid is attached has been shown to be in all cases adenylic–cytidylic–cytidylic (A–C–C). The amino acid–adenylic acid from the activating reaction interacts with tRNA, and the amino acid is transferred to the second or third carbon of the ribose molecule in the terminal adenylic acid of tRNA (Fig. 12.12).

Although all tRNA appears to be similar in gross structure, there is much evidence which indicates some structural characteristics that are unique to each class and specific for a given species of activating enzyme. As discussed later, the tRNA molecules must contain a special short sequence, the anticondon, that allows them to seek out the proper complementary series of bases in the special RNA (messenger) found at the protein-forming site.

Ribosomes. The protein-forming system uses the amino acid–tRNA complexes as the raw material for making a polypeptide chain. The biochemical events in protein synthesis take place on small granules (ribosomes) that are located in the cytoplasm. Ribosomes contain about 60 per cent RNA and 10 per cent protein. In bacteria ribosomes are found throughout the cytoplasm, whereas in higher organisms they are often found attached to the surface of tubular lipoprotein membranes found in the cytoplasm. This membrane system is called collectively the endoplasmic reticulum. Ribosomes have been shown to be the site for protein synthesis in a number of indirect ways; however, much stronger support has been obtained recently by the isolation of the ribosomal particles, freed from other cellular components. Such isolated ribosomes, under appropriate conditions, can incorporate amino acids from the tRNA–AA complex into a polypeptide chain. Ribosomes are rather complicated structures that can dissociate into smaller inactive subunits. The function of the structural RNA of the ribosomes is not known, although it seems reasonably clear that it

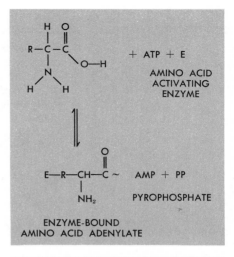

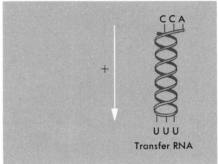

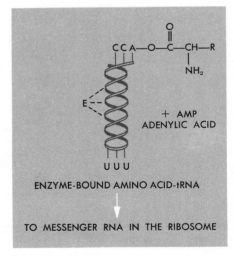

Figure 12.12 The activation of amino acids for protein synthesis.

is not directly involved in the coding of the sequence of amino acids into protein.

mRNA—the Messenger. Research during the past few years indicates that a special class of RNA molecules acts as the template upon which proteins are built. These are called messenger RNA (mRNA) since, as recent evidence indicates, they are ribopolynucleotides that are direct copies of the genetic messages from DNA.

A number of workers have shown that an enzyme, RNA polymerase, catalyzes the synthesis of mRNA and that DNA is required as the primer. Additional evidence indicates that the enzyme catalyzes a preferential copy of the base sequence in DNA rather than that it joins nucleotides in a random fashion. Not all messenger RNA is made with DNA as the template. As indicated previously, certain viruses contain RNA instead of DNA as the genetic material, and this RNA acts directly as messenger for the synthesis of a specific **RNA replicase.** This enzyme is very specific for the virus RNA and explains how RNA viruses are specifically made in bacterial cells where other RNA molecules are present. Using this enzyme, it has been possible to synthesize for the first time a biologically active nucleic acid, that is, an infectious RNA virus.

Figure 12.13 The first break in genetic code was the discovery that a synthetic messenger RNA containing only uracil (poly U) directed the manufacture of a synthetic protein containing only one amino acid, phenylalanine (PHE). The X's in transfer RNA signify that the bases that respond to code words in messenger RNA are not known.

After Nirenberg

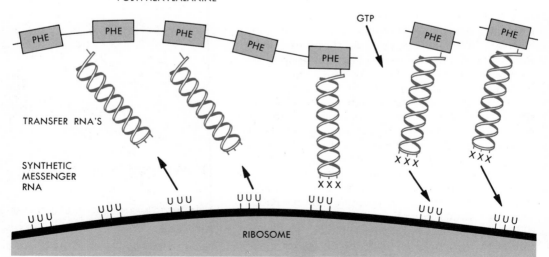

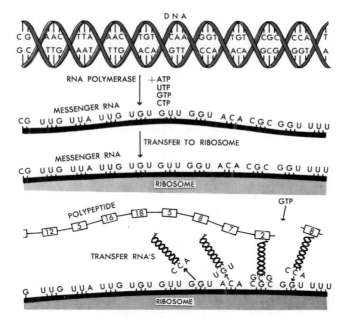

Figure 12.14 The sequence of events from DNA to protein is shown here. Only one of the two polynucleotide strands of DNA transcribes and forms messenger RNA. The mRNA becomes attached to the ribosomes where the RNA's, each carrying a particular amino acid, match up in complementary fashion with the successive triplet codes of mRNA. As this occurs, the amino acids are linked and form a polypeptide chain, and the tRNA's float free.

After mRNA is synthesized, it leaves the template and is adsorbed onto the ribosome. This mRNA–ribosome complex is the active unit for protein synthesis. If we add to this complex the 20 amino acids, the various tRNA's, ATP, and the activating enzymes, we observe the synthesis of a polypeptide chain. In the formation of the peptide, additional enzymes are required as well as guanosine triphosphate (GTP). See Fig. 12.14.

There are numerous experiments which indicate that the mRNA is the specific template for the synthesis of specific polypeptides and that activating enzymes and tRNA are used over and over again in the synthesis of a number of different proteins. Probably the most striking example of this is furnished by the experiments of Nirenberg in which he used synthetic messengers. Synthetic ribopolynucleotides can be made by using an enzyme, polynucleotide phosphorylase (which probably serves normally to degrade RNA), and the appropriate nucleotide triphosphates. The composition of the product depends almost entirely on the concentration of the nucleotides. It has been possible, for example, to synthesize a polymer containing only uracil (polyuridylic acid or poly U). When poly U was added to a ribosomal system along with the other components essential for protein synthesis, it was found that only phenylalanine was incorporated into the polypeptide. Thus, it would appear that the triplet code for this amino acid was UUU (Fig. 12.13). This system has proved very useful in translating the genetic code for other amino acids. Poly C (CCC) appears to be the code for the amino acid proline, CCG for alanine, GUA for aspartic acid, ACC for histidine, and so on.

The genetic code for all the amino acids is shown in Table 12.2. It is clear that more than one codon can code for a single amino acid. For example, UUU as well as UUC can code for phenylalanine. This "degeneracy" seems to be essential, for if we use only two bases, there would be only 16 (4^2) possible sequences for coding the 20 amino acids. The three-letter code gives 64 (4^3) sequences for 20 amino acids, thus allowing for some degeneracy.

Thus the specific organization and alignment of amino acids in the polypeptide chain of a protein are dictated by the base sequence in the mRNA, which in turn is a direct reflection of the base sequence in the gene (DNA). The specific adaptor molecule, tRNA, is responsible for carrying a specific amino acid to the site for protein synthesis. The growth of the polypeptide chain (that is, peptide bond formation) requires GTP and specific enzymes. When the polypeptide is complete, it is released from the

TABLE 12.2. Genetic Code for Amino Acids

First Position	Second Position				Third Position
	U	C	A	G	
U	PHE	SER	TYR	CYS	U
	PHE	SER	TYR	CYS	C
	LEU	SER	OCHRE†	AMBER†	A
	LEU	SER	AMBER†	TRP	G
C	LEU	PRO	HIS	ARG	U
	LEU	PRO	HIS	ARG	C
	LEU	PRO	GLN	ARG	A
	LEU	PRO	GLN	ARG	G
A	ILE	THR	ASN	SER	U
	ILE	THR	ASN	SER	C
	ILE	THR	LYS	ARG	A
	MET*	THR	LYS	ARG	G
G	VAL	ALA	ASP	GLY	U
	VAL	ALA	ASP	GLY	C
	Val	ALA	GLU	GLY	A
	VAL*	ALA	GLU	GLY	G

* Chain-initiating codon.
† Chain-terminating codon.

ribosome and becomes folded in the active configuration. Much remains to be discovered about this latter process, and indeed many important questions still remain to be answered in regard to this general scheme of cell-free protein synthesis. Even so, this model has been very useful in explaining a number of genetic, or heritable, changes **(mutations)**. A mutation appears to be due to a specific change in one of the bases in DNA, and this seems to be eventually reflected in the mRNA and finally the protein. One of the best-known examples of this takes place in the hemoglobin of a person with sickle-cell anemia. In this disease, which is inherited as a single gene difference, the hemoglobin is electrophoretically different from the normal. Analysis of the amino acid sequence of the altered hemoglobin reveals that one particular glutamic acid residue is replaced with valine, and this alteration is sufficient to account for the change in charge.

Hemoglobin is an iron-containing red pigment, a protein, contained in the red blood cells. Its function is to carry oxygen and carbon dioxide in the blood stream. The kind of hemoglobin in most individuals is **hemoglobin A,** and it is made up of a total of about 600 amino acids. (Since there are only 20 different amino acids, each one is represented many times in the hemoglobin molecule.) **Hemoglobin S** is an abnormal type, and anyone having it develops a disease known as **sickle-cell anemia,** the name being derived from the sickle shape of the red blood cells. The difference between hemoglobin *A* and hemoglobin *S* is caused by a mutation and affects only one amino acid out of the several hundred in the whole molecule (Fig. 12.15), yet the change profoundly affects the individual and his health. The actual nucleotide change that took place in the DNA at the time of mutation is now known, but once we know how mRNA codes for the hemoglobin molecule, we can then work backward and say what actually happened in the DNA molecule. It is apparent, then, that our future knowledge of how cellular activity is controlled lies in a better understanding of the interactions and interrelations of DNA, RNA, and protein. Until these are fully known, our insight into the behavior of cells will continue to be fragmentary. A few examples of hereditary disorders in man are shown in Table 12.3.

There are some unusual problems in connection with the initiation of the synthesis of a polypeptide chain. In *E. Coli* it has been observed that the most frequent *N*-terminal amino acid found in proteins is methionine. In isolated systems *N*-formyl methionine is the actual initiator of polypeptide chain formation and if this occurs *in vivo* the formyl group is removed after the chain is released from the ribosome. As indicated in Table 12.2, the codon for methionine is AUG. However, there are two known tRNA's specific for methionine. When methionine is attached to one of these tRNA molecules, it can be rapidly formylated at the amino group. Blocking the amino group prevents it from reacting with carboxyl groups of other amino acids to form a

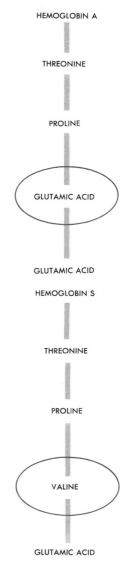

HEMOGLOBIN A

THREONINE

PROLINE

GLUTAMIC ACID

GLUTAMIC ACID

HEMOGLOBIN S

THREONINE

PROLINE

VALINE

GLUTAMIC ACID

Figure 12.15 A small part of the polypeptide chain of hemoglobin A and hemoglobin S is shown here. A change in the nucleotide sequence of the DNA responsible for forming hemoglobin A led to the insertion of valine instead of glutamic acid, and resulted in hemoglobin S. This change of one amino acid out of several hundred greatly affects the function of the hemoglobin molecule.

TABLE 12.3. Examples of Hereditary Disorders in Man

Disorder	*Affected Enzyme*
Albinism	Tyrosinase
Fructose intolerance	Fructose-1-phosphate adolase
Fructosuria	Fructokinase
Goiter (familial)	Iodotyrosine dehalogenase
Hemolytic anemia	Pyruvate kinase
Histidinemia	Histidase
Hypophosphatasia	Alkaline phosphatase
Maple syrup urine disease	Amino acid decarboxylase
Phenylketonuria (PKU)	Phenylalanine hydroxylase
Xanthinuria	Xanthine oxidase

* After White, Handler, and Smith (1968).

peptide bond. Thus *N*-formyl methionine cannot be incorporated at internal positions in the polypeptide, but could function to initiate chain formation.

N-formyl methionine does not seem to play this same role in animals or higher plants. Acetylserine may function in this capacity for some systems in animals. Much remains to be done before we completely understand the process of protein synthesis.

As indicated in Table 12.2, there are three codons, UAA (ochre), UAG (amber), and UGA (umber), that do not code amino acid incorporation into a polypeptide chain. Recent evidence indicates that these are chain-terminating codons. The biochemical mechanism of this process is not clear. It may be a simple passive process in which there is no tRNA for these codons. Therefore, when the growing polypeptide chain reaches one of the triplets, peptide bond formation ceases, and the attached polypeptide is hydrolyzed from the last tRNA used.

Other aspects concerning the control of protein synthesis will be considered in the next chapter.

RNA-DIRECTED DNA SYNTHESIS As indicated in the previous section, it is now well established that in most organisms genetic information is transferred from DNA to RNA and then to protein. The process from DNA to RNA is called **transcription,** and the process from RNA to protein is called **translation.** In addition, genetic information can be transmitted from DNA to DNA in the process of **replication.** There are, however, certain interesting exceptions to the above concept. As discussed previously, there are certain RNA viruses in which the RNA acts as a template to make more RNA, and there is a highly specific enzyme that carries out this process.

Recently, a third type of virus has been shown to be able to carry out a "reverse flow" in genetic information; that is, RNA acts as the template for making specific DNA, which in turn acts as the genetic template for making more virus. There is a great deal of interest in this type of virus since many have been shown to cause tumors in chickens, mice, monkeys, and other species of animals. Quite recently similar viruses have been isolated from human mammary tumors.

Based on the facts briefly described above, Howard Temin suggests that viruses can be grouped into three major classes insofar as genetic information flow is concerned (Fig. 12.16).

Recently, Temin and, independently, David Baltimore have been able to demonstrate a viral enzyme that is capable of catalyzing the synthesis of DNA, using the viral RNA as a template. The enzyme is known as the RNA-directed DNA polymerase, but many workers refer to it as the "reverse transcriptase." In addition to RNA, the enzyme can also use DNA as a template for making more DNA; thus, in many ways, it is similar to the normal DNA polymerase that we discussed previously. These observations are of great interest, because they not only offer a possible explanation how tumor viruses produce a cancerous transformation, but may also explain how normal cells can carry the tumor virus DNA for years and then, by some unknown activation process (inducer), begin to give rise to infectious virus particles (Fig. 12.17). It is clear that a great deal of exciting research remains to be done in this important area. Not only do the present results offer some rational approach to the study of tumor-causing viruses, but they also may offer some new insights into the control and regulation of normal cellular differentiation during embryonic development.

After H. Temin

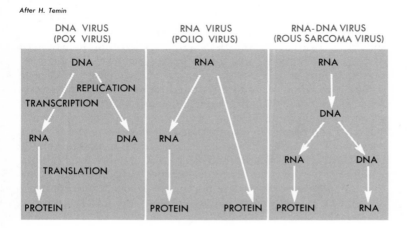

Figure 12.16 The three major classes of viruses based on genetic information.

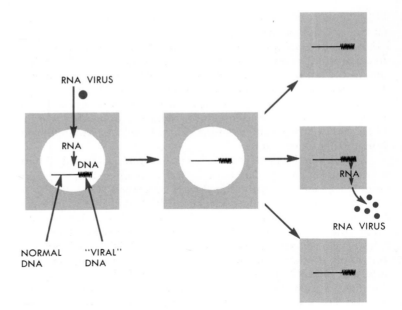

RNA VIRUS

RNA

DNA

NORMAL
DNA

"VIRAL"
DNA

RNA

RNA VIRUS

Figure 12.17 Scheme showing how RNA viruses can incorporate the genetic information into the host DNA and remain latent for long periods before making more RNA viruses.

genes and enzymes

DNA is a molecule having certain physical and chemical properties. It also has hereditary function. A hereditary unit, or gene, can therefore be defined in a preliminary way as a segment of DNA ultimately responsible for the formation and character of a protein. Only rarely do we recognize the change that takes place in the protein.

More commonly we recognize a change, or mutation, in DNA by a change in a **trait.** Brown versus blue eyes is one such variable trait that is inherited. In a shorthand way of speaking, we say that the gene is responsible for determining the trait. We should be aware, however, that heredity is an inherited pattern of chemical reactions, and that a process of development is required before a trait can be recognized. A change in a gene through mutation leads to an altered pattern of chemical reactions. If we now recall the statement made earlier that specific reactions in the cell are controlled by specific enzymes, then it is logical and correct to assume that genes control enzymes, since enzymes are proteins. In fact, one very fruitful hypothesis, made in 1941 by G. W. Beadle and E. L. Tatum, then at Stanford University, is the "one-gene–one-enzyme" hypothesis. It states that a single gene acts by determining the specificity of a particular enzyme (or one polypeptide). In turn, the enzyme governs a particular chemical reaction.

A change in the gene controlling an enzyme can have two consequences. Either the enzyme is no longer formed, or it is altered structurally so that it no longer functions in a normal way. In either case, the chemical reaction stops or is changed. We already know that the product of a chemical reaction can change, for we have seen how a mutation can alter both the structure and function of normal hemoglobin to give a trait known as sickle-cell anemia. The first consequence can also occur, for it can be shown in some instances that when a chemical reaction is missing, so too is the enzyme.

To illustrate the control of single chemical reactions by genes, let us consider the synthesis of **arginine,** one of the amino acids. Our knowledge originally came from a study of the red bread mold, *Neurospora,* but the results are generally applicable to virtually all organisms.

Ordinarily, *Neurospora* can manufacture its own arginine from a simple medium containing mitrate, a vitamin (biotin), and glucose (one of the sugars), the latter providing a source of energy as well as of carbon. By exposing spores of *Neurospora* to x-rays or ultraviolet light, we can cause various genes to mutate. These mutated spores will grow, and among them will be some that are unable to synthesize their own arginine. To keep such strains alive, it is necessary to supply them with arginine, since it is an essential amino acid.

There are many steps in the synthesis of arginine from glucose, but we need be concerned only with the three end steps (Fig. 12.19), each governed by an enzyme. If any one of these steps is blocked by lack of a functional enzyme, arginine will not be synthesized. It must be added to the growth medium as an essential nutrient. Let us now test our original hypothesis. We

Museo Del Prado

The Granger Collection

Figure 12.18 These portraits of two members of the Hapsburg royal family of Austria (top: Philip IV; bottom: his son, Charles II) show the typical long jaw and protruding lower lip. These inherited traits occurred in most members of this family.

Figure 12.19 This diagram shows the last three steps in the sequence of reactions leading to the formation of arginine. Each step is enzyme-controlled and would be blocked if the enzyme were missing or altered to a nonfunctional state.

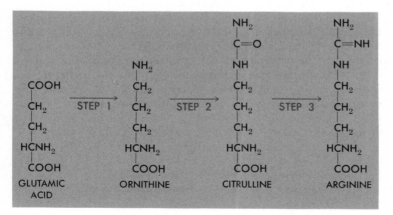

are assuming that genes control specific chemical reactions through their control of the specific enzymes. But we must make a further assumption. We must also assume that an arginine-less mutant (a strain that cannot synthesize arginine) can result from a loss or impairment of three different enzyme systems. As a consequence, not all arginine-less mutants need be similar in character. We can now prove that this is the correct assumption.

Refer again to Fig. 12.19. Assume that step 1 has been blocked, and that no arginine has been formed. We know that the strain will grow if supplied with arginine, but will it also grow if it is supplied with the amino acid **ornithine** or **citrulline**? It turns out that it will, since steps 2 and 3 are not impaired, and the block is earlier in the sequence of reactions. Similarly, if step 2 is blocked, the strain can use either citrulline or arginine, but the addition of only ornithine would be inadequate for growth, since ornithine cannot be converted to citrulline. If step 3 is blocked, the organism can grow only if arginine is added. Further work has also revealed that the particular enzymes are either missing or are altered so that they no longer function properly.

To demonstrate that each step is controlled by a single gene, it is necessary to use the inheritance test, which involves controlled matings and careful observation of the offspring. This is the subject of Chapter 14.

summary

DNA is a macromolecule consisting of two complementary strands of repeating units wrapped around each other in the form of a double helix. The repeating units are of four kinds, each consisting of a base, a sugar, and a phosphoric acid. Each one is a nucleotide, and the four bases—thymine, adenine, guanine, and cytosine—give infinite variety to the DNA molecule. They also supply the genetic information to the cell and the organism.

DNA does two things: (1) it makes copies of itself by replication; and (2) it makes several kinds of RNA. These RNA's pass to the cytoplasm where they participate in the formation of proteins. These proteins form structures such as membranes, spindles, or ribosomes, act as enzymes that control the chemical reactions of the cell, or perform both kinds of functions. The uniqueness of a cell or an organism, therefore, results from the DNA it contains and the kinds of proteins formed as a result of the action of DNA and its derivative RNA molecules.

for thought and discussion

1   The triplet code of the mRNA is called the **codon,** the complementary on the tRNA's, the **anticodon.** Suppose that a

piece of transcribing DNA has the following arrangement of nucleotides:

A T C G G A C T T A C A C C T A G G

What will be the nucleotide sequence of the mRNA? What tRNA's will be necessary for protein formation? Using Table 12.2, determine the nature of the polypeptide to be formed.

2 In the nucleotide sequence of DNA given in Problem 1, suppose that the second nucleotide from the left (T) is lost, but that the process of transcription is unimpaired. What change will be found in the mRNA? Will the tRNA's be the same as before, and used in the same sequence? What will be the nature of the polypeptide that is formed?

3 DNA is a double helix, but all evidence indicates that for any given gene only one of the two polypeptide strands can transcribe. Can you suggest why? Can you suggest a mechanism that permits one to be transcribed, but not the other?

selected readings

BEADLE, G., and M. BEADLE. *The Language of Life.* Garden City, N.Y.: Doubleday & Company, Inc., 1966. An introduction to the science of genetics, including an account of the role of DNA.

KENDREW, J. *The Thread of Life.* Cambridge, Mass.: Harvard University Press, 1966. An introduction to molecular biology, with an emphasis on proteins and nucleic acids.

STANBURY, J. B., J. B. WYNGAARDEN, and D. S. FREDRICKSON. *The Metabolic Basis of Inherited Disease* (2nd ed.). New York: McGraw-Hill Book Company, 1966.

WATSON, J. D. *Molecular Biology of the Gene.* Menlo Park, Calif.: W. A. Benjamin Inc., 1965. An excellent account of molecular genetics.

control
of cellular
metabolism
chapter 13

The DNA–RNA–polypeptide chain of synthesis provides proteins for the building of cellular structures. It also forms enzymes, which catalyze the chemical reactions taking place within a cell. As we found earlier, there are many different kinds of cellular structures; also, each cell is a complex, but orderly, mixture of hundreds of different enzymes. Although the cells of all living organisms are remarkably alike in some ways, in other ways they are different. A liver cell, for example, looks and behaves differently from a cell in a muscle, the brain, or skin. We also know that each kind of cell is stable chemically, even though it is chemically active. It performs its own special task regularly, doing a kind of work not done by other cells. Each cell is a closely regulated system; and because it is regulated, some kind of control is constantly guiding its activity. We do not yet have answers to many of the problems relating to cellular control, but as we gain more and more insight into the cell, we find that control usually involves enzyme synthesis and action.

214

THE EFFECT OF NUTRIENTS ON ENZYME SYNTHESIS The ability of an organism to make a particular enzyme depends on the nature of the organism's DNA. If it does not have the proper DNA to make a given enzyme, it cannot, of course, make that enzyme. However, even though the cell has the proper kind of DNA, it may not maintain a constant supply of the enzyme. The nutritional environment of a cell seems to control the kind and amount of enzyme a cell has.

If a culture of bacteria is being grown on a medium containing glucose as the only source of carbon and energy, the bacteria will have the enzymes necessary to metabolize glucose. The bacteria can also grow on lactose, a disaccharide sugar found in milk. Lactose contains two hexoses—glucose and galactose. The enzymes in the cells that have been grown on glucose cannot break (hydrolyze) the lactose into the two single sugars. If we replace glucose with lactose as a nutrient source, we find that there is a delay—a lag period—before the cells begin to metabolize lactose. After a while, however, the cells make use of lactose with no difficulty. During the lag period the cells manufacture the enzyme necessary for the hydrolysis of lactose, which makes the simple sugars available. When enough of the enzyme is formed, the cells grow as well on the disaccharide lactose as they did on glucose. In this instance, the lactose *induced* the formation of the enzyme needed to metabolize it. Such an enzyme is said to be

Richard F. Trump

Figure 13.1 The tiger salamander is an animal that is unable to keep up the production of enough heat to remain active in cold weather. It survives by hibernating in winter.

inducible in contrast to the **constitutive** enzyme normally present in cells grown on glucose.

Clearly, the bacteria must have the proper DNA for making the inducible enzyme. In the absence of glucose, however, this DNA is inactive. With the addition of lactose, the DNA is activated and the proper enzyme is made. This means, of course, that the formation of a specific enzyme requires the prior formation of a specific messenger RNA. This takes time, and accounts for the lag period; mRNA is being formed, and it, in turn, forms the enzyme. This information, on the other hand, does not tell us how lactose can induce the formation of a specific mRNA. Although we are still not certain, some experiments have given us a clue. It appears that there are at least two kinds of DNA. One kind is concerned directly with the formation of mRNA, and consequently with the synthesis of the enzyme. This kind of DNA is called the **structural** gene. The other kind of DNA, which controls the structural gene by turning it on or off, is called the **regulator** gene. The situation is similar to the difference between a light bulb and a light switch. It is the light bulb that does the work, but the switch determines whether it will be on or off. It appears that the regulator gene produces a repressor substance which can inhibit the action of the structural gene. The inducer—lactose in this instance—combines either with the regulator gene and prevents the formation of repressor substance, or it combines with the repressor substance itself. In either case, the structural gene proceeds with the formation of mRNA, and the enzyme can be made (Fig. 13.2).

Figure 13.2 Schematic representation showing induced enzyme formation.

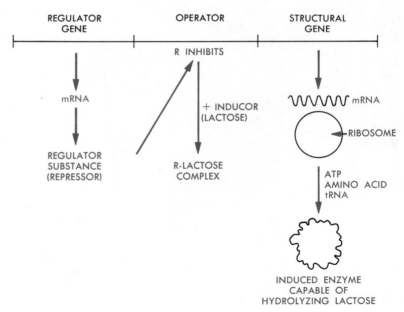

The rate of synthesis, as well as the kind of enzyme, is also controlled in part by the inducer substance. The greater the amount of inducer added, the faster the rate of enzyme induction. At very high concentrations of substrate, the rate tends to level off (Fig. 13.3), so the relation holds only over a limited range of concentrations. Also, the induction of enzymes is quite specific. In fact, some inducers cause the formation of enzymes even though the enzyme that is formed cannot metabolize the inducer. This indicates that DNA can be activated—or the regulator gene repressed—thus stimulating enzyme formation, but the enzyme is not used. The cell, therefore, can make "mistakes," just as an organism can.

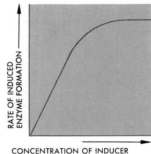

Figure 13.3 Rate of induced enzyme formation depends upon the concentration of the inducer.

ENZYME REPRESSION AND FEEDBACK INHIBITION If an enzyme and its substrate are present in a cell, a chemical reaction will take place. The reaction can be controlled in two ways: (1) by controlling the activity of the enzyme; or (2) by controlling the amount of enzyme formed. For example, if product A is converted to B by the action of an enzyme, the rate at which A goes to B can be shown to be controlled by the amount of B present in the cell. The pathway of synthesis of the amino acid arginine, shown in Fig. 13.4, illustrates this point. Bacterial cells grown in a medium containing a high arginine concentration have low amounts of the enzyme that converts ornithine to citrul-

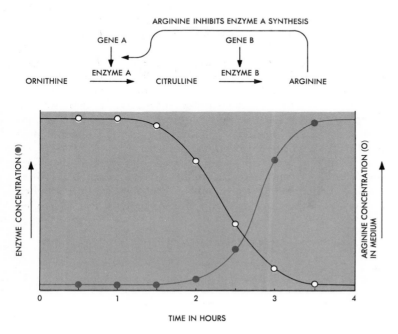

Figure 13.4 Enzyme repression—see text for details.

line. As the arginine is used up during growth, the amount of enzyme rises, and the rate of formation of arginine will consequently increase. A careful analysis of this and similar systems reveals that arginine actually represses, and thereby controls, the formation of this enzyme. The enzyme that converts citrulline to arginine, in this system at least, is not subject to the same control. Only one enzyme in the system is, therefore, affected, but by this the whole system is controlled. The whole system illustrates what is called **enzyme repression.**

At the present time we know of many instances in which enzyme activity can be demonstrated in a cell-free extract only after an inhibitor has been removed. Some of these inhibitors are proteins. The products of an enzyme-catalyzed reaction can also act as inhibitors if they are tightly bound to the enzyme; in this case the enzyme is inactive until the product is removed. In addition, the product may regulate a metabolic pathway by inhibiting an enzyme associated with an earlier step. This mechanism of control, called **feedback inhibition,** may be of considerable importance in regulating the amount of a particular product formed in a biosynthetic pathway (Fig. 13.5). For example, in the biosynthesis of histidine in certain bacteria there are at least 10 different steps requiring eight different enzymes. If we add histidine to the culture in which the cells are growing, they stop making the particular amino acid and use the outside source of histidine for protein synthesis. This blockage of histidine synthesis in the cell is due to the inhibition by histidine of the enzyme that catalyzes the first step in the biosynthetic pathway. Thus, by this mechanism of feedback inhibition the cell's machinery and energy supply are relieved of an additional duty. Since the first step in the pathway is inhibited, wasteful intermediates are not accumulated.

Positive feedback systems also operate in a cell. For example, energy is provided for the cell when ATP is broken down and ADP is formed. ADP stimulates the enzymes involved in energy release. When ATP is reconstituted from ADP, it depresses enzyme action as it increases in amount. By maintaining this positive and negative feedback relation, the energy metabolism of a cell can be effectively regulated.

Induction, repression, and feedback controls are, therefore, important factors in the regulation of cell metabolism. They allow a cell to be efficient, yet economical. Enzymes are formed when needed, and their amount and rate of activity are controlled

Figure 13.5 Feedback inhibition of enzyme activity.

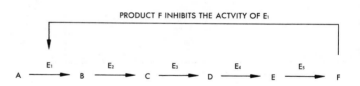

according to the needs of the cell. The advantages to a cell are obvious. We can only assume that these control mechanisms made their appearance during the course of evolution and were retained because of their selective value to the organism.

The special function of an enzyme is due not only to the sequence of amino acids in the molecule, but to the shape or configuration of the molecule. Lactic acid dehydrogenase (LDH) is not a single polypeptide chain, but is made up of four associated chains, or monomers. It is, therefore, a tetramer, and the monomers are of two kinds, called A and B. The number of A and B monomers in a tetramer can vary, and as a result there are five different LDH's. LDH^1 consists only of B monomers, and is given the symbol A^0B^4, which means there are no A and 4 B monomers in the active enzyme;

$$LDH^2 = A^1B^3 \qquad LDH^3 = A^2B^1$$

$$LDH^4 = A^3B^1 \qquad LDH^5 = A^4B^0$$

It is possible to isolate LDH^1 and LDH^5 in a pure state and to break them down into their monomers. A solution of A monomers and B monomers can then be mixed in a test tube. If the conditions are right, they will reaggregate and form tetramers. All five tetramers are formed; the proportions of each depend on the initial concentration of A and B monomers.

The LDH tetramers are called **isozymes.** Interest in them arises from the fact that different isozymes occur in different kinds of cells and act in slightly different ways. LDH^1, for example, is found predominately in heart muscle, while LDH^5 is more prominent in skeletal muscle. LDH^5 functions best when the oxygen concentration is low and lactic acid tends to accumulate, as happens in skeletal muscles. LDH^1, on the other hand, functions best under high oxygen conditions, as is characteristic for heart muscle. The various cells of the body have different internal environments, and respond to varying conditions by modifying the number and kinds of enzyme present. This is a more efficient practice than one requiring a different enzyme for each change in the cellular environment. We do not yet know how this enzyme modification takes place.

The monomers of LDH do not act as enzymes. However, glutamic acid dehydrogenase (GDH), a tetramer similar to LDH, behaves differently. As a tetramer, it acts on glutamic acid; but when it is dissociated into its monomers, the monomers act on the amino acid alanine. The monomers, therefore, are molecules of alanine dehydrogenase. We can see, as a result, that the enzyme's function varies, depending upon the state of aggregation. This is under the control of steroid hormones. The presence or absence of a hormone provides a means for a rapid shift in function without the necessity of forming new enzymes. The gain in efficiency by this kind of a control mechanism gives a cell the capability of meeting new situations rapidly.

THE MECHANISM OF HORMONE ACTION Not all hormones act in the manner indicated above. The hormone **adrenalin,** for example, increases the activity of an organism. When you are excited or frightened you can mobilize and metabolize glycogen much faster than normal, thus supplying energy. During such times of excitement one uses up more energy than when one is at rest. The mechanism of the action of adrenalin is now fairly well understood. As discussed previously, the hormone stimulates the function of a membrane-bound enzyme, **adenylate cyclase,** which in turn synthesizes intracellularly the important cyclic adenylate from ATP. The cyclic AMP in turn acts indirectly to increase the activity of the enzyme, phosphorylase, which catalyzes the breakdown of glycogen to glucose-1-phosphate. These and similar studies have led many workers to conclude that cyclic adenylate is the "universal" intracellular hormone, that is, the second messenger in hormone action. The scheme presented in Fig. 13.6 represents this concept, as developed by Earl Sutherland, who recently won the Nobel Prize for his important discoveries relating to the function and synthesis of cyclic adenylate. This hypothesis suggests that different cells contain on the cell surface or in the membrane specific receptors for different hormones. When the hormone and receptor interact, the adenyl cyclase is stimulated, leading to the production of cyclic AMP inside the cell where it acts. Because of the variation of enzymatic composition of

Figure 13.6 Schematic representation of the second messenger concept in mechanism of hormone action.

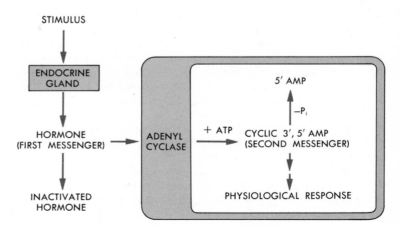

TABLE 13.1. Examples of Hormonal Activity that Appear to Be Mediated by an Increase in the Level of Cyclic AMP

Tissue	Hormone	Response
Bone	Parathyroid hormone	Calcium reabsorption
Muscle	Adrenalin	Glycogen breakdown
Fat	Adrenocorticotrophic hormone	Lipid breakdown
Thyroid	Thyroid-stimulating hormone	Thyroid secretion
Heart	Adrenalin	Increased contractility
Kidney	Vasopressin	Water reabsorption
Ovary	Luteinizing hormone	Progesterone secretion

Modified from I. Pastan, Scientific American, August 1972.

different cells, the response will vary from one cell or tissue to another. For example, in the case of inactive phosphorylase there is a specific enzyme, **protein kinase,** that catalyzes the formation of an active phosphorylase. The inactive protein kinase is composed of two subunits. When cyclic AMP is present, it combines with one of the subunits (regulatory unit) in a way that leads to the release of a catalytic subunit, which is only active in the free form. Thus, cyclic AMP activates an enzyme (a kinase), which in turn catalyzes the production of an active phosphorylase. This important concept has been tested by a wide variety of experiments on different tissues. A summary of some of the results is shown in Table 13.1. Recently, cyclic AMP along with a specific receptor protein has been shown to be important in the transcription process for certain genes. RNA polymerase in some way is activated to form messenger molecules, whereas in the absence of the cyclic AMP-receptor protein, it cannot do so even in the absence of the specific repressor. Not all genes are regulated by this mechanism.

ENZYME COMPLEXES Fatty acids are long-chain molecules formed by linking smaller molecules, such as those of acetic acid, in sequence. The fatty acid synthetase system is responsible for this biosynthesis, and at least seven different chemical reactions are known to be involved. According to our usual definition, each chemical step is catalyzed by a different enzyme, but attempts to separate the synthetase complex into distinct enzymes have been unsuccessful. It would appear that the enzymes are inactive alone, perhaps because they do not have the proper shape; but when they are aggregated into a **multienzyme complex,** they can carry out all the necessary reactions in sequence. Not all single enzymes in a multienzyme complex are inactive when isolated by themselves, but it is clear that their close association in a complex makes for a more efficient system.

There are obvious advantages to such a system. If we are dealing with the chemical series

$$A \xrightarrow{e_1} B \xrightarrow{e_2} C \xrightarrow{e_3} D \xrightarrow{e_4} E,$$

and the product of one reaction, say B, is the substrate for the next enzyme, e_2, then the enzyme complex becomes a sort of cellular assembly line. This would be immensely more efficient than a system in which all the enzymes and substrates are in solution, and must haphazardly find each other. Also, control is more effectively exercised, since blockage of one step blocks the entire system.

The mitochondrion and the chloroplast in the cell are examples of multienzyme complexes. Here one can visualize how such enzymes may be structured. A large number of enzymes, for example, are known to be associated with mitochondria, and they fall into two major groups: (1) the respiratory, or citric acid cycle, enzymes; and (2) those enzymes engaged in oxidative phosphorylation. ATP is one of the principal end products formed by the mitochondrion, and both enzyme groups join together in converting the energy in carbohydrates, fats, and proteins into the useful energy of ATP. Some of the mitochondrial enzymes appear to be in solution in the matrix (interior); others are definitely bound to the walls of the mitochondrion (Fig. 13.7).

It seems that even the soluble enzymes are bound in an organized array, but free themselves from the membranes during

Figure 13.7 Schematic representation of a "respiratory assembly" which is embedded in the protein layer of the mitochondrial membrane at regular intervals (after Lehninger). The respiratory assembly contains one molecule of each of the separate catalysts required for complete oxidation and energy conversion, presumably located adjacent to each other and forming a "molecular machine" as indicated at right. When the electron moves from the high-energy toward the low-energy level, a high-energy intermediate is formed, making use of a coupling enzyme (M), which takes up inorganic phosphate, P_i, to form M∼P. This intermediate reacts with a phosphate transfer enzyme (E) and ADP, forming ATP. This is a suggested scheme to account for a number of observations, but it should be emphasized that much more information is needed before it can be accepted.

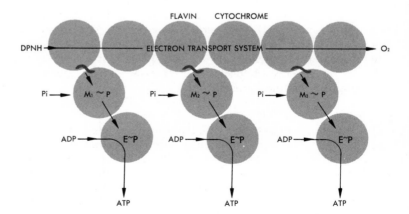

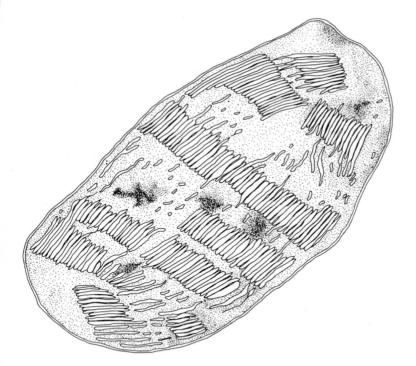

Figure 13.8 This chloroplast from the leaf cell of tobacco shows the stacked lamellae of the grana and the more open regions of the stroma.

the course of extraction. The structure of a chloroplast (Fig. 13.8) also leads us to believe that the enzymes for the conversion of CO_2 and H_2O into carbohydrates, and for photophosphorylation, are grouped into an assembly-line array for greater economy of action. The difficulty is that, while we can see, by means of electron microscopy, the elaborate structure of the mitochondrion or the chloroplast, we still cannot distinguish structural protein from enzyme protein in the membranes.

biological membranes and control systems

Membranes are an important structural feature of cells. They enclose the cell and the nucleus; they separate mitochondria, Golgi materials, lysosomes, and plastids from the rest of the cytoplasm; and they form a rich network as the endoplasmic reticulum in the cytoplasm. Membranes, therefore, are intracellular partitions. When we realize that most enzymes are also membrane-bound within organelles, we have some concept of how reactions in part of a cell can go on independently of reactions in another part. Thus, the breakdown of food materials occurs in lysosomes and mitochondria with the conversion and conservation of energy in the form of ATP, while the buildup of other chemicals, utilizing ATP, can occur in the endoplasmic reticulum and nucleus. Since membranes are made up of protein and lipid materials (Fig. 13.9), water in the cell is also controlled in its movement. It is also possible to immobilize substances in the cell by tying these materials into membranes. For example, cholesterol, a fatty substance associated with hardening of the arteries and with aging, is often found in the smooth endoplasmic reticulum of liver cells. The cell is, therefore, a compartmentalized system, both structurally and functionally, and membranes are the compartmentalizing agents.

One of the most interesting phenomena in cellular studies is how membranes act as "pumps"—how they govern the flow of substances in a given direction. The process is called **active transport** and requires energy. The result is the movement of soluble substances *against a concentration gradient,* quite the opposite of what occurs during diffusion or osmosis.

Let us consider active transport in terms of the cell membrane. This membrane has an ATPase associated with it. When this enzyme breaks down ATP, the energy released forces sodium ions to move in one direction, potassium ions in the other (Fig. 13.10). When these ions are absent, the ATPase is inactive, so the ion movement appears to be definitely related to the splitting of ATP and the release of energy. The enzyme acts as a traffic policeman as well, giving direction to the particular ions. We can visualize the ATPase as an active site, or crossroads, in the membrane, with the enzyme being able to "look" in two directions at once and direct the flow of ionic traffic. The ATPase in this

Figure 13.9 Proposed bimolecular lipid–protein structure for membranes. The basic structure of cell membranes appears to be composed of fatty material sandwiched between inner and outer layers of protein. This structure is not completely uniform, however, for we know that there are special proteins and other molecules attached to the surface of membranes.

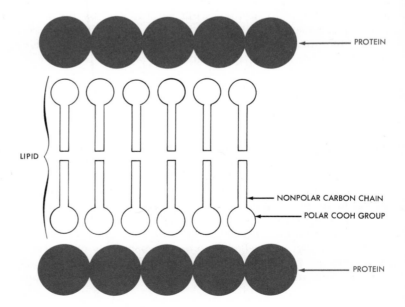

PROTEIN

LIPID

NONPOLAR CARBON CHAIN

POLAR COOH GROUP

PROTEIN

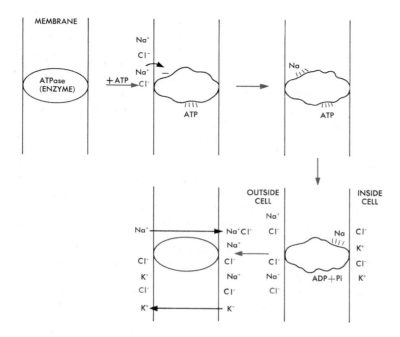

Figure 13.10 Proposed mechanism for the sodium and potassium pump in the cell membrane. A special membrane bound ATPase makes use of ATP to transport sodium ion across the membrane. When ATP is bound to the enzyme a negative charge is exposed on the protein, which attracts a sodium ion from outside the cell. When ATP is hydrolyzed, the energy release allows the movement of sodium to the inner surface of the membrane. The unbalanced negative charge on the outside attracts a potassium ion from the inside. When an unbalanced charge is maintained by the pumps, there develops an electrical potential difference between the inside and the outside of the cell.

manner acts as a **pump.** Other pumps at other active sites direct the flow of other molecules. Again, this means that the cell, through its enzymes, governs and compartmentalizes the activities going on within it.

Other metabolic pumps are also known. One in the mitochondria governs the flow of calcium and magnesium ions. One in muscle cells determines the direction of flow and the amount of calcium, an important aspect of muscular contraction and relaxation. In all such systems, the protein–lipid structure of the cellular membranes plays a key role. It is the boundary between one environment and another, and permits the development of little "islands" of activity both within and between cells. It is unlikely that effective control systems could develop in the absence of such membranes.

summary

Each cell is a highly controlled and regulated metabolic machine. The regulation in most cases involves enzyme synthesis and function. The nutrients surrounding the cell have an effect on the enzyme composition of the cell. In some cases synthesis of enzymes can be induced by an appropriate substrate, provided the cell has the right kind of DNA. **Induced enzyme formation,** therefore, is an important factor that determines the enzymatic activity of cells.

Enzyme synthesis can also be inhibited or repressed by certain nutrients. For example, if a cell is capable of making an essential nutrient (such as an amino acid), it will stop doing so if the nutrient is added to the medium in which the cells are growing. Analysis of this situation indicates that the nutrient inhibits the synthesis of the enzyme, which catalyzes the first step in the biosynthetic pathway. This enzyme repression conserves cell energy and specific chemicals.

The product of a biosynthetic pathway can also inhibit an enzyme that catalyzes an earlier step. This is called **feedback inhibition.**

The structure of enzymes is an important factor in their control and regulation. For example, some enzymes are made up of several different polypeptides or a multiple combination of two or more polypeptides. Dissociation of the polymers either by specific hormones or other agents will alter their catalytic activity.

Enzyme activity is also affected by being bound to membrane-like structures in the cell. The **multienzyme complexes** are important in the mitochondria, where the aerobic metabolism of fatty acids and carbohydrates occurs. Enzymes bound to the cell membrane are also important in that they are capable of making use of metabolic energy (ATP) to move substances across the membrane against a concentration gradient. These "metabolic pumps" are essential for regulating the concentration of nutrients inside the cell.

for thought and discussion

1 Can you think of various mechanisms that would lead to a different enzymatic makeup for a liver cell and a heart cell during the time a fertilized egg is developing into an adult organism?

2 In studying the scheme for aerobic metabolism of carbohydrates, can you think of a way that enzyme repression or feedback inhibition could function to regulate oxygen consumption?

3 Why do you think there are different types of isozymes in different cells? What are some of the factors leading to the synthesis of one type over another?

4 What is the value of having enzymes associated with one overall chemical process bound onto a structure in a multienzyme complex?

selected readings

INGRAM, V. M. *The Biosynthesis of Macromolecules.* Menlo Park, Calif.: W. A. Benjamin, Inc., 1965.

PTASHNE, M., and W. GILBERT, "Genetic Repressors," *Scientific American,* June 1970.

TEMIN, H. M., "RNA-Directed DNA Synthesis," *Scientific American,* January 1972.

WOESE, C. R. *The Genetic Code: The Molecular Basis for Genetic Expression.* New York: Harper & Row, Publishers, 1967.

inheritance
of
a trait
chapter 14

In an earlier chapter we discussed DNA as the principal molecule of heredity. Of the 3 billion human beings on Earth, no two are exactly alike (except, possibly, identical twins); it follows that their DNA's are also dissimilar, which, in turn, means that their proteins must be dissimilar. It is impossible to pinpoint this dissimilarity by determining the number and distribution of the 8 billion nucleotide pairs in the DNA of a human cell. But, if we want to know something of the details involved in the inheritance of similarities and dissimilarities from one generation to another, we must have some way of tracing the passage of DNA from one generation to another. The simplest way is to concentrate on a character or trait that is determined by DNA.

We should make it clear that patterns of inheritance can be studied *only* when variation exists. We would not learn very much about inheritance if we studied a population of organisms all alike. Without "fingerprints" of some sort, there would be no way of distinguishing one individual organism from another. In a way, genetics has developed by focusing attention on differences between individuals rather than on their similarities. Fortunately, variations of all kinds exist. The problem is to select that variation best suited to the situation being investigated. Whenever possible, the variation should consist of a pair of easily distinguishable, sharply contrasting, and mutually exclusive characters. For ex-

Ewing Galloway

Figure 14.1 These identical twins have strikingly similar facial features. What do we mean when we say that identical twins are "identical?"

ample, red versus white flower color; the presence or absence of an enzyme; black as opposed to white spores; straight versus curly hair; and so on. When studying a pair of contrasting characters, we disregard all other characters so that we do not complicate our results.

The basic tool in studying the inheritance of a trait is the **inheritance test.** It was used long before the role of DNA was understood, and it is still used today. The inheritance test is relatively simple. It consists of controlled matings and the study and tabulation of the offspring. For these reasons, man is not the most suitable organism to use. His mating cannot be controlled because of social restrictions, the offspring are too few in number, and their generation time is too long. Consequently, the choice of organism is important, and we find that the greater part of our knowledge of inheritance is not from the study of man, but rather from peas, fruit flies, corn, bacteria, and even viruses. The knowledge gained, however, inevitably sheds light on inheritance in man himself.

THE LIFE CYCLE *Neurospora,* which is easy to grow,
OF NEUROSPORA handle, and observe, is a fungus
(mold) and an **ascomycete.** It gets
this name from one of its main fruiting structures, the **ascus.**
Figure 14.2 outlines the life cycle of this plant.

The normal growth of *Neurospora* is by means of a spreading
mat of fine threads, the **mycelium.** The individual threads, or
hyphae, are not broken up into cells with single nuclei; rather
there are many nuclei in a common mass of cytoplasm. When
the mycelium is broken up, each hyphal piece forms a new
mycelium (Fig. 14.2). This represents a form of asexual reproduction. The mycelium can also form small asexual spores called
conidia (singular, **conidium**), each having one or more nuclei in
them. These germinate and form mycelial masses similar to the
parental mycelium.

Neurospora also has a form of sexual reproduction, even
though it is not possible visually to distinguish male from female
mycelia. The different kinds of mycelia are classified into mating
types *A* and *a. A* will not mate with *A,* nor *a* with *a. A* will mate
only with *a.* These are distinct, mutually exclusive traits.

When mating types *A* and *a* are brought together, the
mycelial strands fuse, or a conidium from one type unites with
a **protoperithecium** from the other, and two nuclei—one of each
type—move into a single cell that is cut off by a wall. The two
nuclei fuse and then undergo a series of three divisions, producing
eight nuclei. A wall forms around the mass of cytoplasm surrounding each nucleus, and eight ascospores are formed. When
the ascospores germinate, each one produces a mycelium. The
mycelium from any single ascospore will be of mating type *A*

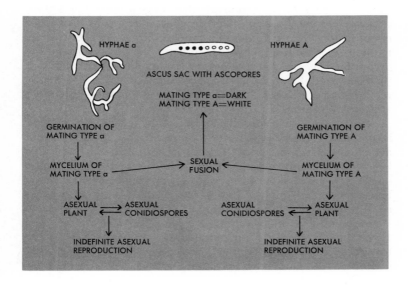

Figure 14.2 The life cycle of
the ascomycete, *Neurospora,*
has both asexual and sexual
aspects.

or *a,* never a mixture of the two. Of the eight ascospores, four will produce *A* mycelia, and the other four, *a* mycelia. That is, the ratio of *A* to *a* is 1 : 1.

THE INHERITANCE TEST Let us now consider in detail the meaning of the 1 : 1 ratio. A portion of the DNA—how much we do not know—is responsible for determining traits *A* and *a*. This portion of DNA is a gene, and we can speak of genes *A* and *a*, which have the function of determining mating types. How they do this is unknown, but it is probably through some kind of a protein structure or reaction. Since genes *A* and *a* are also in chromosomes, the behavior of chromosomes should tell us something about the behavior of genes; since genes determine traits, the inheritance of a trait should follow the inheritance of a chromosome.

Genes *A* and *a* are never found in the same mycelium. They are, therefore, mutually exclusive, and we can think of them as alternate states of a single character. The term **allele** is used to designate such contrasting genes. If we assume that *A* represented the original state, then *a* is a mutation of *A*. (It could, however, be the other way around, with *A* a mutation of *a*.) If *A* were the original allele, then *a* probably arose by some change in the portion of DNA controlling mating type. Our present guess is that this change reflects an alteration of nucleotide pairs.

With the information above, let us now consider chromosome behavior in our inheritance test. The nuclei in the mycelium of *Neurospora* contain seven chromosomes. These can be numbered 1 through 7. From other evidence we know that gene *A* (or *a*) is located on chromosome 1. This group of seven chromosomes, each of which is different from all others in gene content, is known as a **haploid set.** Since we tend to characterize cells, tissues, or organisms by their chromosome content, *Neurospora* is essentially haploid; only the fusion nucleus is diploid. We shall see that other organisms may differ in the degree of haploidy and diploidy of their cells, and their features of inheritance also differ. All sexually reproducing organisms, however, show the alternation of haploid and diploid phases.

Figure 14.3 shows that the mating of the two types of mycelia is soon followed by fusion of the two nuclei. This is the principal act of sexual reproduction, the act of fertilization. The single fusion nucleus now contains two haploid sets of chromosomes, and is said to be **diploid.** Instead of 14 different chromosomes, however, we can think of the nucleus as containing seven pairs of chromosomes. Thus, mating type *A* contributed chromosome 1 containing gene *A;* mating type *a* contributed its chromosome 1 containing gene *a.* For our purposes, we can consider that these two chromosomes are identical, except for the particular allele carried at the mating-type position or **locus.** The two chromosomes are said to be **homologous** with each other.

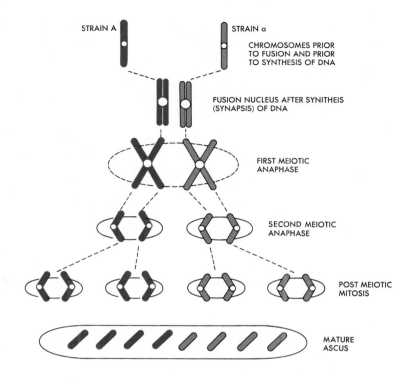

STRAIN A   STRAIN a

CHROMOSOMES PRIOR
TO FUSION AND PRIOR
TO SYNTHESIS OF DNA

FUSION NUCLEUS AFTER SYNITHEIS
(SYNAPSIS) OF DNA

FIRST MEIOTIC
ANAPHASE

SECOND MEIOTIC
ANAPHASE

POST MEIOTIC
MITOSIS

MATURE
ASCUS

Figure 14.3 The sequence of events taking place in the sexual cycle of *Neurospora*. Chromosome I from each strain, or mating type, joins with its homologue soon after formation of the fusion nucleus, and then separates from it at the first meiotic anaphase. The two chromatids of each chromosome separate at the second meiotic anaphase, and each again divides at the postmeiotic mitosis. Each of the eight chromosomes is now enclosed in a separate ascospore in the mature ascus. The behavior of the chromosomes, like the behavior of genes, illustrates the law of segregation discovered by Mendel.

The fusion diploid nucleus now goes through a series of divisions called **meiosis.** The behavior of the chromosomes differs significantly from that occurring in prophase of the earlier described mitotic division. In mitosis each chromosome entered prophase with the chromosomes longitudinally double. At anaphase the chromatids separated from each other and formed two nuclei having the same chromosome number as the mother cell from which they arose. In prophase of meiosis the homologous chromosomes pair with each other—number 1 from type *A* pairs with number 1 from type *a*, and so on. They arrive at metaphase as pairs of chromosomes, called **bivalents,** instead of single chromosomes longitudinally double. The homologues separate from each other at anaphase, and the newly formed nuclei now have seven instead of 14 chromosomes. The diploid fusion nucleus has now been reduced to two haploid nuclei. In terms of chromosomes, therefore, the first division is a **reduction division.**

The second division merely separates the two chromatids of each chromosome, producing four haploid nuclei. The third division is a normal mitotic division, so the ascus now has eight haploid nuclei within it. These now are enclosed with a mass of cytoplasm inside of a spore wall, and eight ascospores result.

A summary diagram of these events is shown in Fig. 14.3 for one pair of chromosomes. It is clear that sexual reproduction involves the union of haploid nuclei to give a diploid state, and

231

that meiosis is the reverse of this process. The life cycle of a sexually reproducing organism is, therefore, an alternation of haploid and diploid states. In *Neurospora*, however, the diploid state is a brief one and is confined to the fusion nucleus. *Neurospora*, consequently, is primarily a haploid organism.

The events we have just described follow what is known as **Mendel's first law of inheritance.** Gregor Mendel, an Austrian monk, formulated this law in 1865 when he was studying inheritance in garden peas. It is also called the **law of segregation.** In terms of the *Neurospora* life cycle, the law states that within sexual organisms there are pairs of factors (we now call them genes) which unite at the time of fertilization (when the two nuclei fuse), and which segregate at the time of meiosis. Stated another way, genes of sexually reproducing organisms come together in a single cell and then segregate during any given life cycle. The genes do this without losing their identity.

random assortment of genes

The remaining six haploid chromosomes of *Neurospora*, numbered 2 to 7, also contain genes. On chromosome 4 is a mutant gene *c*, which causes a colonial type of growth. The normal gene *C*, of which *c* is an allele, governs the spreading type of growth. Suppose that we cross a colonial strain of mating type *A* with a spreading strain of mating type *a*. These strains can be designated, respectively, as *Ac* and *aC* (Fig. 14.4). From such a cross, ascospores will be produced. These can be isolated and grown, and the mycelial colonies produced can be examined for traits. We would find that 25 per cent of them would be *Ac,* 25 per cent *aC,* 25 percent *AC,* and 25 per cent *ac.* However, what if we took individual asci, isolated the ascospores one by one, grew them into colonies, and then examined their genes? We would find that half of the asci produced four *Ac* spores, and four *aC* spores, while the other half produced four *AC* spores and four *ac* spores.

Let us consider this in terms of chromosomes 1 and 4, which contain these genes (Fig. 14.5). The fusion nucleus would contain two of chromosome 1 and two of chromosome 4, each with their respective genes. At meiosis the homologous chromosomes would pair and then segregate.

At the first meiotic division, chromosome 1(*A*) goes to one pole, chromosome 1(*a*) to the other. There is an equal chance of chromosome 4(*C*) going with 1(*A*) or with 1(*a*). If the former, then four spores in the ascus will be *AC*, the other four *ac*. If 4(*C*) goes to the same pole with 1(*a*), then four spores will be *aC*, the other four *Ac*.

The combinations *aC* and *Ac* are parental types; that is, they

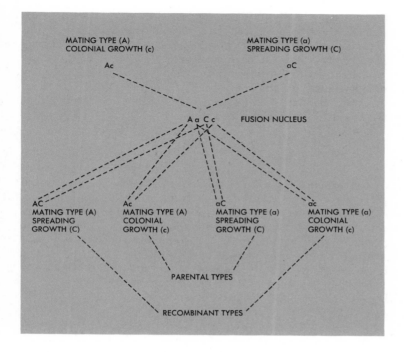

Figure 14.4 Random assortment at meiosis of two pairs of genes found on two different and nonhomologous pairs of chromosomes. Their behavior illustrates the Mendelian law of independent assortment.

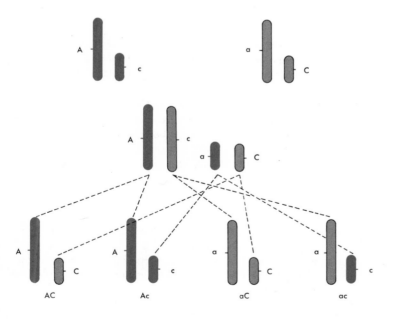

Figure 14.5 The independent assortment of two pairs of nonhomologous chromosomes produces the same results as those obtained by following the genes in Fig. 14.4.

are the same as the original strains with which the cross was started. The *AC* and *ac* spores are **recombinant** types. The results obtained indicate that the chromosomes assort at random during meiosis, and so do the individual genes.

Mendel also discovered this principle, now known as the **law of independent assortment** of genes. We shall see in a later chapter, however, that when two genes are on the same chromosome, the law does not always run true.

summary

The **inheritance test** is the basic tool of genetics. Its use requires pairs of contrasting characters or traits, which are controlled by contrasting genes or alleles. The inheritance test also requires controlled matings and a quantitative record of the offspring produced from such matings. In a haploid organism, using trait *A* in one parent as contrasted to trait *a* in the other, the inheritance test shows that the ratio of *A* to *a* among the offspring is 1:1. When two traits are studied simultaneously, they are inherited independently of each other to give ratios of 1*AB*:1*Ab* and 1*aB*:1*ab*.

These two ratios provide us with information needed to formulate the two basic laws of genetics; (1) the **law of segregation of genes;** and (2) the **law of independent assortment of genes.** The first law is used when only one pair of contrasting traits is being studied, the second law when two or more pairs of traits are being studied. The physical basis of genetic inheritance lies in the behavior of chromosomes at meiosis.

for thought and discussion

1 List a group of contrasting characters in man. Select characters that you think might be relatively easy to follow in inheritance studies. Discuss them and try to decide which ones would serve your purpose and which would not.

2 Some genetically determined characters are said to show "qualitative" differences, others "quantitative" differences. What do these terms mean? Which would be the easier to study by the inheritance test?

3 Why is the use of the inheritance test in man impractical? Can you think of any way to get around this problem so that genetic studies in man can be made?

4 Can you study the inheritance of a trait in an asexually reproducing organism? Give reasons for your answer.

5 It is thought that the average size of a gene is about 1,000 nucleotide pairs long. If this is so, how many genes can man have in a haploid set of chromosomes? Have you any reason for thinking this number to be reasonable or not?

6 If a gene of average size controls mating types A and a in

Neurospora, and the difference between the two alleles is due to differences in nucleotide pairs, would you expect more than two kinds of mating types? If only two are present in nature, can you think of any reason why this should be so?

7 Genes X and x are found on the members of one pair of homologous chromosomes, and genes Y and y on another homologous pair. A cross is made between the XY strain and an xy strain, and the ascospores are removed in serial order from a number of asci. How many different combinations can you expect to get from individual asci? Diagram your results to show the distribution of chromosomes in each case.

8 Yeast is like *Neurospora* in that it is haploid and is an ascomycete, but it is unicellular and does not form a mycelium. After the fusion of two haploid yeast cells, meiosis occurs, but only four spores instead of eight are formed (the final mitotic division is missing). One strain of cells cannot ferment glucose (glu⁻) and the other strain cannot ferment the sugar maltose (mal⁻) because of defective genes. These genes segregate independently of each other. The two strains are crossed, and the spores are isolated from individual asci. What kinds of spores will you get from each ascus? Will all asci be alike in this respect? How are you to test for the genotype of each spore? If you group the results from many asci, what kinds of ratios would you obtain?

selected readings

AUERBACH, CHARLOTTE. *Genetics in the Atomic Age.* New York: Essential Books, Inc., 1956. An elementary treatment of genetics and the effects of radiation on genetic systems.

AUERBACH, CHARLOTTE. *The Science of Genetics.* New York: Harper & Row, Publishers, 1961. Covers inheritance in plants and animals.

BONNER, D. M. and S. E. MILLS. *Heredity.* Englewood Cliffs, N.J.: Prentice-Hall, Inc., 1964. The emphasis in this small book is on the microbial and biochemical aspects of inheritance.

COOK, S. A. *Reproduction, Heredity and Sexuality.* Belmont, Calif.: Wadsworth Publishing Company, Inc., 1964. Deals broadly with the many variations of reproduction and inheritance in the plant world.

HERSKOWITZ, I. H. *Genetics.* Boston: Little, Brown and Company, 1965. A college textbook that includes as a supplement some of the classical papers of genetics.

LEVINE, R. P. and U. GOODENOUGH. *Genetics,* 2nd ed. New York: Holt, Rinehart and Winston, Inc., 1974. Covers most aspects of genetics including haploid and diploid inheritance and its chromosomal relationships.

PETERS, J. A. (ed.). *Classic Papers in Genetics.* Englewood Cliffs, N.J.: Prentice-Hall, Inc., 1959. Contains 28 classic papers that laid the foundations of genetics from its early beginnings to today.

SRB, A. M., R. D. OWEN, and R. EDGAR. *General Genetics.* San Francisco: W. H. Freeman and Company, 1965. One of the most broadly comprehensive textbooks on genetics at the college level, including both classical and modern aspects of the subject.

meiosis and its relation to sexual reproduction
chapter 15

In the life history of a sexual organism such as *Neurospora*, two events are of particular significance. One is fertilization. In *Neurospora*, fertilization occurs when the haploid nuclei of the two mating types fuse and form a diploid nucleus. In higher plants and animals, fertilization occurs when the sperm penetrates the egg and their two nuclei fuse, giving rise to a **zygote,** or new individual, which is diploid.

If you stop for a moment and think of the consequences of fertilization, it should soon become apparent that there always must be some other process as well, an accompanying counterpart. Remember that when two nuclei fuse the chromosome number doubles. In each succeeding generation, therefore, the chromosome number would continue to double. If no event occurred to offset this trend, it would not take long to arrive at unmanageable numbers of chromosomes. Furthermore, we know that the haploid number of chromosomes in *Neurospora* remains at seven, generation after generation, just as the diploid number of man remains 46.

The answer, of course, is that meiosis is the opposite of fertilization; it is a form of nuclear division that halves the number of chromosomes at some stage in the life cycle. In *Neurospora*, meiosis follows fertilization immediately. In man, however, the zygote is formed when fertilization occurs. Afterward the

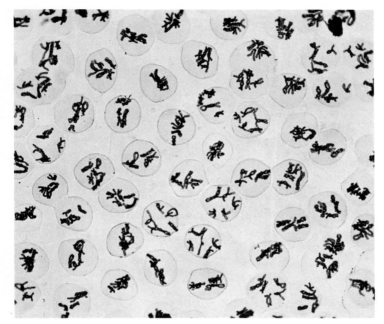

Brookhaven National Laboratory

Figure 15.1 These cells at the first meiotic metaphase in the anther of the wake robin, *Trillium erectum*, were taken at low magnification. The cells are synchronized in division so that most of them arrive simultaneously at metaphase. A single anther contains hundreds of such cells.

zygote develops first into an embryo, and then gradually over the years into an adult. Meiosis occurs when the individual passes through puberty to gain sexual maturity, and eggs and sperm are produced. Only these two kinds of human cells are haploid. Man, like all higher animals and plants, is essentially a diploid organism during most of his life cycle. Here we shall consider meiosis as a variation of mitosis in which a number of interesting innovations appear.

THE STAGES OF MEIOSIS Neither *Neurospora* nor man is a favorable organism in which to study meiosis. Maize (corn), grasshoppers, amphibians, and lilies are much more suitable because of their large chromosomes and the ease with which the cells can be fixed and stained. Prophase is the first stage of meiosis, as it is in mitosis, but in meiosis it lasts longer and is more complicated.

In the early stage of meiotic prophase the chromosomes of the nucleus are seen as delicate threads within the nuclear membrane. Because of their great length, the chromosomes are not individually identifiable. The nucleoli are quite large and easily visible. This stage is not strikingly different from the early prophase of mitosis, except that along the extended length of the

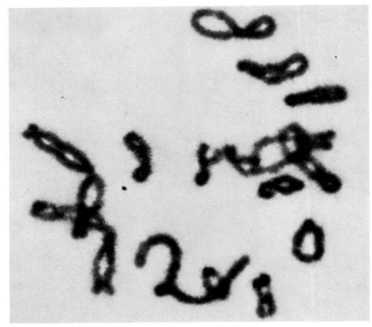

Figure 15.2 In this meiotic
prophase cell from the testes of
an amphibian the chromosomes
have formed homologous pairs.
Can you identify all 15 pairs?

Brookhaven National Laboratory

Brookhaven National Laboratory

Figure 15.3 An early meiotic
prophase stage in a testicular
cell of an amphibian. The
nucleolus is not visible because
the stain used is absorbed only
by chromatin, but the
chromomeres are visible along
the lengths of the elongated
chromosomes. Some indications
of pairing are evident.

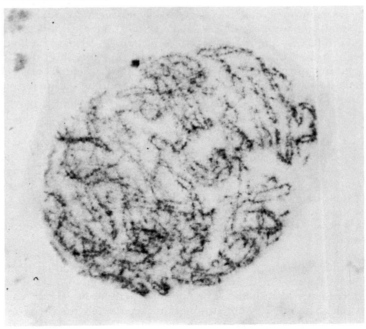

chromosomes are dense granules of irregular size. These are the **chromomeres.** Any given chromosome has chromomeres of characteristic size and position. It is thought that the chromomeres arise as a result of the chromosome contracting unevenly along its length. However they arise, the number, size, and position of chromomeres are constant features of each chromosome.

The chromomeres gradually become more visible, and the individual chromosomes become thicker as the result of contraction. The nucleoli can be seen to be attached to particular chromosomes at special points called the **nucleolar organizers.** This is the stage when the homologous chromosomes begin to pair. The beginning of this stage is shown in Fig. 15.4. The homologues contact each other at one or more points. Then, like a zipper, the two chromosomes pair, chromomere by chromomere. When pairing is complete, it appears as if only a haploid number of chromosomes were present in the nucleus. With the aid of a high-power microscope, however, each closely associated pair can be seen as two. Each pair of homologues is called a **bivalent.**

Contraction continues to shorten the chromosomes and makes them more readily identifiable. Figure 15.6 shows a few

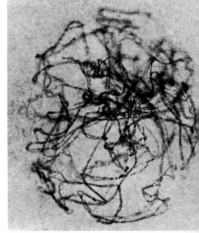

Courtesy Dr. J. MacLeish

Figure 15.4 A meiotic prophase cell from the anther of the regal lily, *Lilium regale.* The pairing of homologous chromosomes can be seen at various points; at other regions pairing is not yet complete.

Figure 15.5 Mid-prophase stage of meiosis in an amphibian: the chromosomes are nearly paired along their entire lengths, the chromomeres are large and coarse, and considerable shortening has taken place (compare with Fig. 15.3).

Brookhaven National Laboratory

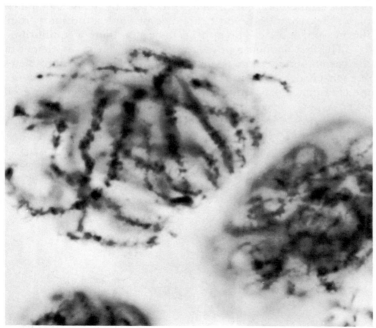

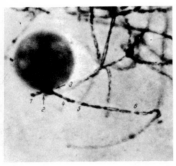

Figure 15.6 A mid-prophase stage in meiosis of maize, showing the paired homologues and the attachment of one pair to the nucleolus. Recognizable regions of the bivalent are numbered. This figure is comparable to Fig. 15.5 from an amphibian, but the chromosomes are clearer as to intimate details of pairing and structure. There are 10 pairs of chromosomes in maize, but only chromosome 6 is shown in its entirety.

of the chromosomes of maize (the haploid number is 10). Remember that what appears to be a single chromosome consists of two homologues closely paired. This is shown clearly in Fig. 15.6. Here chromosome number 6 of maize has the nucleolus attached to it, the individual chromomeres are visible, and the paired nature of the strands can be seen in several places.

An abrupt change in the appearance of the bivalents now takes place. The paired homologues fall apart, showing that each consists of two chromatids. The homologues do not separate completely, however. They are held together at various points called **chiasmata** (singular, **chiasma**) (Figs. 15.2 and 15.7). The chiasmata can be seen as points where chromatids cross over from one homologue to another. We shall refer to these chiasmata in a later chapter.

The chromosomes are more contracted in this stage, and in some organisms small coils can be seen. It is by means of a coiling process that a long thin chromosome is converted into a shorter, more maneuverable one.

At the end of prophase the chromosomes are shorter, the nucleoli disappear, and the nuclear membrane disintegrates. As these events occur, the spindle forms and the bivalents orient themselves on it halfway between the poles. The **first meiotic metaphase** stage has thus begun.

You will recall that at *mitotic* metaphase the centromere of each chromosome occupied a position on the metaphase plate. At *meiotic* metaphase, however, each bivalent has two centromeres, one for each homologue. Instead of both being at the metaphase plate, they come to lie on either side of, and equidistant from, the metaphase plate. This appears to be a position of equilibrium.

The **first anaphase** of meiosis begins with the movement of the chromosomes to the pole (Fig. 15.8). The two centromeres of each bivalent remain undivided, and their movement causes the chiasmata to be undone, freeing the homologues from each other. When movement ceases, a reduced or haploid number of chromosomes is located at each pole. Unlike mitotic anaphase,

Figure 15.7 A photograph of a single bivalent from a male salamander revealing the individual chromatids of the homologous chromosomes, and the two chiasmata where the chromatids cross over from one homologue to the other. The centromeres of the two homologues are seen as the dark circular areas at the left of the bivalent.

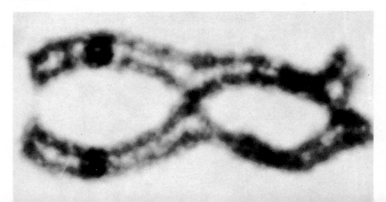

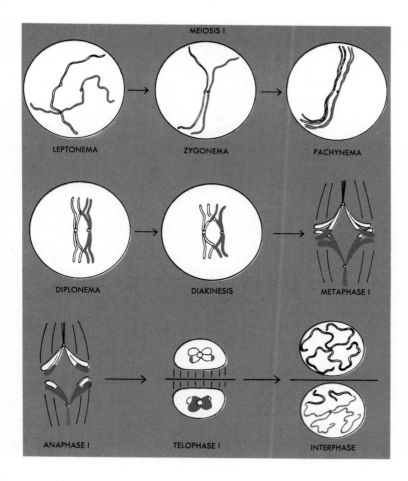

MEIOSIS I

LEPTONEMA

ZYGONEMA

PACHYNEMA

DIPLONEMA

DIAKINESIS

METAPHASE I

ANAPHASE I

TELOPHASE I

INTERPHASE

Figure 15.8 The sequence of stages that occurs in the first meiotic division of both plants and animals is shown here. Only a single pair of chromosomes is represented. Prophase includes the first five stages shown in the diagram, and distinctive terms have been given to each recognizable stage. The progression of prophase, however, is continuous, and one stage leads directly into the following one without delay. The line drawn between the telophase and interphase nuclei represents the cell membrane, which cuts the cell in two. The important thing to remember is that *chromosomes,* not chromatids, segregate in the first meiotic division.

in which the chromosomes appeared longitudinally single, each chromosome now consists of two distinctly separate chromatids held together only at their centromeres.

A nuclear membrane forms around the chromosomes, and the chromosomes uncoil. A wall or membrane divides the cell in two in some kinds of meiotic cells; in others it does not. This is the **first telophase** of meiosis.

The second meiotic division follows the first without appreciable delay (Fig. 15.9). It appears to be quite similar to a mitotic division, but there is one important distinction. The second meiotic prophase chromosomes have the same structure as those in the previous first meiotic anaphase. In other words, *no chromosomal replication took place during meiotic interphase,* and the centromere of each chromosome remains undivided.

A spindle forms in each cell, initiating the second meiotic metaphase, and at the second anaphase the centromeres divide

MEIOSIS II

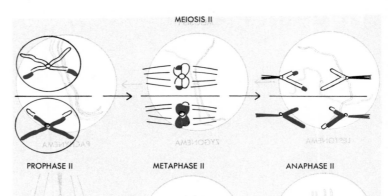

PROPHASE II   METAPHASE II   ANAPHASE II

Figure 15.9 The second meiotic division follows the first with little delay. Each chromosome now divides and its two chromatids segregate at anaphase II. As a result, the four chromatids of a first-division bivalent become segregated into four separate cells.

and the chromatids separate and pass to the poles. The four haploid nuclei are reorganized in the second telophase. Depending on the species and the kind of cells, the nuclei may or may not be segregated into individual cells by segmentation of the cytoplasm.

Neurospora is representative of those organisms, which include many algae and fungi in which the two meiotic divisions are followed by a mitotic division resulting in eight haploid nuclei and then eight ascospores. This mitotic division does not occur in animal species.

If you look back over the events of meiosis, you will find that the structure of each chromosome remained unchanged. From mid-prophase to the second anaphase, the chromosomes maintained their longitudinal subdivisions. The replication of each chromosome took place in the interphase preceding meiosis. Then two divisions occurred: during the first the homologues separated, thus reducing the chromosome number; during the second the two chromatids of each chromosome were separated. It might be argued that a reduction in chromosome number could have been accomplished by only one division. But when viewed in terms of gene recombination, the two divisions make good genetic sense. This point will be discussed in the following chapter.

reproduction in animals

Sexual reproduction in most animals differs from that in *Neurospora* in two major respects. First, the sexual cells, or gametes, in animals are produced in specialized structures: **eggs** are produced in **ovaries,** and **sperm** are produced in **testes.** The gametes are visibly different, both in size and in character. In *Neurospora* there are no special organs in which sexual cells are produced, and the nuclei differ only in mating-type genes. Second, the body of an animal is a diploid structure in contrast to the haploid mycelium of *Neurospora.* Let us consider what takes place in man (Fig. 15.10).

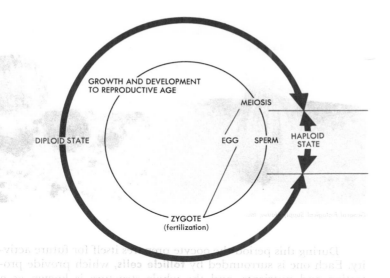

meiosis and its relation to sexual reproduction

Figure 15.10 This diagram demonstrates the relation of meiosis and fertilization to the haploid and diploid states of a sexually reproducing species such as man. The only haploid cells are the eggs and sperm cells after meiosis has been completed. All other cells of the body at all other stages of the life cycle are diploid.

THE EGG The ovaries of a female begin development early in fetal life, and all the eggs that an individual will produce are formed by about the sixth month of fetal life, that is, three months before the child is born. Figure 15.11 shows a section of an ovary with eggs in the course of development. On the outside is the **germinal epithelium,** which gives rise to **oogonial** cells. These cells then develop into **oocytes.** The oocytes are those cells in which meiosis takes place. The first stages of meiosis in all oocytes begin before birth, stopping, however, at a stage just prior to the first metaphase. Since a female does not shed eggs until 12 to 14 years of age, and may continue to do so up to 50 years of age, the oocytes remain retarded in division for many years.

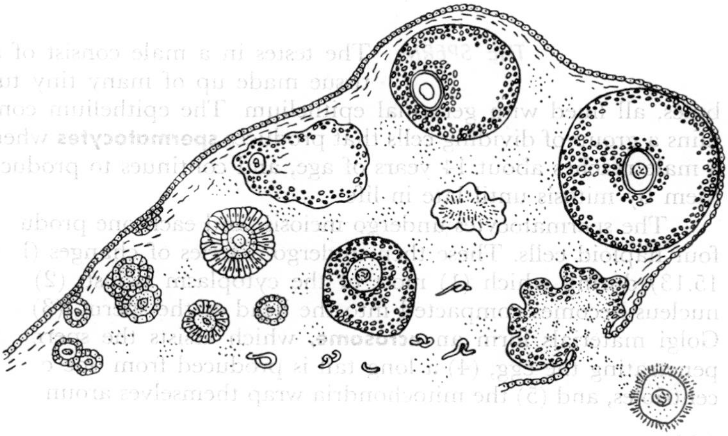

Figure 15.11 A section through a mammalian ovary would show the progressive enlargement of the eggs, and the eventual shedding of one egg or ovum (lower right). The human female, after the age of puberty, sheds one egg per month, usually.

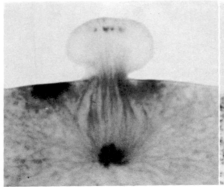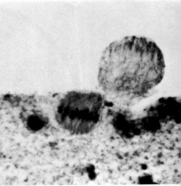

Figure 15.12 Polar body formation in the egg of a fish: (left) late anaphase of the first meiotic division, with the first polar body being pinched off; (right) metaphase of the second meiotic division, with the first polar body resting on the cell membrane above and to the right. This division will eventually pinch off a second polar body. In the meantime, the first polar body may divide again, producing a total of three polar bodies and one functional egg.

General Biological Supply House, Inc.

During this period the oocyte prepares itself for future activity. Each one is surrounded by **follicle cells,** which provide protection and nutrients, and the whole structure is known as a **Graafian follicle.** Yolk materials, which the young embryo will use for food, are stored inside. In a chicken egg the yolk can be seen as a yellow mass, but in human egg the yolk is much less prominent and is scattered throughout the cytoplasm of the egg.

When the egg is released from the ovary and then fertilized by a sperm, meiosis starts up again. The first meiotic division gives rise to two cells. One, very minute, is called a **polar body.** This is difficult to detect in humans, but the results can be seen in a fish egg (Fig. 15.12). The division, however, brings about a reduction in chromosome number.

The second meiotic division is similar, and another polar body is produced. In the meantime, the first polar body has divided. Thus, as a result of meiosis, four haploid cells are produced: three small polar bodies, which eventually disintegrate, and one large egg, which will be functional.

THE SPERM The testes in a male consist of a tissue made up of many tiny tubules, all lined with germinal epithelium. The epithelium contains a group of dividing cells that produces **spermatocytes** when a male reaches about 14 years of age, and continues to produce them by mitosis until late in life.

The spermatocytes undergo meiosis, and each one produces four haploid cells. These then undergo a series of changes (Fig. 15.13) during which (1) most of the cytoplasm is lost, (2) the nucleus becomes compacted into the head of the sperm, (3) the Golgi materials form an **acrosome,** which assists the sperm in penetrating the egg, (4) a long tail is produced from one of the centrioles, and (5) the mitochondria wrap themselves around the

244

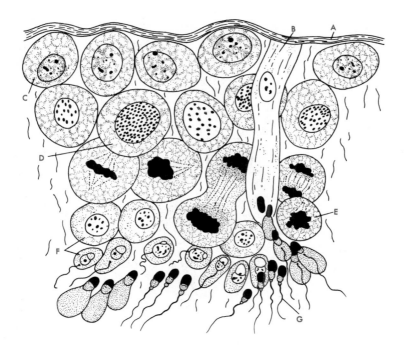

Figure 15.13 A section through a mammalian testes would show the development of mature sperm cells: (A) outer membrane of the testis tubule; (B) accessory cell; (C) early prophase in a spermatocyte; (D) later prophase stage (metaphase I and anaphase I stages are also present but not labeled); (E) second spermatocyte ready to begin the second meiotic division; (F) spermatids at two stages of development; (G) mature sperm cells.

Figure 15.14 Mature human sperm cells. The acrosome caps the nucleus, with the two parts making up the head of the sperm cell. The midpiece contains the mitochondria, and the tail extends beyond the midpiece, providing a means of locomotion.

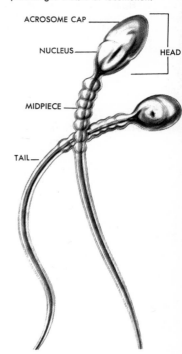

tail just below the attachment to the nucleus, presumably supplying the sperm with energy. By the whipping action of its tail, the sperm swims to the egg, which has no power of movement. The sperm, therefore, can be thought of as a haploid nucleus with an outboard motor.

FERTILIZATION The mature egg and sperm must unite in a limited period of time, since their life span is limited. Fertilization begins when the sperm enters the egg. This occurs in the **Fallopian tube,** a structure that connects the ovary to the uterus, where the embryo will develop. Once one sperm has entered, other sperm are blocked from entering by the quick development of an impenetrable membrane. The sperm loses its tail, and its nucleus fuses with the nucleus of the egg. The sperm also contributes a **centriole** to the egg. The centriole aids in the formation of a spindle, and the fertilized egg, or zygote, prepares to divide and become an embryo.

In summary, then, the sperm does several things:

1. It adds a haploid set of chromosomes to form a diploid zygote.
2. It contributes a centriole, which enables the zygote to divide.
3. It provides a stimulus for division of the egg.

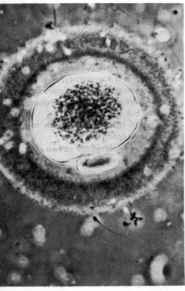

Figure 15.15 Photograph of
a human egg at the moment of
fertilization by a sperm cell.

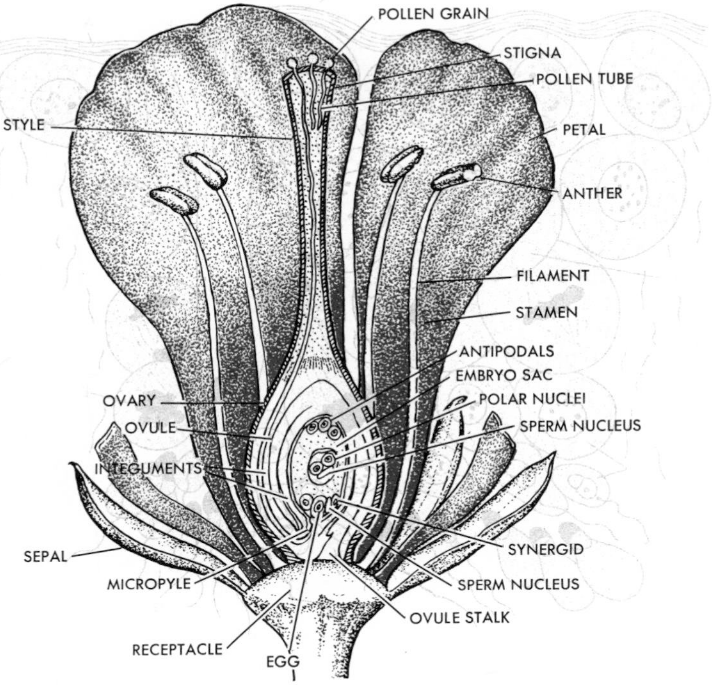

Figure 15.16 A lily flower at a point in its life cycle when the ovule has just
been fertilized. The stages leading up to this point are shown in Fig. 15.17.

Labels in figure 15.16:
STYLE, POLLEN GRAIN, STIGNA, POLLEN TUBE, PETAL, ANTHER, FILAMENT, STAMEN, ANTIPODALS, EMBRYO SAC, POLAR NUCLEI, SPERM NUCLEUS, OVARY, OVULE, INTEGUMENTS, SEPAL, MICROPYLE, RECEPTACLE, EGG, SYNERGID, SPERM NUCLEUS, OVULE STALK

reproduction in plants

If we were to consider reproduction in all kinds of plants—algae,
fungi, mosses, ferns, pine trees, and flowering plants—we would
find a bewildering array of life cycles. Here we shall consider
only one, that in the lily.

The simplicity of the lily flower makes it ideal for an examination of the reproductive organs and cells. The lily plant is a
diploid structure, having developed from a fertilized egg. The
lily flower is part of the diploid plant and contains both male
and female organs in the same flower.

The female part of the flower is the **pistil**, consisting of **ovary,
style**, and **stigma**. The seeds develop in the ovary and each seed,
or **ovule**, forms a single egg. The egg develops from a cell that
enlarges and becomes the **megasporocyte.** The megasporocyte
goes through two meiotic divisions, forming four haploid nuclei

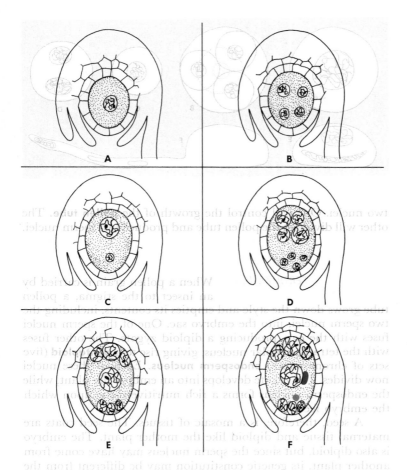

Figure 15.17 Formation of an embryo sac of lily: (A) the megasporocyte. (B) Four-nuclei stage at the end of the two meiotic divisions. (C) Two-nuclei state, one nucleus (the smaller one) being haploid, and the other triploid. (D) Eight-nuclei stage resulting from two successive mitotic divisions; four nuclei are haploid and four triploid. (E) Rearrangement of the eight nuclei in the embryo sac. (F) Fertilization, with one sperm (shown in solid color) uniting with the egg and forming a zygote. The other sperm unites with the polar nucleus and forms a pentaploid endosperm nucleus which will produce a nutritive tissue for the developing zygote.

(Fig. 15.17), after which three of the nuclei fuse and form a triploid nucleus (it contains three sets o chromosomes). The megasporocyte, therefore, contains only two nuclei, each different from the other. These two nuclei now go through two successive divisions, giving a megasporocyte with eight nuclei, four of which are haploid, and four triploid. These nuclei now rearrange themselves. Three triploid nuclei go to the basal part of the megasporocyte; three haploid nuclei cluster at the micropylar end (with the middle one becoming the egg); and the remaining two nuclei (one triploid and one haploid) fuse in the center, forming a tetraploid **polar nucleus.** The whole structure is now called an **embryo sac** and is ready to be fertilized.

The male portion of the flower is the **anther,** in the center of which are many cells called **microsporocytes.** These undergo two meiotic divisions, producing four haploid cells that will develop into mature pollen grains. First, however, the single nucleus in each of these cells divides by mitosis, providing a cell with

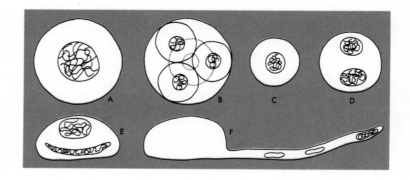

two nuclei. One will control the growth of the **pollen tube.** The other will divide in the pollen tube and produce two sperm nuclei.

FERTILIZATION   When a pollen grain is carried by an insect to the stigma, a pollen tube grows down the style and empties its contents, including the two sperm nuclei, into the embryo sac. One of the sperm nuclei fuses with the egg, producing a diploid zygote. The other fuses with the tetraploid polar nucleus, giving rise to a **pentaploid** (five sets of chromosomes) **endosperm nucleus.** Each of these nuclei now divides. The zygote develops into an embryo lily plant, while the endosperm nucleus forms a rich nutritive tissue upon which the embryo feeds.

A seed, therefore, is a mosaic of tissues. The seed coats are maternal tissue and diploid like the mother plant. The embryo is also diploid, but since the sperm nucleus may have come from another plant, its genetic constitution may be different from the seed coat. The endosperm is pentaploid tissue, and is used up when the seed germinates and draws upon it for nutrient.

summary

Fertilization and **meiosis** are of particular significance in the life history of a sexually reproducing organism. Fertilization doubles the chromosome number; meiosis halves it. In the two processes, genes are recombined and segregated, giving a physical basis for the inheritance of traits and a means for producing offspring with varying combinations of traits.

The key to an understanding of meiosis is the fact that chromosomes occur in pairs. Two members of each pair come together and then segregate from each other during the first meiotic division. At the second meiotic division, the chromatids of each chromosome separate from each other, resulting in four haploid cells from each diploid cell entering meiosis.

The cells resulting from meiosis can be eggs, sperm, or spores, depending on the kind of organism and the kind of tissue in which meiosis takes place.

1 Does a haploid gamete contain a full set of genes? Explain. What is meant by a "full set" of genes? A "full set" of chromosomes?

2 What are the basic differences that distinguish meiosis from mitosis? In what ways are they similar? Consider your answers in terms of both genes and chromosomes.

3 Can you think of any reason why it is of advantage to form only one functional egg instead of four from each oocyte?

4 What would be the result if an oocyte failed to go through meiosis but was fertilized by a sperm?

5 What would a zygote be like if the second division of meiosis did not take place, but fertilization occurred?

6 What do you think would happen if no chiasmata formed between homologues?

7 Why should the replication of DNA be lacking between the two meiotic divisions? Is the lack simply an economical measure, or is the lack necessary for the success of meiosis?

8 Why, in both plants and animals, should more male than female gametes be produced?

9 Can you think of any reason why it is of advantage to have the endosperm of lily pentaploid?

10 Let us assume that the maternal member of a pair of homologous chromosomes is white, the male member red. If a particular meiotic cell has two pairs of homologues, how many kinds of gametes are possible in terms of red and white chromosomes? How many kinds if three pairs of homologues are present? Four pairs? Twenty-three pairs, as in man? Can you diagram this for four pairs? Can you think of a mathematical formula to describe the number possible for any given number of pairs of homologues? What law of Mendel's does this behavior of chromosomes support?

11 By combining knowledge of the inheritance of genes and the behavior of chromosomes in meiosis, the *chromosome theory of inheritance* was developed. How would you state this theory in your own words?

12 In what cells does meiosis take place in lily? In man?

13 What would happen if meiosis took place in other cells of the body?

See the books listed at the end of the previous chapter.

inheritance
in
a diploid
organism
chapter 16

In our discussion of *Neurospora* we made the point that it is a haploid organism; that is, each nucleus has but a single set of unpaired chromosomes. The only diploid phase (each nucleus having paired chromosomes) is that brief period of time after the two nuclei fuse and each contributes its single set of chromosomes. For any single pair of segregating alleles, a ratio of $1A:1a$ is to be expected. With two pairs of independently assorting genes, ratios of $1AB:1Ab:1aB:1ab$ are the rule.

However, most of the organisms with which we are familiar are diploid. The fertilized egg divides by mitosis, and continued cell division brings about a mass of cells that differentiates into an adult organism. The genetic constitution of an individual, therefore, is expressed by cells containing two sets of chromosomes and, consequently, two sets of genes. For example, a given gene could exist in cells or in organisms such as *AA*, *Aa*, or *aa*. When both *A* and *a* are present, do both genes express themselves? Does *A* mask *a*? Or does some new expression make itself evident? In the haploid mycelium of *Neurospora*, any gene generally expresses itself because no other allele is present. As we shall see, several different conditions are possible in diploid organisms. We can consider them all under the general term of **diploid inheritance.**

THE 3:1 RATIO Let us describe here the classical experiments performed by Mendel, the results of which enabled him to formulate his laws of inheritance (Fig. 16.1). Mendel studied inheritance in the garden pea, and he selected a number of pairs of contrasting, mutually exclusive characters. One of these was *tall* versus *dwarf* plants (Fig. 16.2).

The garden pea is ordinarily a self-fertilized plant. That is, the eggs it produces in its ovaries are fertilized by sperm brought in by its own pollen. With care, however, the anthers in a given flower can be removed, and pollen from another plant can be substituted. It is thus possible to control all matings, and to make certain which sperm fertilizes each egg.

The tall and dwarf plants used by Mendel bred true when they were allowed to self-pollinate. They showed no segregation of the tall versus dwarf character in that tall plants produced tall offspring, dwarf plants dwarf offspring. It can be assumed that the genes controlling height were in a pure allelic state, even though it can be further assumed that the genes responsible for

Figure 16.1 Gregor Mendel, an Austrian monk and science teacher, carried out his studies on the inheritance of characters in the garden pea in the monastery garden. From the results obtained, he discovered the basic laws of biparental inheritance operating in diploid organisms.

Figure 16.2 Two varieties of peas which were planted at the same time. They have about the same number of leaves, but the internodes (that portion of the stem between successive leaves) of the taller variety grow much faster than those of the dwarf variety. At maturity, the dwarf variety will reach a height of only 20 inches. The taller variety will reach a height of about 60 inches.

251

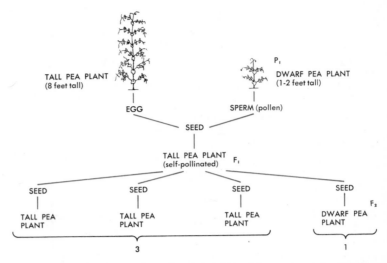

Figure 16.3 This diagram shows the results of Mendel's experiment, in which he crossed two distinct strains of peas—one tall, the other dwarf. The members of the F_1 generation were all tall; but when these plants were self-pollinated, the seeds developed into both tall and dwarf plants, in a ratio of 3:1.

the tall character were different from those governing the dwarf character. Mendel then crossed the two strains of plants. The cross is shown in Fig. 16.3, with the parent generation labeled P_1. The seeds were harvested, sown, and produced the F_1 **(first filial)** generation. *The F_1 plants were all tall; the dwarf character was not expressed in any way.*

Mendel then allowed the F_1 plants to self-pollinate themselves. Seeds again were harvested and sown, and the F_2 generation plants grew. The population consisted of 1,064 plants. Of these, 787 were tall, and 277 were dwarf. This is essentially a 3:1 ratio of tall to dwarf progeny.

Mendel carried the experiment further by allowing each F_2 plant to self-pollinate and produce seed. From these seeds he discovered that the dwarf plants bred true; that is, they produced only dwarf progeny. However, only one third of the tall plants bred true for the tall character. The other two thirds segregated for tall versus dwarf, again in a 3 tall:1 dwarf ratio. These segregating tall plants behaved, therefore, exactly like the F_1 tall plants.

Remember that Mendel published his work in 1865. He knew about the necessity of pollen as an agent for fertilizing the egg, and he had a good mathematical training, but he knew nothing about chromosomes, meiosis, or the concepts of diploidy and haploidy. Mendel's analysis of his data is all the more remarkable when we consider the meager foundation of knowledge he had to build on. Let us now list the facts that emerge from his experiments, and then list the assumptions he made to explain the facts.

HERE ARE THE FACTS:

1. The characters **tall** (6 to 7 feet) and **dwarf** ($\frac{3}{4}$ to $1\frac{1}{2}$ feet) are constant. There are no plants of intermediate height.
2. The dwarf character does not appear in the F_1 generation. But it appears unchanged in the F_2 generation; and when it does, it reappears with a predictable frequency.
3. Part (one-third) of the tall F_2 generation breeds true for tallness. The remainder (two-thirds) continues to show segregation in the F_3 generation.
4. The dwarf plants always produce dwarf progeny when self-pollinated.

The fact that predictable ratios were achieved suggested to Mendel that factors were being passed from generation to generation in a constant manner; also that the factor for dwarfness was present in the F_1 generation even when not expressed.

HERE ARE MENDEL'S ASSUMPTIONS:

1. Assume that T stands for the tall factor (we would now call this a gene), t for dwarfness.
2. Assume that one factor is contributed by the egg and one by the sperm (pollen). The zygote and the plant itself would have two factors existing together in each cell. These would segregate from each other when the eggs and pollen were formed. (Meiosis was discovered and understood in the 1880s, so Mendel formed his assumptions without this knowledge.)
3. Assume that T dominates t in such a way that when the two are together only T is expressed. These assumptions can be diagrammed as follows:

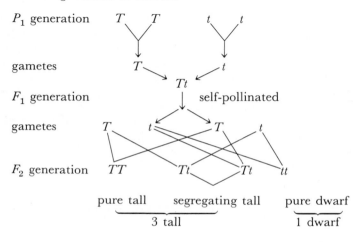

The assumptions made by Mendel fitted his data very well, and they have been fully supported since by hundreds of similar crosses in a wide variety of organisms. The law of segregation, therefore, applies not only to garden peas, but to all sexually reproducing diploid organisms.

Mendel, however, was not fully satisfied. He devised another way of testing his hypothesis that the F_2 tall plants were of two kinds. He made what is known as the **backcross** or **testcross.** The testcross involves the crossing of the individual being tested to another individual so that its genetic constitution will be revealed. The testcross made by Mendel was to cross the tall F_2 plants to any dwarf plant. When TT was testcrossed to tt, only Tt tall progeny resulted. If Tt were testcrossed to tt, then two kinds of progeny resulted in a 1:1 ratio.

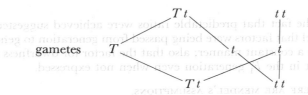

The testcross, therefore, fully supported his contention that a dwarf factor could be obscured when in the presence of T, but that it could be recovered unchanged in the next generation when combined with another t gene.

Some terms, commonly used, should now be defined: T is said to be **dominant** over the **recessive** t; TT and tt are, respectively, **homozygous dominant** and **homozygous recessive;** Tt is a **heterozygous** state. The terms tall and dwarf describe the **phenotype** (appearance) of the pea plant. However, because of dominance and recessiveness, the phenotype tall is the same even when the **genotype** (genetic constitution) is different, that is, Tt or TT.

THE 9:3:3:1 RATIO The breeding experiment just described is known as a **monohybrid cross.** It involves only one pair of characters. All other characters are disregarded. Mendel also made **dihybrid crosses** in which two pairs of independent characters were tested simultaneously. For example, in one of his crosses he used seed characters: yellow (Y) versus green (y) and round (R) versus wrinkled (r). Mendel knew from previous studies that yellow was dominant over green and round dominant over wrinkled. Using the symbols as before, the cross can be expressed as follows:

$$YYRR \times yyrr$$

By using the Punnett square method of demonstration (Fig.

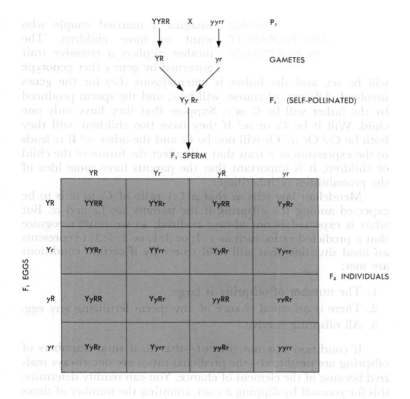

Figure 16.4 The Punnett square method for demonstrating the F_1 gametes and the F_2 progeny resulting from a dihybrid cross. Among the F_2 progeny, Mendel obtained 315 plants that had yellow, round seeds (Y-R-); 108 plants that had yellow, wrinkled seeds (Y-rr); 101 plants that had green, round seeds (yyR-); and 32 plants that had green, wrinkled seeds ($yyrr$). The dash (-) indicates that the missing symbol can be either dominant or recessive. For example, Y- can be either YY or Yy, but the phenotype will be yellow. The ratio obtained is basically a $9:3:3:1$.

16.4) to show the F_1 gametes and the F_2 individuals, we can demonstrate that the expected phenotypic ratio will be 9 yellow round : 3 yellow wrinkled : 3 green round : 1 green wrinkled. In the legend of Fig. 16.4 the exact values obtained by Mendel are shown. The pairs of genes behaved independently of each other. The **law of independent assortment** of genes, therefore, applies to diploid inheritance as well as to haploid inheritance. The difference between the $1AB:1Ab:1aB:1ab$ ratio in haploid organisms, such as *Neurospora*, and the $9AB:3Ab:3aB:1ab$ ratio in the diploid garden pea is due to single versus double doses of genes in the individual organisms, and to the phenomena of dominance and recessiveness. Otherwise, the same rules of inheritance apply. One need only remember that in a haploid organism *every* gene has the opportunity of expressing itself since there is no other allele to interfere with it. Its phenotype directly reveals its genotype. This, of course, is not necessarily true in a diploid organism.

In Chapter 14 the segregation of genes during meiosis in *Neurospora* was explained in terms of chromosome segregation. Those same rules of chromosome behavior operate in the garden pea, man, or any other diploid organism having normal meiosis.

Imagine a married couple who
want to have children. The
mother displays a recessive trait
governed by gene *c* (her genotype
will be *cc*), and the father is heterozygous (*Cc*) for the genes
involved. All eggs, of course, will be *c*, and the sperm produced
by the father will be *C* or *c*. Suppose that they have only one
child. Will it be *Cc* or *cc?* If they have two children, will they
both be *Cc?* Or *cc?* Or will one be *Cc* and the other *cc?* If *cc* leads
to the expression of a trait that will affect the future of the child
or children, it is important that the parents have some idea of
the probabilities of inheritance.

Mendelian law tells us that a 1:1 ratio of *Cc* to *cc* is to be
expected among the offspring if the parents are *Cc* and *cc*. But
what is expected is not always realized; so we must recognize
that a predicted ratio, such as 1:1, or 3:1, or 9:3:3:1 represents
an ideal situation that will hold true only if certain conditions
are met:

1. The number of offspring is large.
2. There is an equal chance of any sperm fertilizing any egg.
3. All offspring survive.

If condition 1 is not present—that is, if small numbers of
offspring are produced—the predicted ratios are not always realized because of the element of chance. You can readily determine
this for yourself by flipping a coin, counting the number of times
heads or tails turns up, and comparing the ratio when the number
of flips is small as opposed to when it is large. If conditions 2
and 3 are not realized, the expected ratios are bound to be
distorted.

If condition 2 holds in our *Cc* × *cc* mating, the probability,
or chance, that a *C* sperm will fertilize the *c* egg is one half. The
probability of the *c* sperm fertilizing the egg is also one half. There
is, in other words, a 50:50 chance that the first offspring will
be *cc*. Let us assume that this is so. If the family has two children,
what, then, is the probability that both will have the *cc* genotype?
Here we must make one assumption—the occurrence of the first
event does not influence the probability of the second event. In
terms of coins, if the first flip shows heads, it does not influence
the next flip so that tails is any more likely to turn up than heads.
Each flip is still 50:50 heads or tails.

Therefore, if the probability of the first child being *cc* is one
half, the same probability holds for the second child. The answer
to our question, then, is $\frac{1}{2} \times \frac{1}{2}$, or $\frac{1}{4}$; or, to phrase it differently,
there is one chance out of four that both children will be *cc*. On
the same basis, there is an equal probability that both children
will be *Cc,* or that one will be *Cc* and the other *cc*. The probability
of three *cc* children will be $\frac{1}{2} \times \frac{1}{2} \times \frac{1}{2}$, or $\frac{1}{8}$.

The Punnett square previously described is a pictorial representation of Mendelian segregation and recombination to give ideal ratios. When dealing with small numbers, however, it is often easier to make use of a simple algebraic representation of the genetic information. This is done by expanding the binomial $(a + b)^n$, where a is the probability of one genotype, b the probability of the other, and n the number of offspring. In our $Cc \times cc$ family, and with three children, the binomial would be written $(\frac{1}{2}Cc \times \frac{1}{2}cc)^3$, which becomes

$$\frac{1}{8}(Cc)^3 + \frac{3}{8}(Cc)^2 \times cc + \frac{3}{8}Cc \times (cc)^2 + \frac{1}{8}(cc)^3$$

The exponents assigned to the genotypes indicate the character of the three children: $(Cc)^3$ means three Cc children, $(Cc)^2 \times cc$ means two Cc children and one cc child, and so on. The fractions indicate the probabilities for the occurrence of the particular families; that is, in many families of three children, arising from a $Cc \times cc$ mating, you would expect all children to have a cc only one eighth of the time.

If one is dealing with a 3:1 ratio, the same binomial can also be used, but in this instance the formula would be $(\frac{3}{4}a + \frac{1}{4}b)^n$, with a being the dominant phenotype and b the recessive phenotype. With two pairs of alleles, we can write the usual 9:3:3:1 ratio as $\frac{9}{16}A-B- + \frac{3}{16}aaB- + \frac{3}{16}A-bb + \frac{1}{16}aabb$, the dash (–) indicating that the second allele can be either dominant or recessive without altering the phenotype. Let us now ask: What is the probability in a family with three children that the offspring will be two $aaB-$ and one $aabb$? The probability is $\frac{3}{16} \times \frac{3}{16} \times \frac{1}{16}$, or $\frac{9}{4096}$. Therefore, in families of three children whose parents were heterozygous for both pairs of alleles, we should expect the above genotypes to appear once in about 455 times.

If we assume in the problem above that A is normal and a is albino, and that B is normal and b is dwarf, we can tabulate all aspects of the problem in the following way:

	Phenotype			Phenotypic
Genotype	**Color**	**Height**	**Probability**	**Frequency**
$A-B-$	Normal	Normal	$(\frac{3}{4})(\frac{3}{4})$	$\frac{9}{16}$
$aaB-$	Albino	Normal	$(\frac{1}{4})(\frac{3}{4})$	$\frac{3}{16}$
$A-bb$	Normal	Dwarf	$(\frac{3}{4})(\frac{1}{4})$	$\frac{3}{16}$
$aabb$	Albino	Dwarf	$(\frac{1}{4})(\frac{1}{4})$	$\frac{1}{16}$

summary

In a **haploid** organism, all genes express themselves without hindrance. In a **diploid** organism, each gene is represented twice in every cell, and problems of gene expression (dominant versus recessive) are raised. The ratios resulting among the offspring differ accordingly, and monohybrid ratios of 3:1 and dihybrid ratios of 9:3:3:1 are encountered. The laws of inheritance and the relation of meiosis to gene segregation, however, are the same for diploid and haploid organisms; only gene dosage and expression of genes differ.

for thought and discussion

1   In a cross between pure breeding red-flowered and white-flowered plants, the F_1 offspring are pink-flowered. In the F_2 generation, the offspring appear in a ratio of 1 red:2 pink:1 white. From these data what can you conclude about the action of the genes concerned with color formation? What ratios would you expect if you crossed an F_2 red with an F_2 pink? F_2 pink with F_2 white? Pink with pink? In the original cross, diagram the genotypes of all individuals from parental types to F_2 offspring.

2   If two yellow mice are crossed, they will produce offspring with an average ratio of 1 black:2 yellow. Can you explain these results? How would you test your conclusions?

3   The major blood types in human beings, due to a chemical substance in the blood and determined by several alleles of a given gene, are of four kinds: A, B, AB, and O. The pattern of inheritance is indicated in the table. Can you give an explanation

Parents	Offspring	Parents	Offspring
A × A	A or O	B × O	B or O
B × B	B or O	AB × O	A or B
A × B	A, B, AB, or O	AB × A	A, B, or AB
A × O	A or O	AB × AB	A, B, or AB

for these patterns of inheritance? What combinations of alleles show dominance and recessiveness, and which reveal a lack of dominance? If a child has an AB blood type, what possible genotypes can its parents have? What use is the above information in determining the parentage of a disputed child?

4   In chickens, assume that the gene *I* inhibits the formation of feather color; *i* permits color formation. Assume also that gene C, which is independent of I, governs the formation of black pigment; gene c inhibits its formation. In a cross of *IiCc* × *IiCc*, what is the expected phenotypic ratio of black to white offspring? Explain your results.

5   In rabbits, the genotypes C^hC^h or C^hc produce the Himalayan phenotype in which the extremities (nose, ears, feet) are black and the remainder of the body white. The black pigment, melanin, is formed by enzyme action. From this information and your

knowledge of enzyme action, what explanations can you offer for the Himalayan phenotype? How could you test your hypothesis?

6   The parents of two offspring are heterozygous for the genes A and B. What are the probabilities that both offspring will show only the dominant phenotype? One with the double recessive phenotype and one with the double dominant phenotype? Both with an aaB– phenotype? One with aaB– and the other with A–bb? If there are four offspring, what is the probability that all will have the double recessive phenotype? That all will have the dominant phenotype and all will be girls? (Assume that being a boy or girl has a 50:50 probability.)

7   The ability to taste PTC (phenylthiocarbamide) is inherited as a dominant trait. Using the tester materials provided by your instructor, test yourself and your family. Do the results agree with the hypothesis that it is a dominant trait?

8   Perform the following experiment yourself, using two pennies and two nickels. Put them in a container, shake them, and then toss them onto a table. Score the results for 100 such tosses as follows: H = heads, T = tails.

Class	Pennies	Nickels	Observed	Expected
1	HH	HH		
2	HH	HT		
3	HH	TT		
4	HT	HH		
5	HT	HT		
6	HT	TT		
7	TT	HH		
8	TT	HT		
9	TT	TT		
			Total 100	100

9   Before beginning the experiment, calculate your expected results for 100 tosses. How do your observed results agree with the expected? If there is a difference, explain it. Compare your observed results with those of another student, and then with the pooled results of all students. Is the difference between observed and expected greater for a single comparison or for the pooled results? Why?

10   Assume heads to be dominant, tails recessive, and pennies and nickels to be two different traits. What is the expected genotypic ratio? Phenotypic ratio? What kind of a genetic experiment is this comparable to? How do your observed results agree with what you expect? What accounts for the difference, if it exists?

selected readings

AUERBACH, CHARLOTTE. *Genetics in the Atomic Age.* New York: Essential Books, Inc., 1956. An elementary discussion of chromosomes and genes, and how radiation can affect them

AUERBACH, CHARLOTTE. *The Science of Genetics.* New York: Harper & Row, Publishers, 1961. An excellent and clear description of the basic principles of genetics.

BAKER, W. K. *Genetic Analysis.* Boston: Houghton Mifflin Company, 1965. A difficult but excellent book concerned with the analysis of genetic results.

BONNER, D. M., and S. E. MILLS. *Heredity.* Englewood Cliffs, N.J.: Prentice-Hall, Inc., 1964. More concerned with the physical nature of the gene, but containing much good material on inheritance.

BREWBAKER, J. L. *Agricultural Genetics.* Englewood Cliffs, N.J.: Prentice-Hall, Inc., 1964. Contains a good deal of information on basic Mendelian inheritance as well as data relating to inheritance in domesticated plants and animals.

MCKUSICK, V. A. *Human Genetics,* 2nd ed. Englewood Cliffs, N.J.: Prentice-Hall, Inc., 1969. A readable volume on the methods and results of genetic studies in humans.

linkage, crossing-over, and gene maps
chapter 17

Our understanding of genetic systems has been built up largely from information gained through inheritance tests. Controlled matings and accurate recording of offspring are required in such tests (Fig. 17.1). Fortunately, the behavior of those genes which we cannot see is reflected in the behavior of the chromosomes, which are visible. Therefore, when we find genetic ratios unlike those we expect, we can often turn to the chromosomes for possible explanations.

We have already pointed out that the law of segregation applies to most organisms. There are exceptions, however, to the law of independent assortment. We can understand the reasons for many of the exceptions if we relate them to the behavior of chromosomes.

The garden pea, for example, has seven pairs of homologous chromosomes. When two different pairs of genes are on different pairs of chromosomes, they segregate independently of each other, giving rise to the familiar 9:3:3:1 dihybrid ratio characteristic of a diploid organism. *Neurospora* also has seven chromosomes, but being haploid the dihybrid ratio is 1:1:1:1. But both the garden pea and *Neurospora* have many more than seven pairs of genes. It follows, then, that any given pair of chromosomes contains many genes. If the two pairs of genes being studied happen

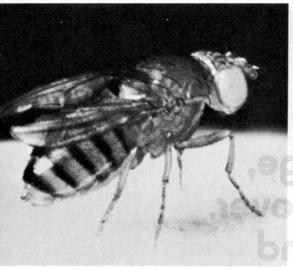

Figure 17.1 Photographs of two females of the fruit fly *Drosophila melanogaster.* The one on the right is a normal individual, commonly referred to as *wild type.* That on the left exhibits two mutant characters: white eye instead of the wild type (red eye), and miniature wing instead of the longer wing of the wild type.

to be on the same chromosome, the chances are great that these genes will not show independence of assortment. This tendency for the genes in a given chromosome to segregate together is called **linkage.**

Let us think of genes as beads on a string. Each bead may be separate from the others and is individually distinct (Fig. 17.2). But if one bead is moved the others follow, simply because of the string that holds them together. In the same way, genes show linkage because they are on the same chromosome. If the genes on any chromosome are always segregated as a group, these are said to show **complete linkage.** Complete linkage, however, is rare, and groups of linked genes can recombine with other groups within homologous chromosomes by a process of **crossing-over.** A study of meiotic chromosomes indicates this at the cell level by the formation of chiasmata.

LINKAGE AND CROSSING-OVER IN DROSOPHILA Two easily recognized variations in *Drosophila melanogaster,* the fruit fly, are black body (*b*) and vestigial wings (*vg*). Both are recessive to the normal gray body color and to normal wing size. If a cross is made between a normal fly homozygous for these traits ($BBVgVg$) and a black one with vestigial wings ($bbvgvg$), the F_1 is normal in appearance, but heterozygous for the two pairs of genes ($BbVgvg$). A testcross of an F_1 female with a double recessive ($bbvgvg$) male would be expected to produce offspring in a ratio

Figure 17.2 The genes in a single chromosome can be represented by the different symbols shown here. It is not known, however, that genes are separated from each other by nongenic material, as the diagram would suggest. The chromosome is probably a continuous piece of DNA, in which case the genes would be adjacent to each other and distinguishable only by variations in the nucleotide sequence.

of $1BVg : 1Bvg : 1bVg : 1bvg$ if the genes were on different chromosomes. However, the actual results are quite different in two ways.

Notice in Fig. 17.3 that the genes are designated somewhat differently than before: BVg/BVg instead of $BBVgVg$. This is the geneticist's way of indicating that the genes are linked. The genes BVg are located on one chromosome, while the other set of BVg genes, indicated below the line, are located on its homologue. The F_1 individuals are, of course, heterozygous: BVg/bvg. After the F_1 females are mated to the double recessive male in the testcross and 1,000 of the offspring are singled out, the results might be as indicated in Fig. 17.3. The same results would be obtained if the experiment were repeated, although the actual numerical figures might vary somewhat from one experiment to another.

The results clearly indicate that a marked deviation from a $1:1:1:1$ phenotypic ratio has been obtained. The parental phenotypic combinations, (BVg/bvg) and (bvg/bvg), account for 81.5 per cent of the total population, with the two parental combinations being represented almost equally. The new combinations, (Bvg/bvg) and (bVg/bvg), add up to 18.5 per cent of the population. Again each new combination of genes is represented equally. From these data we can say that these two pairs of genes do not segregate independently of each other. The BVg and bvg parental combinations show a strong tendency to be inherited together. The new arrangements of genes, Bvg and bVg, called **recombinations,** show up relatively infrequently. The strength of the tendency to be inherited as a unit, calculated as a percentage $[(407 + 408)/1,000]$ of the total population, is a measure of the strength of linkage. Conversely, the frequency of recombinations, again calculated as a percentage $[(92 + 93)/1,000]$ of the total population, is a measure of the frequency of crossing-over between these two genes.

If we do the same experiment, but cross F_1 males to double recessive females this time, we obtain the results indicated in Fig. 17.4. Again, if 1,000 flies were counted, we would find that they would be about equally divided into (BVg/bvg) and (bvg/bvg) individuals. Therefore, only the parental combinations were recovered, and we must assume that linkage is complete.

The percentages obtained from these two experiments are not haphazard. When *Drosophila* F_1 males are used, complete

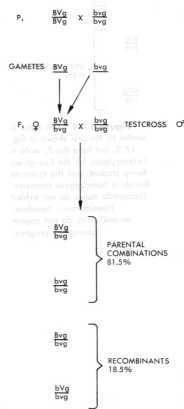

Figure 17.3 Diagram of the results of a dihybrid cross: the genes being studied are part of the same chromosome rather than being on separate chromosomes. In this particular cross, the F_1 female was the heterozygous parent, and the testcross male was homozygous recessive. Compare these results with those in Fig. 17.4

$P_1 \quad \dfrac{BVg}{BVg} \times \dfrac{bvg}{bvg}$

GAMETES $\quad BVg \qquad bvg$

$F_1 \ ♀ \quad \dfrac{BVg}{bvg} \times \dfrac{bvg}{bvg}$ TESTCROSS $♂$

$\dfrac{BVg}{bvg}$

$\dfrac{bvg}{bvg}$

PARENTAL COMBINATIONS 81.5%

$\dfrac{Bvg}{bvg}$

$\dfrac{bVg}{bvg}$

RECOMBINANTS 18.5%

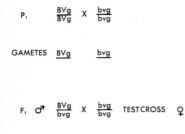

P₁
$$\frac{BVg}{BVg} \times \frac{bvg}{bvg}$$

GAMETES   BVg   bvg

F₁ ♂ $$\frac{BVg}{bvg} \times \frac{bvg}{bvg}$$ TESTCROSS ♀

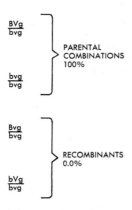

$$\frac{BVg}{bvg}$$
$$\frac{bvg}{bvg}$$

} PARENTAL COMBINATIONS 100%

$$\frac{Bvg}{bvg}$$
$$\frac{bVg}{bvg}$$

} RECOMBINANTS 0.0%

Figure 17.4 This cross is similar to the one shown in Fig. 17.3, but here the F₁ male is heterozygous for the two genes being studied, and the testcross female is homozygous recessive. *Drosophila* males do not exhibit crossing-over; therefore, recombinants do not appear among the progeny.

linkage is the rule. When F_1 females are used, parental combinations will be found in about 80 per cent of the population, and recombinations in about 20 per cent. Clearly, there must be some physical reason for this. Since we have already discussed the fact that gene segregation is mirrored in the behavior of chromosomes, it is natural that we should turn to the behavior of meiotic chromosomes. It is in meiosis that segregation occurs.

If you look back to Fig. 15.7, you will notice that in meiotic prophase the homologous chromosomes, which were paired, fall apart except at those points along their length where they are held together by chiasmata. *Notice carefully that a chiasma is an exchange of chromatin between only two of the four chromatids,* one from each of the two homologous chromosomes. This exchange results from the process of crossing-over, with the chiasma being visible evidence of it at the four-strand stage.

Assume now that the genes for body color and wing size are located on this pair of chromosomes. Gene B and its allele b occupy corresponding places in the homologues, and the same is true for gene Vg and its allele vg. A chiasma forms between these two pairs of genes, and a cross-over results (Fig. 17.5). Two of the chromatids are unaffected; they retain the parental gene combinations, BVg or bvg. The other two have new gene combinations, Bvg and bVg. These are **cross-over chromatids.** Therefore, every cross-over yields only 50 per cent cross-over chromatids and 50 per cent parental, or non-cross-over, chromatids. If a cross-over occurred in every meiotic cell between these two genes, no more than 50 per cent recombination of genes could result. When crossing-over occurs less frequently, more non-cross-over chromatids are found.

Let us now view our data in terms of crossing-over. In the experiment where the F_1 individual was a female, 18.5 per cent of the offspring had new gene combinations and 81.5 per cent had the parental combinations. With the knowledge that only two of the four chromatids are involved in a single cross-over, we can state that 37 per cent (18.5 × 2) of all eggs had a cross-over taking place between the genes b and vg.

In the experiment where the F_1 individual was a male, no

Figure 17.5 This diagram shows a cross-over that has taken place between the chromatids of homologous chromosomes. The chromosomes are marked with the mutant genes shown in Figs. 17.3 and 17.4.

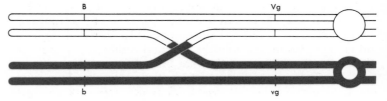

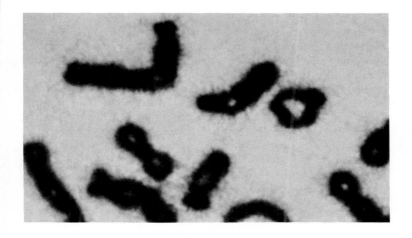

Figure 17.6 In this meiotic cell of a male amphibian you can see the chiasmata holding the homologous chromosomes together. The shorter bivalents have two chiasmata while several of the longer ones have as many as four or possibly five.

cross-over chromatids were recovered in the offspring. Only the parental combinations appeared. Therefore, no cross-overs take place in *Drosophila* males. This is not true for males of all species. In fact, *Drosophila* males are exceptional in this regard, but they are useful in demonstrating complete as opposed to partial linkage.

If we look at meiotic chromosomes in a number of species, we find in general that the longer the chromosomes, the more numerous the chiasmata. Small pairs of homologues generally have one chiasma holding them together. Larger ones may have two or more (Fig. 17.6). Furthermore, the position of a chiasma is not constant. It may be any place along the length of a chromosome. A general rule therefore is that the farther apart two genes are on a chromosome, the greater the opportunity for a cross-over to occur between them. The frequency of recombination will, of course, also be greater. When the percentage of recombination is low, the genes can be physically close to each other, or crossing-over for some reason does not occur, as in male *Drosophila*.

PROOFS OF
CROSSING-OVER

Those of you who have a skeptical turn of mind probably have been asking a few questions as you have been reading this chapter: How do we *know* that crossing-over involves an exchange of chromatin? How do we *know* that crossing-over takes place in the four-strand stage and between only *two* of the four chromatids? How can we be *certain* that a chiasma is visible evidence of a cross-over between genes? These questions need answering if we are to take the position that the behavior of genes is reflected in the behavior of chromosomes, because this assumption lies at the heart of genetics.

The first question was answered by a simple yet elegant experiment carried out by Harriett Creighton and Barbara McClintock, then at Cornell University. They took advantage of the fact that some chromosomes in maize have visibly distinct knobs while others with which they are homologous do not. They crossed two strains of maize, producing a hybrid with the chromosomal composition shown in Fig. 17.7. The knobbed chromosome also carried an additional cell marker indicated by the dashed portion in the diagram. In addition, the chromosomes were marked genetically with two pairs of heterozygous alleles: *C*, colored kernel; *c*, colorless kernel; *Wx*, starchy kernel; *wx*, waxy kernel. This hybrid plant, heterozygous now for the two genes, was crossed with one carrying knobless chromosomes, and with *c* and *wx* in a homozygous state. The testcross progeny was then examined genetically and cytologically.

Since *C* was very close to the knob, little or no crossing-over took place in this region. When a cross-over does occur between the two genes, however, and if an exchange of chromatin accompanies crossing-over, then all *CWx* plants should have the knobbed chromosome. The *cwx* plants should not have them. However, the *cwx* chromosomes should have the dashed portion at the other end of the chromosome. In all cases studied there was complete agreement between the genetical and cytological information. Whenever a cross-over occurred, an exchange of chromatin resulted. A comparable study by Curt Stern in *Drosophila* confirmed the maize work in all respects.

To answer our second question about crossing-over in the four-strand stage, we turn to *Neurospora,* because all the chromatids from a single meiotic cell can be recovered. Each one is contained within a single ascospore. Assume that genes *A* and *B* are linked, and that strain *AB* is crossed with *ab*. If crossing-over occurs, two alternatives exist: either crossing-over occurs at the two-strand or the four-strand stage. Figure 17.8 shows these two possibilities. If crossing-over occurs at the two-strand stage, the ascospores will contain *only* cross-over chromatids. If no crossing-

Figure 17.7 An experiment carried out in maize demonstrates that when a cross-over occurs there is also an exchange of a block of chromatin. See text for further explanation.

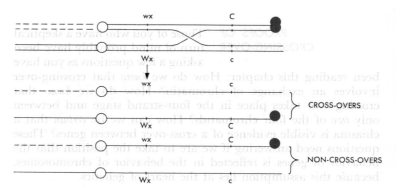

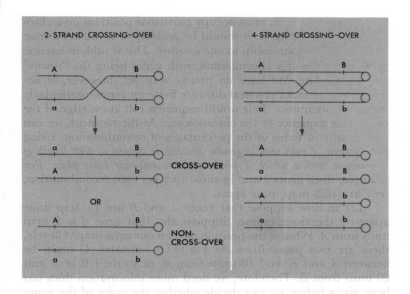

A   B

a   b

a   B

A   b   CROSS-OVER

OR

A   B   NON-CROSS-OVER

a   b

A   B

a   b

A   B

a   B

A   b

a   b

Figure 17.8 If crossing over in *Neurospora* occurs in the two-strand stage—that is, between chromosomes—each ascus would show all crossover products, but not a mixture of parental and crossover products. The latter would occur if crossing over took place in the four-strand stage—that is, between chromatids.

over occurs, only the parental chromatids will be found. *No ascus will contain both parental and cross-over chromatids.* If crossing-over occurs at the four-strand stage, however, both parental and cross-over chromatids are to be expected in the same ascus. The fact that the latter situation prevails is proof of the fact that crossing-over does take place at the four-strand stage, and that only two of the four chromatids are involved in any given cross-over.

Proof that all chiasmata represent points of crossing-over has not been established with certainty. The kinds of cross-overs we have been discussing should be visible in good preparations of meiotic cells as chiasmata. However, we are faced at the present time with the fact that meiotic cells in male *Drosophila*, the homologous chromosomes, are sometimes held together by chiasmata, but yield no evidence of having had any recombination of genes. The best we can say, therefore, is that, in the majority of organisms, a chiasma is very probably the cytological equivalent of a genetic cross-over. To go beyond this is to go beyond the evidence at hand.

CHROMOSOME MAPS You will recall that linked genes are on the same chromosome, and that the frequency of crossing-over between two genes is constant and characteristic. For example, in *Drosophila*, *b* and *vg* always show about 18.5 per cent recombination in females if the environment is kept constant. Another pair of genes might show a different rate of crossing-over, but it, too, would be constant for these two genes.

Genes, therefore, must occupy particular positions on a chromosome. This being so, it should be possible to construct a map showing their relationship to one another. This would, in essence, be a road map of a chromosome, with genes being the "towns" along the route. We have no precise way yet, however, of determining the exact physical distance between genes, particularly in higher organisms. This would require a full knowledge of the nucleotide sequence of the chromosome. At the moment, we can do this only in terms of the percentages of recombination. Using this system, *1 map unit of genetic distance is defined as that length of chromosome within which 1 per cent of crossing-over takes place.* For example, since genes *b* and *vg* show 18.5 per cent recombination, they are 18.5 map units apart.

Let us now suppose that genes *A* and *B* are 15 map units apart on the chromosome. Suppose also that gene *C* is 5 map units from *A*. What is the position of *C* on our gene map? Clearly, there are two possibilities. As Fig. 17.9 indicates, *C* could be between *A* and *B*, and 10 units from *B*, or to the left of *A* and 20 units from *B*. Therefore, we need more information than has been given before we can decide whether the order of the genes is *A C B* or *C A B*.

A standard test procedure, developed by A. H. Sturtevant, then at Columbia University, and known as the **three-point test-cross,** is now commonly used in determining gene order. The basis of the test is as follows: suppose that the correct serial order of three genes is *a b c*. Suppose also that the distance between genes is reflected in the frequency with which crossing-over takes place between them. If these two assumptions are correct, then a testcross of the heterozygote *A B C/a b c* to the triple recessive *a b c/a b c* can yield the following phenotypes, the complementary cross-over and non-cross-over types being grouped (in all instances, each individual would also have the *a b c* chromosome, but it is omitted for the sake of convenience):

Parental or non-cross-overs	*A B C*
	a b c
Single cross-overs	*A b c*
type I	*a B c*

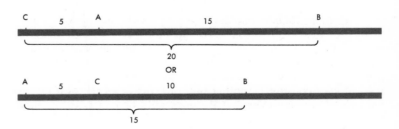

Figure 17.9 Gene C has two possible positions when genes A and B are 15 map units apart, and when gene C is five map units from gene A. The only way to resolve the position of gene C in respect to the other two genes is by use of the three-point testcross described in the text.

Single cross-overs type II	$A\ B\ c$ $a\ b\ C$
Double cross-overs	$A\ b\ C$ $a\ B\ c$

On the basis of the mechanism of crossing-over described in Fig. 17.7, the non-cross-over types are the most frequent. Type I single cross-overs result from a single cross-over between genes *a* and *b*. Type II single cross-overs result from a cross-over between genes *b* and *c*.

If the Type I cross-overs show a frequency of 10 per cent and the Type II cross-overs a frequency of 20 per cent, then *the double cross-overs would be expected to occur at a rate much lower than either alone.* This is because the simultaneous occurrence of both Type I and Type II in the same cell is the product of the two frequencies (10 per cent of 20 per cent), and the double cross-overs should occur in approximately 2 per cent of the cells.

An experiment carried out with maize will illustrate the use of the three-point cross. The genes brown midrib (*bm*), red seed color (*pr*), and virescent (light green) seedling (*v*) are located in one chromosome and, of course, are linked. The crosses and the data obtained are as follows:

$$P_1: \qquad \frac{Bm\ Pr\ V}{BM\ Pr\ V} \times \frac{bm\ or\ v}{bm\ pr\ v}$$

$$F_1: \qquad \frac{Bm\ Pr\ V}{bm\ pr\ v}$$

$$\text{Testcross: } F_1 \times \frac{bm\ pr\ v}{bm\ pr\ v}$$

Testcross progeny (the $\overline{bm\ pr\ v}$ is omitted for all progeny but assumed to be present):

Bm Pr V	232	non-cross-overs = 42.1 per cent
bm pr v	235	
Bm pr v	84	single cross-overs between
bm Pr V	77	*bm* and *pr* = 14.5 per cent
Bm Pr v	201	single cross-overs between
bm pr V	194	*pr* and *v* = 35.6 per cent
Bm pr V	40	double cross-overs between *bm* and *pr*,
bm Pr v	46	and *pr* and *v* = 7.8 per cent

Among this progeny we can recognize the non-cross-over parental types. They are the most numerous. We can also recognize the double cross-overs. They are the two least frequent classes of progeny. The double cross-overs also tell us the order of the genes. That is, *pr* has shifted position with *Pr,* while *Bm* and *V,* and *bm* and *v* are still linked. Therefore, the *Pr* (or *pr*) gene lies between the other two genes.

In calculating map distances, all cross-overs that have taken place between any two genes must be considered. Thus, the distance between *bm* and *pr* is not 14.5 map units, but 14.5 plus 7.8, or 22.3. In other words, the single cross-over frequency between *bm* and *pr* does not represent the total frequency. The double cross-overs must be added, since they also represent cross-overs in this same region. Similarly, the total cross-over frequency between *pr* and *v* is 35.6 plus 7.8, or 43.4. Therefore, the total distance from *bm* to *v* is 65.7 map units.

There is one more point to consider. If the cross-over frequency between *bm* and *pr* is 22.3 per cent, and 43.4 per cent between *pr* and *v,* then we should expect the double cross-over frequency to be 22.3 per cent of 43.4 per cent, or 9.7 per cent. Actually, however, we find that the frequency is only 7.8 per cent, less than expected. Many studies have shown that this is due to something called **interference.** It turns out that two adjacent cross-overs in a single pair of chromosomes are not generally independent of each other; one tends to interfere with the other and reduce the frequency. In general, and in such organisms as maize and *Drosophila,* double cross-overs do not occur between two genes if the genes being tested are 10 map units or less apart. Interference is complete. If the genes are more than 45 map units apart, interference is difficult to detect, or is nonoperative. For reasons that we do not yet understand, the cross-overs on one side of the centromere of a chromosome do not interfere with those on the other side.

On the basis of many three-point tests, it is possible to build up the gene maps of organisms. Figure 17.10 shows the gene maps of the four chromosomes of *Drosophila melanogaster.* The left end of the chromosome is indicated as zero (0) map distance. Successive distances are added to give the location of individual genes.

summary

When two genes are on the same chromosome, the law of independent assortment does not hold. The genes are linked to each other and tend to be inherited together. **Crossing-over** breaks up linkage groups, and the frequency of crossing-over gives a measure of the distance one gene is from a neighboring gene. A chiasma formed between homologous chromosomes in the four-strand state is physical evidence of crossing-over between genes.

Since crossing-over is a measure of distance between genes, gene maps of chromosomes can be, and have been, constructed. In such maps, 1 per cent crossing-over is equal to 1 map unit.

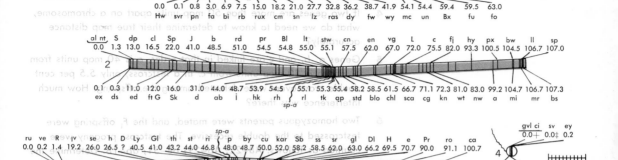

Figure 17.10 Genetic maps of the four chromosomes of *Drosophilia melanogaster*: each gene has a lettered symbol, with lower case letters indicating recessiveness, capital letters indicating dominance in relation to the normal, or wild type, gene. Each gene also has a given position on the chromosome. For example, in chromosome 1 at the top of the illustration, the gene yellow body color (*y*) is at the extreme left at position 0.0; the gene white eye (*w*) is close to it at position 1.5, or 1.5 map units distance from yellow. Bobbed bristle (*bb*) is at the other end of the chromosome, 66.0 map units from yellow. Chromosomes 2 and 3 are the longest in size and in map distance. Chromosome 4 is very small, rarely undergoing crossing-over.

Map distances calculated between genes close together are more accurate than map distances for distant genes, because of the occurrence of double cross-overs.

The position of one gene in respect to two other genes on the same chromosome can be determined through the **three-point testcross.** In such a cross, the double cross-overs form the least frequent class and provide a clue about the order of the three genes.

for thought and discussion

1 Why do we use testcross offspring rather than F_2 offspring in calculating cross-over frequencies? What do we mean by the term "frequency"?

2 If a double cross-over occurred between the linked genes *A* and *B*, could it always be detected? In considering this question, remember that at any position of crossing-over, any chromatid can cross-over with any nonsister chromatid. Can crossing-over be detected if it occurs between sister chromatids? Explain.

3 An organism has five pairs of chromosomes as a diploid number, and an examination of many meiotic cells reveals that these pairs, from longest to shortest, have the following chiasma frequencies:

3.72, 2.60, 2.10, 1.50, and 1.00. Assuming that a chiasma always represents a cross-over between two of the four chromatids, what is the maximum genetic map length of these chromosomes?

4   If two genes are more than 50 map units apart on a chromosome, what do we need to know to determine their true map distance accurately?

5   Genes *a*, *b*, and *c* are linked in that order; *a* is 40 map units from *b*, and *b* is 20 map units from *c*. In a testcross, only 5.5 per cent double cross-overs were detected among the offspring. How much interference was there?

6   Two homozygous parents were mated, and the F_1 offspring were testcrossed to the double recessive. The testcross progeny were counted as follows: *AB*–22; *Ab*–29; *aB*–31; *ab*–18. Determine the genotypes of the parents, and the amount of crossing-over that has occurred.

7   Consider meiosis in *Neurospora*, and suppose that a gene *m* is not too far from the centromere on one of the pairs of chromosomes. You have mated a strain with *M* to one with *m*. How can you determine the genetic map distance from the centromere to the gene? Diagram your results.

8   The genes *x*, *y*, and *z* are linked on the same chromosome, but not necessarily in that order. The homozygous parents were mated to produce an F_1 offspring, and this F_1 was then testcrossed to a homozygous triple recessive. The following progeny were obtained:

XYZ–230   What are the genotypes of the parents?
XYz–26   What is the order of the genes on the chromosome?
XyZ–145   What are the map distances between the genes?
Xyz–98   How much interference was there?
xYZ–102   Why do we list only part of the genotype of the testcross progeny?
xYz–155
xyZ–24
xyz–220

selected readings   See the books listed at the end of the previous chapter.

sex as
an inherited
trait
chapter 18

The most obvious difference among human beings is that of sex. Maleness and femaleness represent a pair of contrasting, mutually exclusive characters (Fig. 18.1), and they serve well to illustrate the basic mode of inheritance and its physical, cellular basis.

Before the principles of inheritance were firmly established, biologists generally viewed heredity as some mechanism that operated in the offspring and blended the characters of the two parents. Common experience tells us that this is true for such characters as weight, height, and skin color. For example, the child of a white–Negro parentage is colored (intermediate shade), and it makes no difference which parent is white and which is Negro. Sex, however, is not a "blended" character. Except in rare instances, human offspring are either male or female.

The hereditary basis of sex was experimentally established about 1900. By this time microscopic techniques had been developed, and the behavior of chromosomes was well understood. It was around this time that biologists noticed that the males and females of certain insects differed in their chromosome number. The male generally had one less chromosome in its somatic cells than did the female. This odd chromosome in the male became known as the **X-chromosome,** simply because its role was initially poorly understood. Figure 18.2 shows the X-chromosome as it appears in a grasshopper.

273

Figure 18.1 The markings that distinguish a male mallard drake (lower left) and the female duck (upper right) are obvious in this photograph of a pair of mallards. Not all animals show such clear-cut sexual differences. The horns of a ram, the tail of the peacock, and the mane of a lion are distinguishing features of maleness; but bears, rabbits, and horses, for example, do not show such striking differences.

During meiosis the paired chromosomes are equally distributed to the gametes—except for the X-chromosome. Because it is without a pairing partner, it is found in only half the total number of sperm cells. When this was first worked out, biologists thought that here, obviously, was a mechanism that could account for the equality of numbers of males and females in sexually reproducing populations. If so, then the female must have a pair of X-chromosomes, while the male has but a single one. It also follows that all eggs must have an X-chromosome in their haploid set, and that sex is determined at the time of fertilization:

Figure 18.2 The single X-chromosome in this photograph of a mid-prophase meiotic cell from the testis of a male grasshopper can be seen as the dark-staining rod in the lower left. Since there is only one X-chromosome, the grasshopper males have an XO constitution, the females an XX constitution.

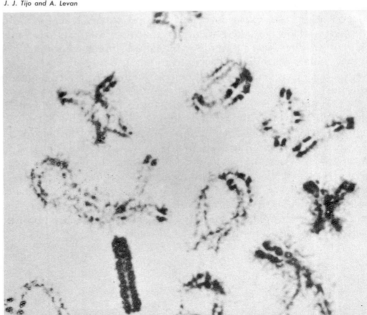

an *X*-bearing sperm combines with an *X*-bearing egg to produce a female zygote; a sperm lacking an *X*-chromosome combines with an *X*-bearing egg to produce a male zygote.

These assumptions have been fully borne out. Female organisms bearing the *X*-chromosome can be designated **XX,** while the males are labeled **XO** (the **O** stands for the lack of an *X*). Figure 18.3 shows the inheritance of *X*-chromosomes. Notice that the male offspring always receive their *X*-chromosomes from their mother. Female offspring receive an *X*-chromosome from each parent.

In many species of organisms, including man, both males and females have the same number of somatic chromosomes. In these males, the *X*-chromosome has a pairing partner, which may be either similar to it in shape and size or strikingly dissimilar. This is known as the **Y-chromosome** (Fig. 18.4). In the male the chromosomal composition is designated **XY** instead of **XO,** while the female remains *XX* as before. The *Y*-chromosome may or may not exert a sex-determining influence; in either event, two types of sperm are produced—those bearing an *X*- and those bearing a *Y*-chromosome.

Proof that sperm of two kinds is produced in equal numbers is provided by *Protentor,* an insect of the *XX–XO* type. At the end of meiosis, the **spermatids** (those cells which will be transformed into motile sperm) are of two kinds. One shows very little nuclear structure; the other has a deep staining body within it (Fig. 18.5). This body has been identified as the *X*-chromosome, which has a tendency to stain darkly during interphase.

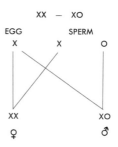

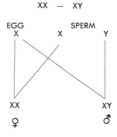

Figure 18.3 Diagrams of the inheritance of *XX-XO* and *XX-XY* species.

Figure 18.4 The chromosomes of a normal human male. A cell was photographed and the chromosomes were cut out, arranged in homologous pairs, and numbered according to size. The human male has an *XY* sex-determining system. The tiny *Y*-chromosome is shown at the bottom right. The *X*-chromosome, which is difficult to identify positively, is one of those in the second row.

Courtesy Dr. Barbara Migeon

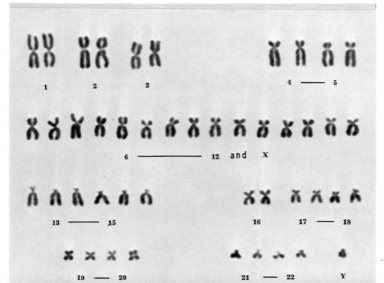

Figure 18.5 This section cut through the testis of the bug, *Protentor,* shows the haploid spermatid cells before they have differentiated into motile spermatozoa. About one-half the cells show a dark-staining body, which has been identified as the *X*-chromosome, the other half do not.

The remainder of the chromosomes in a cell are known as **autosomes.** Of the 46 chromosomes in humans, females have 22 pairs of autosomes plus one pair of *X*-chromosomes. Males have the same number of autosomes plus one *X*-chromosome and one *Y*-chromosome. We shall see that the autosomes may also exert a sex-determining influence. Not all sexually reproducing species show the female to be *XX* and the male to be *XY* or *XO*. Birds, moths, butterflies, fishes, and some reptiles and amphibia are the reverse; that is, the female is *XY* and the male is *XX*.

SEX DIFFERENTIATION The *X*- and *Y*-chromosomes provide us with only part of the genetic basis for the development that transforms a fertilized egg into either a male or a female. Sex is the **phenotypic** character, and like all phenotypes it is dependent upon a genotype. Sex, as a distinct character, becomes a reality only after a long and rather complex period of development. The early embryo in human beings, which arises from the zygote through cell division, is neither male nor female. Essentially it is neutral, and it remains neutral even after the early development of **gonads,** or sex organs. This is so because the embryonic gonad consists of two parts: (1) an external layer of tissue, the **cortex,** which can develop into an ovary; and (2) an internal mass, the **medulla,** which can develop into a testis. In addition, two pairs of sex ducts are present in the neutral embryo. One of these is the **Müllerian duct,** which persists if the individual becomes a female. The other, the **Wolffian duct,** persists if the individual becomes a male (Fig. 18.6). There are additional embryonic parts, which are later transformed into the external genitalia. But at an early stage in embryogeny these parts are in such a primitive state that they can be transformed into either male or female organs, depending on genotypic influences.

It is only after the neutral stage of gonadal development that the specific sex determiners take over and stimulate the development of sex organs that are distinctly male or female (Fig. 18.7). In an embryo whose chromosomal constitution is *XX,* and, therefore, potentially female, the cortical part of the gonad begins to develop, while the medullary part becomes inconspicuous and may even degenerate. In this way the neutral gonad is transformed into an ovary. If the embryo, on the other hand, happens to be *XY,* the reverse situation takes place. The medullary part of the gonad enlarges and begins to differentiate, while the cortical portion disappears; the neutral gonad is transformed into testes.

It is at a particular point in normal development, therefore, that the genetic constitution of the individual becomes decisive. In a morphological sense, this is the time when either the cortex or the medulla enlarges and the other part degenerates. In a

Figure 18.6 In early embryonic development the gonads appear the same in both sexes. Males and females alike have comparable ducts. In females, the Müllerian duct will persist, and the Wolffian duct degenerate. The reverse occurs if the individual is to be a male.

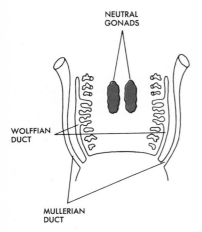

NEUTRAL
GONADS

WOLFFIAN
DUCT

MULLERIAN
DUCT

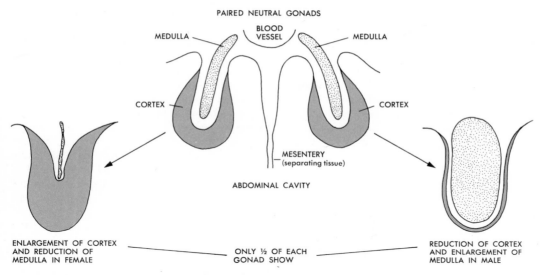

Figure 18.7 This diagram shows the developmental change that takes place in each gonad as sexual differentiation occurs.

genetic sense the "decision" was probably made much earlier, since a whole chain of biochemical changes must precede any obvious structural change. The chromosomes, consequently, act as a trigger mechanism to start off one or the other of a set of developmental reactions that leads either to maleness or femaleness. Any upset or delay in these reactions leads to an individual who is undeveloped sexually, or who falls into a "twilight" zone of intersexuality. Later on, other influences tend to reinforce the established sexual pattern so that maleness and femaleness can often be recognized visually as well as by the type of gonad and gamete produced. The plumage of some birds, which differs in the two sexes, is an example of a secondary sex character (Fig. 18.1).

In human beings, sexual differentiation is by no means complete at birth. Secondary sex characters develop as beginning adulthood is reached, that is, during the period of puberty. These may involve anatomical differences which change the larynx and consequently change the character of the voice; differences in general body conformation, which influence the structure of the pelvic region and the mammary glands; and differences in hair growth and distribution, the male generally being more hairy than the female. All these are influenced by the **gonadal hormones,** chemical substances that are produced as development proceeds. The important thing to remember, however, is that these influences begin in the individual with the initial genetic constitution present when the fertilized egg begins to grow.

277

sex linkage

The distribution of the *X*- and *Y*-chromosomes in meiosis, and their combinations in the fertilized zygote, account for the fact that there are approximately as many males as females. Sex is but one of many inherited characters, however. Do we need to assume that the sex chromosomes are concerned solely with maleness and femaleness? In many organisms, including man, the answer is "no." They may govern other characters, unrelated to sex. If the factors governing the appearance of other characters are carried on the *X*-chromosome, then these particular characters should show a relationship to sex. We refer to such a relationship as **sex linkage.** Those factors carried on the autosomes should show transmission relationships unrelated to the sex of the individual.

Let us use an example from *Drosophila melanogaster*. In 1910 the American geneticist T. H. Morgan, then at Columbia University, found that one of the fruit flies raised in a bottle culture had white instead of the usual red eyes (Fig. 17.1). Morgan isolated this fly and eventually obtained a strain of true-breeding white-eyed flies. When a white-eyed fly was crossed with a con-

Figure 18.8 A cross in which the mutant gene white (*w*) is in the male parent, and hence in only one *X*-chromosome. Of the F_2 progeny, only one-half of the males, and none of the females, are white-eyed. The *Y*-chromosome, although partially homologous with the *X*-chromosome, lacks the genes found in the *X*-chromosome, and hence does not contribute to sex linkage.

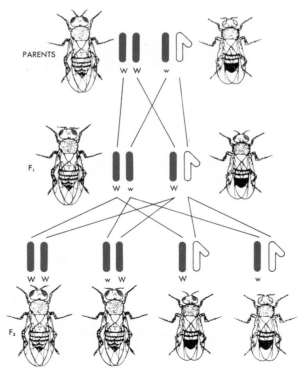

ventional red-eyed fly, the eye color of the offspring depended on whether the white-eyed parent was male or female. The results suggest that eye color is transmitted in much the same way as sex. The details of these crosses are shown in Figs. 18.8 and 18.9.

From the cross of a white-eyed male with a homozygous red-eyed female, the first generation (F_1) cross shows all red-eyed individuals. From this we can assume that whatever is determining red eyes prevails over that factor controlling the development of white eyes—in other words, red is dominant over the recessive white. When these F_1 individuals are bred, white reappears in the F_2 generation, but only in one quarter of the flies. The diagram also shows that all the white-eyed individuals are males. White-eyed females are absent from both F_1 and F_2 generations. If we diagram the cross in such a way (Fig. 18.8) that the red-eye character is designated by a large W, and the white-eye character by a small w, we find that in the F_2 generation the females must be of two different kinds—one homozygous WW, the other heterozygous (Ww). We can further support our assumption by more breeding. One type of F_2 female, when crossed with a white-eyed male, produces only red-eyed offspring. The other type of female produces half white- and half red-eyed offspring. The latter female must, therefore, carry white in a

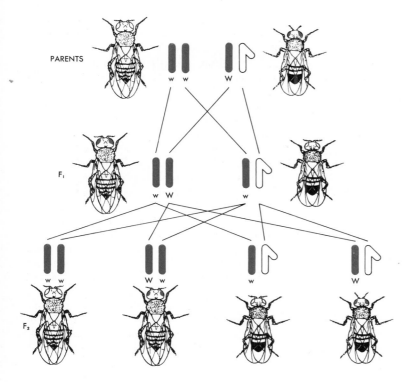

Figure 18.9 The results of a cross in *Drosophila* when the female parent carries the mutant gene *white* (w) in both of her X-chromosomes. In the F_2 generation one-half of the females and one-half of the males are white-eyed. Compare these results with those of Fig. 18.8. (w = white; W = red.)

PARENTS

F_1

F_2

recessive heterozygous condition, but its heterozygosity is revealed only by further breeding trials. The white-eye character is not lost when heterozygous; it is merely suppressed in its action.

If we cross a red-eyed male with a white-eyed female, we get different results (Fig. 18.9). Among the F_1 offspring, all the females are red-eyed and all the males are white-eyed. When these are bred, the F_2 offspring show red- and white-eyed characters in equal proportions, and with an equal distribution among the sexes. If the F_2 females are backcrossed once again to a white-eyed male, the white-eyed female will not produce red-eyed offspring but the red-eyed female produces half red- and half white-eyed offspring. It therefore appears that the white-eyed F_2 female is pure in respect to the characters shown, but that the red-eyed female, again, is heterozygous. The red- and white-eyed males bred back to a white-eyed female would produce either white- or red-eyed offspring, respectively. Therefore, it appears that these F_2 males are pure for the character they display.

Let us analyze what we have just been through. The male transmits the eye character to his grandsons through his daughters, but never directly to his sons. The mother transmits the character to both sons and daughters. A glance at Figs. 18.8 and 18.9 will reveal this, if we assume that the factor for eye color is carried on the X-chromosome and that the Y-chromosome plays no part in eye color. Only the daughters get an X-chromosome from the father, but both sons and daughters receive an X-chromosome from the mother. Therefore, we can conclude, as Morgan did, that the factor for white eye color, as well as red eye color, is carried on the X-chromosome, and that red and white are allelic to each other in a genetic sense. The factor, or gene, exists in two forms, and a single X-chromosome carries only one at a time. A particular genetic factor, then, must be located on a particular chromosome. It remained for C. B. Bridges of Columbia University to prove that this is so.

Bridges made a series of crosses between white-eyed females and red-eyed males of *Drosophila*. Ordinarily, if everything goes as it should, the offspring would be white-eyed males and red-eyed females, as indicated in Fig. 18.9. Out of a large number of offspring, however, he found an occasional red-eyed male or white-eyed female in the F_1 generation. This would seem to suggest that the original hypothesis (that the factors for red and white eye color are carried on the X-chromosome) was incorrect. Instead, however, it turned out that these flies provided proof of the conclusiveness of Morgan's hypothesis. Bridges suspected that what had happened was that the X-chromosomes in the female had failed to segregate from each other during meiosis. This failure resulted in some eggs having two X-chromosomes instead of only one, and other eggs having no X-chromosomes. He called this abnormal process **nondisjunction** (Fig. 18.10).

Bridges could now predict that if a two-X-egg with both of the X-chromosomes carrying the factor for white were fertilized

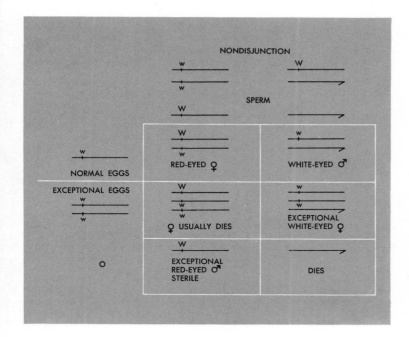

Figure 18.10 This diagram shows what happens when nondisjunction of the X-chromosomes occurs during egg formation in females of *Drosophila* and forms exceptional eggs. The discovery, made by C. B. Bridges, proved that a particular gene is carried on a particular chromosome.

by a *Y*-bearing sperm, which does not carry any eye-color factors, a white-eyed female should be produced. Similarly, a no-*X*-egg fertilized by an *X*-bearing sperm carrying the factor for red should produce a red-eyed male. The transmission in this instance was directly from father to son, whereas in the first case it was the transmission of the white factor from mother to daughter. If this were true, then each of the exceptional white-eyed F_1 females should have, in addition to the two *X*-chromosomes, a *Y*-chromosome in each cell of her body. The red-eyed F_1 exceptional males should not have any *Y*-chromosomes in their cells. To test this, Bridges made a microscopic examination of the cells. The chromosomal composition of these exceptional flies was exactly as he had predicted. Every time there was an irregularity in the pattern of heredity, there was a corresponding irregularity in the distribution of chromosomes.

Bridges' study provided a very convincing demonstration of the close relationship of genes to chromosomes. Additional evidence was supplied by L. V. Morgan, wife of T. H. Morgan. She had experimented with a strain of *Drosophila* in which the two *X*-chromosomes of the mother were always passed on to the daughters, never to the sons. It is difficult to imagine how this could happen unless the two *X*-chromosomes were somehow attached to each other, and always remained together in the egg nucleus, or passed together into the polar body. It is now known that they are actually attached to each other. Females bearing this type of chromosome are known as **attached-X females;** in

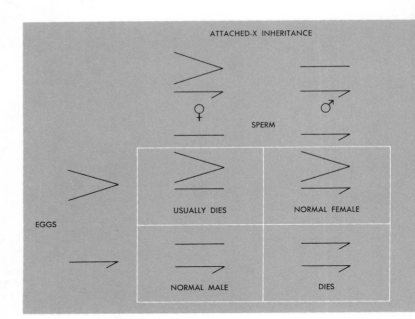

ATTACHED-X INHERITANCE

SPERM

USUALLY DIES | NORMAL FEMALE

EGGS

NORMAL MALE | DIES

Figure 18.11 Inheritance in attached-X females of *Drosophila*: these females also contain a Y-chromosome, but this has no influence on sex determination or sex linkage.

addition, they usually carry a *Y*-chromosome. Figure 18.11 shows the type of transmission that takes place when an attached-*X* female is mated with a normal male.

Of the four kinds of zygotes produced, those with three *X*-chromosomes die, as do those lacking an *X*-chromosome. Thus, the two surviving types are exactly like their parents. Attached-*X* transmission, therefore, stands in contrast to the usual pattern of sex-linked inheritance. Once again the close relationship between gene and chromosome transmission is convincingly demonstrated.

The attached-*X* studies also provide evidence of the nonimportance of the *Y*-chromosome in *Drosophila* sex determination. The *Y*-chromosome can be found in either males or females without affecting sex. However, we cannot generalize and say that the *Y*-chromosome is *always* without such effect. Males lacking a *Y*-chromosome are sterile, so the *Y*-chromosome has a function to perform. As we shall see later, it appears that mammals, including man, have a sex-determining mechanism in which *Y*-chromosomes play a prominent role.

Let us now examine one case of sex-linked inheritance in humans—color blindness. Say that a woman who is heterozygous for color blindness marries a normal man (Fig. 18.12). Their daughters would all be normal, although there is a 50:50 chance that each daughter is heterozygous, and thus able to transmit the color-blindness character to sons in a later generation. On the other hand, one half of all the male children of the marriage are likely to be color blind since one of the *X*-chromosomes carries

282

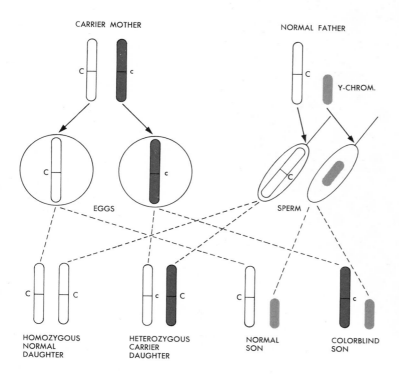

CARRIER MOTHER

NORMAL FATHER

Y-CHROM.

EGGS

SPERM

C — C
HOMOZYGOUS
NORMAL
DAUGHTER

c — C
HETEROZYGOUS
CARRIER
DAUGHTER

C
NORMAL
SON

c
COLORBLIND
SON

Figure 18.12 Sex-linked inheritance of color blindness in man: normal vision is designated by C, color-blindness by c. The Y-chromosome carries no comparable gene, and so is without influence in this respect.

the factor for color blindness. The *Y*-chromosome does not play any part in color blindness.

The number of known characters found on the *X*-chromosome of man that show a sex-linked inheritance of the type just described are relatively few. In *Drosophila* many more characters are known. At the present time several hundred are known, involving the wings, bristles, and eye color, as well as more subtle characters that can be determined only by careful experimentation. Any character, regardless of whether it is involved in the determination of sex, is always sex-linked if it is borne on the *X*-chromosome.

In birds, butterflies, moths, reptiles, some fish, and amphibians, the sex-determining mechanism is the reverse of that found in *Drosophila*. The female is the heterozygous sex (*XY*) whereas the male carries two *X*-chromosomes. The study of certain sex-linked genes has borne this out.

HOW CHROMOSOMES
DETERMINE SEX   We have said that species can have an *XX–XO* or an *XX–XY* sex-determining mechanism. The real situation, however, is not quite so simple. In Fig. 18.13 are the chromosomal conditions observed in *Drosophila* and man. Each haploid set of autosomes is designated by the letter *A*. A glance at the *Drosophila* data indicates that while the principal action

CHROMOSOME CONSTITUTION AND SEX IN DROSOPHILA
AND MAN (A = HAPLOID SET OF AUTOSOMES)

| ——DROSOPHILA—— | | | ——MAN—— | | |
CHROMOSOME CONSTITUTION	SEX	X/A RATIO	CHROMOSOME CONSTITUTION	SEX	X/A RATIO
2A XXX	SUPERFEMALE	1.5	2A X		0.5
2A XX			2A XX	FEMALE	1.0
2A XXY	FEMALE	1.0	2A XXX		1.50
3A XXX			2A XXXX		2.00
3A XX	INTERSEX	0.67	2A XY		0.5
3A XXY			2A XYY	MALE	0.5
			2A XXY		1.0
2A X			2A XXYY		1.0
2A XY	MALE	0.50			
2A XYY					
3A X	SUPERMALE	0.33			

Figure 18.13 The relation of chromosomal constitution to sex in *Drosophila* and man. A haploid set of autosomes is designated by the letter A. In *Drosophila*, the X/A ratio determines maleness and femaleness; in humans, maleness and femaleness are determined by the presence or absence of a Y-chromosome. It is probable that human femaleness is influenced by the autosomes, but we do not know at the present time.

of the *X*-chromosomes is to push development in a female direction, the autosomes push development in a male direction. The sex of an individual is determined, therefore, by a balance between the number of *X*-chromosomes and the number of sets of autosomes. If the *X*:*A* ratio is 1 or greater, femaleness results. If the ratio is 0.5 or less, the individual is a male. However, a ratio of 2*X*:3*A*, or 0.67, is an **intersex.** This is an individual who shows an intermediate kind of maleness or femaleness.

In man the role of the *X*-chromosome in sex determination is not clear. An *XO* individual develops into a female, not into a male as in *Drosophila*. In addition, the *XO* individual is an abnormal female. The ovaries are abortive or missing, the individual is sterile, intelligence may be subnormal, and there are a number of structural abnormalities, particularly of the heart. In the mouse, an *XO* animal is a fertile female and indistinguishable from an *XX* animal. The *X*-chromosome, at least in the mouse, and possibly in man, does not seem to have any pronounced male- or female-determining tendencies.

No *YO* individuals are known to exist in man or in other mammals. The *X*-chromosome is necessary for survival. Any mammal with a *Y*-chromosome, however, is a male. This again is different from the situation in *Drosophila* where the *Y* is without any sex-determining influence. In *Drosophila*, an *XXY* individual is a normal, fertile female. In man, the same chromosomal situation results in a sterile, abnormally underdeveloped male. It is of interest to point out that the silkworm, *Bombyx mori*, carries sex-determining factors in its *Y*-chromosome. In regard to sex determination, man is therefore more like a moth than a fly.

It is, of course, tragic that *XO* and *XXY* individuals exist in a human population. Nondisjunction of chromosomes at

SEX-CHROMOSOME CONSTITUTION	REMARKS ON APPEARANCE OR BEHAVIOR
MALE	
XYY	SOMETIMES SOCIALLY AGGRESSIVE
XXY	KLINEFELTER'S SYNDROME: VARIABLE MENTAL RETARDATION, VARIABLE ABNORMAL PHYSICAL AND BEHAVIORAL ASPECTS
XXXY	LIKE KLINEFELTER'S
XXXXY	SEVERELY RETARDED
XXXYY	LIKE KLINEFELTER'S
XXYY	LIKE KLINEFELTER'S
FEMALE	
XO	TURNER'S SYNDROME: SEE TEXT
XXX	MENTAL RETARDATION
XXXX	MENTAL RETARDATION
XXXX	MENTAL RETARDATION
STRUCTURAL CHANGES	
X + FRAGMENT Y	MAY BE EITHER MALE OR FEMALE; VARIABLE FROM NORMAL TO ABNORMAL
X + FRAGMENT X	FEMALE; NO SEVERE MENTAL DEFECTS
X + LARGER X	LIKE TURNER'S

Figure 18.14 Some sex chromosome variants in human beings (see also Fig. 18.13).

meiosis is undoubtedly the cause, but nothing as yet can be done to alter this condition, once it arises. It is through our understanding of these cases, however, that we gain some of our knowledge of inheritance in the human race.

What we have just described points out the fact that, while sex is an inherited character, it is not determined by a single gene in the same way that eye color is. Sex is determined by whole chromosomes rather than by single genes, at least in diploid organisms. No distinct sex gene has been found in *Drosophila*, man, or mouse. However, the *A* and *a* mating types found in *Neurospora* can be thought of as a primitive kind of sex, not yet differentiated into maleness or femaleness, and these are determined by single genes. Sex probably arose in some early organism as a mutation, and its inheritance later became a chromosomal rather than a genic feature of inheritance.

summary

Sex is an inherited trait, but rather than being determined by a single gene, it is determined by a balance between the sex chromosomes and the autosomes. In *Drosophila*, the *X*-chromosome is female-determining, the autosomes male-determining; the

Y-chromosome is without influence on sex and is concerned only with fertility of the male. In man, the *Y*-chromosome is male-determining, and it would appear as if the autosomes rather than the *X*-chromosome were female-determining. It is probably more accurate, however, to state that the presence of the *Y*-chromosome determines maleness, its absence femaleness.

Genes located on the sex chromosomes follow a type of inheritance linked to sex, and are called **sex-linked.**

for thought and discussion

1   The sex ratio in human populations is not equal; there are 105 boys born to every 100 girls. What reasons can you give to explain this difference? At about age 21, males and females are about equal in number. What genetic explanation can you offer for the shift in ratio?

2   Does the inheritance of *X*-chromosomes suggest that sons or daughters are likely to be more similar to their mothers than their fathers in certain characteristics? Why?

3   The attached-*X* chromosome in *Drosophila* is peculiar in that the order of genes in the two arms is in reverse; that is, ABCDEFG–centromere–GFEDCBA. By folding at the centromere, the two arms can synapse and cross-over in meiosis. Yellow (*yy*) bodied offspring can be obtained from a female known to be heterozygous for this character (*Yy*). Can you show by means of a diagram how this can happen?

4   Color blindness for red–green is due to a sex-linked recessive gene in man. A father is color blind, and the mother is normal with no evidence of color blindness in her ancestry. What is the probability that their son will be color blind? Their daughter? Their grandson, if the grandson is a son of the daughter?

5   In cats, which are *XX–XY*, the sex-linked gene *Y* determines black when homozygous, yellow when homozygous recessive, and tortoiseshell or calico (mixture of black and yellow) when heterozygous. The *Y*-chromosome lacks the gene.
    (a) A calico mother has a litter of six: one yellow male, two black males, one yellow female, and two calico females. What is the genotype of the father?
    (b) A calico mother has a litter of three black females. If the father was black, how often would you expect the same result to happen again?
    (c) A yellow mother has a litter consisting of two yellow and three calico offspring. What is the genotype of the father, and what is the sex of the offspring in relation to color?

6   A convenient way of determining the rate at which genes in *Drosophila* mutate when exposed to x-rays or chemicals is to treat males and mate them to attached-*X* females. What advantage is there to such a technique? What kinds of mutations would be

missed in this sort of experiment? If the treated males were mated to normal *XX* females, what procedures would have to be followed to obtain the same information? Can you think of any advantage of the second technique over the first?

7   In chickens, the barred pattern of feathers of the Plymouth Rock breed is due to a dominant sex-linked gene, *B*. An autosomal gene *F* affects the feathers such that *FF* has brittle, curly feathers, *Ff* has feathers that are slightly curly but otherwise normal, and *ff* has feathers that are fully normal. A number of barred, normal-feathered hens are mated to a nonbarred, slightly curly cock, and 200 hatchable eggs are produced. What is the expected frequency of offspring in terms of feather character and sex?

See books listed at the end of Chapter 16.

selected readings

heredity
and
environment
chapter 19

The inheritance test is the major tool of the geneticist in his study of the transmission of characters from parents to offspring. The sharper the contrast between a particular pair of characters, the easier it is to determine the pattern of inheritance. We have used, for example, such contrasting characters as maleness and femaleness and red and white eyes in *Drosophila*. These characters are

Figure 19.1 Three members of the cat family: the Canada lynx, leopard, and house, or tabby, cat. Among other characteristics, they are readily distinguishable by size, fur color and markings, disposition and native home.

remarkably constant. Because they do not have a wide range of variation, no one has difficulty in identifying the phenotype.

We can extend this idea of contrast in individual characters to groups of characters (Fig. 19.1). An oak tree has an array of characters—leaf, bark, wood, and growth habit—that enables us to distinguish it from a maple or pine tree. Even if the oak tree grew in such widely different environments as a shady forest, a dry open hillside, or a swamp, the characters remain constant enough for us to identify it as an oak. Similarly, a cocker spaniel can be identified not only as a vertebrate, a mammal, and a dog, but a dog different from a setter, a foxhound, or a wirehaired terrier. No change in diet or environment will alter the general character of a cocker so that it will be mistaken for another breed.

Frank Stevens
From National
Audobon Society

Joyce R. Wilson
From National Audubon Society

These facts stress the importance of the role heredity plays in determining individual characteristics. We often remark on this in everyday speech by saying, "He has his mother's eyes," or "She has her father's chin," or "She favors the Smith rather than the Jones side of her family." Rarely do we say that the environment plays an important role in determining the phenotype of an individual. Yet all of us are aware of it. What about the relation of your diet to your weight? We know that some individuals readily gain or lose weight. Others change hardly at all, no matter what they eat.

Every individual organism, of course, inherits certain things from its parents. These things determine its genotype. Every organism also grows and develops *in* an environment, never in the absence of one. The two influences, genotype plus environment, give rise to the individual's phenotype. Beethoven was a musical genius. Would he have been had he not had access to a piano? Here are some examples from biology of how we can separate and then determine the role of each of these two influences.

Some people are born with a kind of diabetes that is hereditary. They lack **insulin,** a hormone that regulates the level of sugar in the blood. Without insulin, which is formed in the pancreas, such individuals would go into diabetic shock and die. The internal environment of these people can be corrected and controlled by periodic injections of insulin, but the genetic deficiency is not altered by this treatment. It is the internal environment—not the genetic deficiency—that determines whether the individual lives or dies.

An equally dramatic example occurs in human females. A single abnormal gene leads to the absence of a certain chemical **(21-keto steroid)** needed for normal development and feminization of the individual. Without medical treatment, a young girl with this genetic deficiency would have a tendency toward masculinity: the ovaries would fail to develop, the mammary glands would not enlarge, facial hair would become prominent. If this deficiency is recognized and treated before adulthood, there can be striking results. The ovaries become fully functional, development of the mammary glands proceeds, facial hair disappears, and the individual can lead a normal life. The treatment, however, must continue for as long as the individual wishes to preserve her feminine traits. Here is another example of how the environment can drastically alter a phenotype. The phenotypic alteration in this instance involves both appearance and reproductive capabilities.

The same problem has been approached experimentally by using plants that can be divided. The subdivisions are then grown in different environments. Remember that the original plant always comes from a single fertilized egg, so all its cells have the same set of genes. This, of course, is true also for the subdivisions.

How do the subdivisions behave in different environments? Which characters are altered, and which remain unchanged?

We will use the cinquefoil *Potentilla glandulosa* as an example. This species is widespread, and in California it grows from sea level to alpine heights (10,000 feet) in the Sierra Nevada Mountains. At sea level the species is active throughout the year, but its primary growth begins in late January. The seeds are mature by mid-July (about 175 days). At sea level, therefore, the plant is timed to a slow rhythm of seasonal development and a long season. At Mather the season is from mid-April to the end of July (about 105 days). At Timberline the season is from July into September (less than 90 days). Plants from each station were selected. Each plant was then subdivided into three parts, and then transplanted at each of the three stations. Figure 19.2 shows growth at each station.

AT STANFORD
EL. 100 FEET

AT MATHER
EL. 4,600 FEET

AT TIMBERLINE
EL. 10,000 FEET

Jens Clausen, Carnegie Institution of Washington

Figure 19.2 Shown here are four altitudinal types of the cinquefoil, *Potentilla glandulosa.* They were taken from four different locations in the Sierra Nevada Mountains and were subdivided into three parts, each part grown in a different environment. The origin and altitude of the four types are as follows: top row, 10,000 feet at Timberline; second row, 4,600 feet at Mather; third row, 2,500 feet in the foothills; bottom row, 900 feet near the coast. The contrasting environments are at Stanford (1,100 feet), Mather (4,600 feet), and Timberline (10,000 feet). Each type, in general, grows best in the region from which it was originally collected, and less well at other locations. This indicates that each type is adapted to a particular locality.

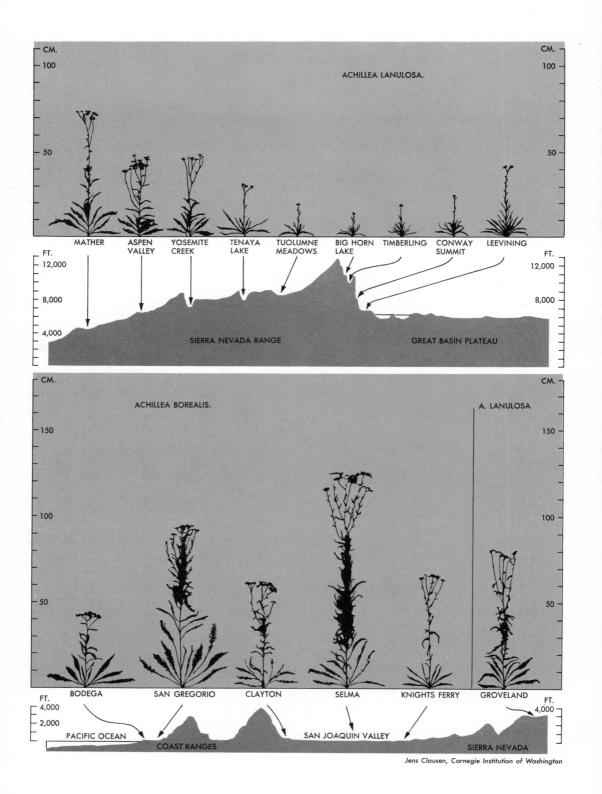

ACHILLEA LANULOSA.

MATHER · ASPEN VALLEY · YOSEMITE CREEK · TENAYA LAKE · TUOLUMNE MEADOWS · BIG HORN LAKE · TIMBERLING · CONWAY SUMMIT · LEEVINING

SIERRA NEVADA RANGE · GREAT BASIN PLATEAU

ACHILLEA BOREALIS. · A. LANULOSA

BODEGA · SAN GREGORIO · CLAYTON · SELMA · KNIGHTS FERRY · GROVELAND

PACIFIC OCEAN · COAST RANGES · SAN JOAQUIN VALLEY · SIERRA NEVADA

Jens Clausen, Carnegie Institution of Washington

Charlie Ott From National Audubon Society

The environment clearly altered the growth habit. The length of the flowering period was also changed, being lengthened at low altitudes, and hastened at high altitudes. However, each of the plants tended to retain certain characters that are constant, regardless of the environment; for instance, arrangement of the leaflets, openness or compactness of the plant, and the shape of the petals. This is also shown for two different species of yarrow growing at different altitudes (Fig. 19.3).

The examples we have considered demonstrate that a single genotype can exhibit different phenotypes when environmental conditions are varied. But, you may wonder, "Can two genotypes exhibit the same phenotype?" The answer is "Yes" when environmental conditions are varied. In *Drosophila,* for example, a mutant gene *y* (yellow) causes the body of the fly to be yellow instead of gray. But a wild-type gray fly, with the dominant *Y* gene, turns yellow if silver nitrate is included in its food. Such a fly is called a **phenocopy,** since it copies the phenotype of a mutant strain. The yellow color in this instance would not be transmitted to any offspring, because the genes are unaffected by the silver nitrate.

The same kind of thing happens in rabbits. Figure 19.4 shows three rabbits of different coat color. Inheritance tests show that the fully colored rabbit (bottom) has a dominant phenotype. Its genotype may be **CC, Cc,** or **CC^h**. The solid white individual is *cc,* while the Himalyan is either $C^h C^h$ or $C^h c$. The order of dominance of these three alleles is $C \rightarrow C^h \rightarrow c$.

Walter Dawn

If we vary the temperature as a means of changing the environment, the phenotypes of the white and colored rabbits stay constant. Temperature seems to have no influence whatever on the genetic expression of *C* or *c.* In the Himalayan form, however, coloring pattern is temperature sensitive. Notice that the black hairs are located at the extremities—feet, nose, and ears. This pattern of distribution of black hair by itself does not necessarily prove anything about temperature sensitivity. Even though the body extremities—feet and ears—have a lower temperature (because of rapid heat loss) than the rest of the body does, we cannot accept this alone as proof of temperature sensitivity. However, if hair is plucked from the back, the new hair coming in

Figure 19.4 Three color patterns in rabbits are shown here. Top: the white phenotype has a cc genotype. Bottom: the black phenotype has one of three possible genotypes, CC C^h or Cc. Middle: the Himalayan phenotype has either a $C^h C^h$ or $C^h c$ genotype.

Walter Dawn

293

is white if the temperature is high, but it is black if the temperature is low. That is the kind of proof needed for us to make a definite statement about temperature sensitivity, which in this instance can be related to the formation of **melanin,** a black pigment. The chemical steps leading to the production of melanin are catalyzed by a series of enzymes. One of the enzymes is temperature sensitive and cannot function at high temperatures. Melanin, therefore, can be produced only when the temperature is lowered, artificially in the case of the plucked hair, naturally in the extremities.

Can we always be so certain of how much of a role the genotype or environment plays in determining the phenotype? For example, the average American soldier of World War II was several inches taller and somewhat heavier than his counterpart in World War I. Amherst College has recorded the weight and height of its students since it was founded. Today's freshman averages 4 inches taller and 20 pounds heavier than the freshman of a previous generation. Is this due to a change in the genotypes of American men, so that there are more genes favoring tallness? Or is it the result of better nutritional standards? How do we find out? We might suspect that height is genetically, not environmentally, determined, but can we be certain? Weight, on the other hand, is a character easily changed by diet. One of the best ways we have of answering such questions is to make use of twin studies.

TWIN STUDIES Twins are either **fraternal** or **identical.** If they are fraternal, they are **dizygotic (DZ),** arising from two separate eggs fertilized by two separate sperm. Despite their birth relationships, they are no more like each other than they would be like their other brothers and sisters born singly, except, of course, that they are of the same age. Identical, or **monozygotic (MZ)** twins, on the other hand, develop from a single fertilized egg. They are, consequently, of the same sex and have the same genotype (Fig. 19.6). Division of the single fertilized egg into two units does not occur at the first cleavage division, as was once believed, but only after the embryo consists of many thousands of cells. Complete separation of the cellular mass into two parts is necessary for twin formation. Incomplete separation leads to Siamese twin formation, with the union between the two being of varying degrees. As Fig. 19.5 shows, the degree of union often interferes with full development, and a variety of abnormalities can result.

Among human twin births, which occur at a rate of about one in 80, the ratio of identicals to fraternals is about $1:2$. Higher multiple births of triplets, quadruplets, and quintuplets also occur, but their occurrence is too rare to provide a ready source of individuals for study.

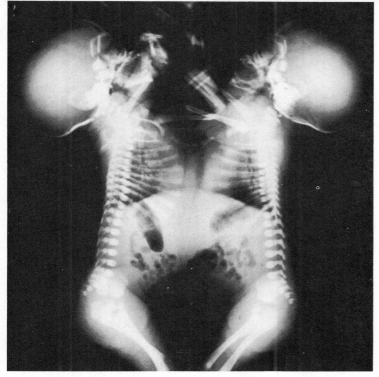

Figure 19.5 An x-ray photograph of Siamese twins at birth. Joined along the chest and abdominal regions of the body, they exhibit various degrees of shared and separate internal organs and bone structures. An operation, in this instance, would not permit complete separation and survival of the twins.

The important thing for us to remember here is that identical twins provide the geneticist with a measure of control over two important variables—inheritance and environment. Consequently, the most important source of information has been studies of identical twins who have been separated from each other early in life, and who have developed in quite different environments. Similarities exhibited by such identical twins would point to the independence of heredity from the environment in determining specific characteristics. Dissimilarities, on the other hand, do not minimize the importance of heredity. Instead, they indicate how the environment and heredity interact. They show us that certain inherited potentialities may be expressed one way in one environment, but quite another way in a different environment.

If you have ever seen identical twins, you know how very much alike they are physically. They are, of course, always of the same sex. Their height, weight, and build are strikingly similar (see Table 19.1). So, too, are other physical features: hair and

| DIZYGOTIC TWINS | MONOZYGOTIC TWINS |

OCCUR WHEN 2 DIFFERENT SPERM FERTILIZE 2 DIFFERENT EGGS BY INDEPENDENT BUT NEARLY SIMULTANEOUS FERTILIZATIONS. THE ZYGOTES ARE GENETICALLY DIFFERENT FROM EACH OTHER.

OCCUR WHEN A SINGLE SPERM FERTILIZES A SINGLE EGG. . .

WHICH THEN DIVIDES SOME TIME DURING EARLY DEVELOPMENT, . . .

EACH FERTILIZED EGG DEVELOPS INTO A SEPARATE INDIVIDUAL, BUT WITHIN THE SAME UTERUS.

GIVING RISE TO TWO GENETICALLY SIMILAR INDIVIDUALS IN THE SAME UTERUS.

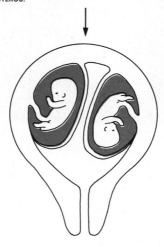

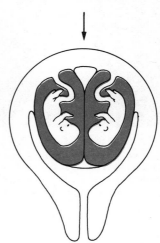

Figure 19.6 This diagram illustrates the origin of monozygotic (identical) and dizygotic (fraternal) twins at the time of fertilization.

eye color, texture and pigmentation of skin, structure and placement of teeth, and the shape and size of nose, eyes, ears, and mouth. These similarities also extend to less obvious features: blood type, fingerprint pattern, pulse and respiration rates, and blood pressure.

You are also well aware of the fact that fraternal twins often look much alike, but no more so than their brothers or sisters born singly. Their likeness, of course, is due to the fact that they have the same parents, and consequently have many of the same genes. However, the segregation of genes during meiosis, in preparation of eggs and sperm formation, makes it highly unlikely that fraternal twins would have the same genotype.

Table 19.1, however, indicates that MZ twins (monozygotic) reared together are more alike than are MZ twins reared apart. Environmental influences, therefore, are exerted on virtually all characteristics. We must remember that the environment is never

TABLE 19.1. Differences, Expressed as Percentages, Between MZ Twins Reared Together, MZ Twins Reared Apart, and DZ Twins

Trait and intrapair difference	MZ twins reared together (44 pairs)	MZ twins reared apart (44 pairs)	DZ twins (39 pairs)
Height			
Under 1 in.	82	58	24
Over 1 in.	18	42	76
Weight			
Under 14 lb	68	64	40
Over 14 lb	32	36	60
Intelligence			
0–4 points	44	32	14
5–9 points	21	30	14
10–14 points	23	14	14
15– points	12	24	57

From J. Shields, *Monozygotic Twins*, New York: Oxford University Press, 1962.

the same for any two individuals, however similar their genotypes may be.

If we now consider such hard-to-define characteristics as intelligence, ability, or personality, the influence of the environment appears to be greater. Let us concentrate on intelligence, since we all are aware of the existence of IQ tests. Intelligence is not a single, clearly defined character. It is a composite of many attributes, all of which provide a basis for rational behavior. But we recognize that differences do exist between individuals, and IQ tests, while not wholly reliable (since it is difficult to rule out the role of experience), can provide usable estimates of these differences.

Table 19.1 provides information about both MZ and DZ twins. It is clear that MZ twins reared together have less variation in their IQ scores than similar twins reared apart. Fraternal twins and paired siblings, however, are still wider apart in score than either of the MZ groups. Additional tests designed to measure personality factors and several kinds of ability show the same trends. It is clear, however, that experience, education, and the general environment play a greater role in determining intellectual and social traits than physical characteristics.

Of some interest is a study made by Johannes Lange, a German professor, on heredity and criminal tendencies in MZ and DZ twins. He selected those twins, one of whom had a criminal record. Among the MZ twins, most of whom had been reared together, 10 of the 13 pairs showed both members with criminal records, and generally for the same kind of crime. Of the DZ twins, only 2 of the 15 pairs revealed tendencies in both members. It is, of course, difficult to determine the relative weight to be given to heredity and environment in these cases. However, it seems clear that certain social and behavioral responses have an inherited basis, and that the environment will determine how these responses will be expressed.

summary

The **phenotype** of an organism is the result of a given **genotype** developing in a given environment. The contribution of the environment versus that of heredity can be assessed through the study of **monozygotic** and **dizygotic** twins. The results show that many traits—physical, behavioral, and intellectual—have a strong genetic basis, but that the environment does play a role. The genotype determines the limits of expression of a given trait, but the environment also determines it to a certain extent.

for thought and discussion

1 Assume that each of the 23 chromosomes a man inherited from his mother is different from the 23 he inherited from his father. How many genetically different kinds of sperm can he form? What is the probability that a sperm will contain all the chromosomes derived from the paternal parent? What does this information tell us about the possible variability found among children in the same family? If a striking likeness shows up among all the children, what conclusions can be drawn?

2 Baldness in human beings is genetically determined. What possible mechanisms or differences in internal or external environment can be responsible for the relative absence of baldness in women?

3 Tuberculosis is an infectious disease caused by a bacterial organism, yet it tends to characterize certain families as though it were inherited. What can you say about the possible environmental and hereditary factors responsible for the expression and distribution of the disease?

4 Some kinds of cancer are hereditary. Does this mean that a gene is directly responsible for the disease?

5 Houseflies and mosquitoes, which were once easily controlled by an insecticide such as DDT, are now often found to be resistant to the insecticide. Would you think that this change in susceptibility is hereditary or environmental? Explain.

6 Parents of a set of girl twins have the following genotypes: *MmNNOoPpRR* and *MmNnooPPRr*. Both twins show the same phenotype for these genes and are very much alike in all other respects. Assuming that the genes are inherited independently of each other, what is the probability that the twins are dizygotic?

7 Diabetes is a lethal trait if left untreated. Is the gene responsible for the lack of insulin a lethal gene? Explain.

8 Inbred strains of animals such as mice have long been used in testing the effects of new drugs. What is meant by "inbred," and what are the genetic consequences of inbreeding? Why should the investigator insist on inbred strains instead of any random group of mice?

9 Do you suppose that Mendel could have derived his laws of segregation from a study of characters that varied widely in different environments? Explain.

10 If a trait is expressed in both members of a pair of monozygotic twins, can one be certain that the trait is genetically determined? Explain.

11 If the changing height and weight of Amherst College freshmen is not due to a changed pattern of inheritance, how can you account for the differences environmentally?

12 The Napoleonic Wars were said to have had a marked effect on the average height of Frenchmen. From your knowledge of history, can you suggest a possible reason for this? In time of war and with heavy loss of life among members of the armed forces, do you think that the present selective service draft procedure—which is an environmental factor of a sort—can alter the genetic structure of a population? If so, in what way?

See books listed at the end of Chapter 16.   **selected readings**

development—
an
inherited
pattern
chapter 20

Our study of identical twins has shown that two individuals with the same genotype, developing in the same environment, will be strikingly similar to each other in appearance and behavior. If you stop to think about this, you should soon realize that it is an extraordinary thing. Consider what it means.

Identical twins are separated from each other at a very early stage in their development. At this time each twin consisted of only a few hundred or a few thousand cells, and all the cells were very much alike. From then on, the two individuals developed independently: from prenatal life, through childhood and adolescence to adulthood. Yet throughout the course of development, a striking similarity was retained. This must mean that both twins follow a precise pattern of development, step by step, as they proceed from the egg to adulthood; the fertilized egg must contain the information that determines this parallel behavior. Development, therefore, is the realization of an inherited pattern: it brings to fulfillment the potentiality present in the egg at the time of fertilization.

There is no resemblance between an adult and the egg that gave rise to the adult. In size the human egg is no larger than a speck of dust. It weighs about one millionth of a gram; and the sperm, when fertilizing it, adds only another five billionths of a gram. If you examine an egg under the light microscope, you will see little visible structure: a clear membrane surrounding the egg, fat droplets (yolk), and a nucleus within. There is no hint among these details of what the egg can become.

The fertilized egg, in addition to being a single cell, is also an organism. Even though it is relatively simple and undeveloped, it is very much alive and completely coordinated. The adult, on the other hand, is one of the most complex structures known, consisting of billions of cells. If the pattern of cell division were one cell to two cells, two cells to four, four to eight, and so on, it would take about 45 cellular divisions to produce a human adult. But a human adult could never be formed by a group of similar cells, no matter how they were organized. As our body cells form, they also become different. This is the process which gives character and uniqueness to us as individuals. Some cells form bone, others muscle, still others skin, kidney, liver, and other organs and tissues of the body. Development is the result of many coordinated processes: **growth,** which leads to an increase in mass and is generally accompanied by cell division; **differentiation,** a process by which similar cells become different in structure and function; and **integration,** or **regulation,** which is really growth and differentiation coordinated in time and space (Fig. 20.1).

The problems of development are, therefore, cellular problems. Before we look into this, however, we might ask a simple question. Why should development begin with a single cell, an egg? The animal could release a highly developed portion of itself, which could then grow into a new adult form. As Figs. 20.2 and 20.3 show, this happens in certain flatworms, and it is a chief means of reproduction in many plants. But the fact that sexual reproduction is so widespread would suggest that it has certain advantages over asexual and vegetative reproduction.

One advantage of sexual reproduction is that, through fertilization, the characters of the two parents can be combined in the offspring. New combinations of genes can be tried. Such a trial would be much more difficult if asexual reproduction were the rule. Furthermore, fertilization could not occur in an organized way if a multicellular body were involved. A single cell (an egg), therefore, is needed. The sperm cell brings to the egg not only a set of genes, but also an activating mechanism that stimulates and initiates the whole process of development.

GROWTH Cells can do one of three things: (1) they can grow and divide; (2) they can differentiate; or (3) they can die. All of these aspects are included in the general term "development." We can think of growth as an increase in mass. Since we are dealing with cells, growth can occur by an enlargement of cells, but is usually accompanied by cell division as well. Sometimes, however, early

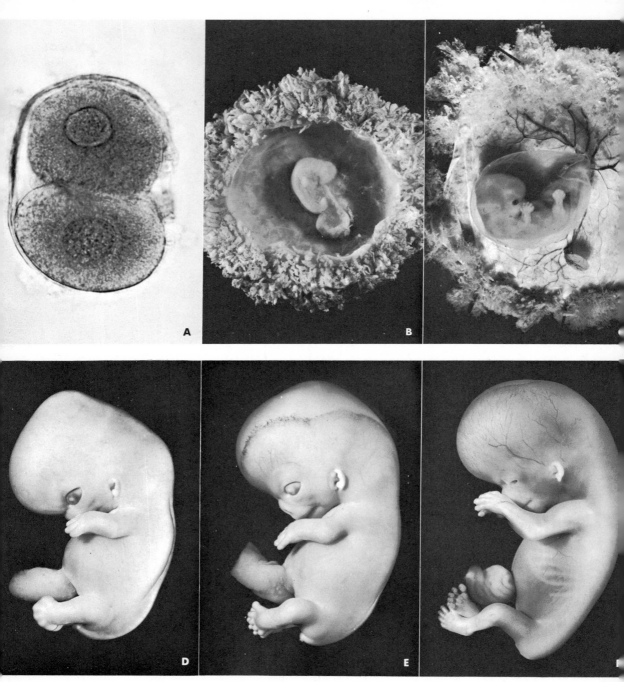

Courtesy C. F. Reather

Figure 20.1 This sequence of photographs shows the development of a human embryo from the two-celled stage to an embryo 56 days old. (A) Two-celled stage; (B) 28-day embryo with chorion removed; (C) 39-day embryo with chorion laid open but with amnion (inner membrane) intact; (D) 40-day embryo with eyes, ears, and limbs evident; (E) 44-day embryo; (F) 56-day embryo. The umbilical cord, or placenta, which attaches the embryo to the wall of the mother's uterus, and through which the embryo obtains its nutrition, is seen in figures (E) and (F).

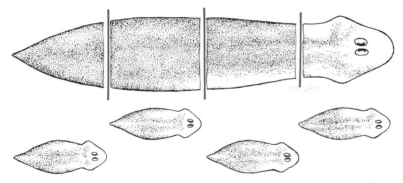

Figure 20.2 The planarian (a flatworm) can be cut into several pieces, each of which develops into a complete individual. This is an example of asexual reproduction.

development may not involve growth as an increase in mass, although cell division is very much a part of the pattern of change. During the early **cleavage** stages in the development of an embryo, there is active cell division, but the mass of the embryo remains the same. This is why it is called "cleavage"; the egg is cleaved into smaller and smaller units by cell division. The individual cells must, therefore, become smaller as their number is increased (Fig. 20.4).

As the number of cells increases, the mass of cells takes the form of a hollow sphere. This is the **blastula.** No particular change in shape or size has occurred, and there is no obvious advance toward an adult form. If we were to cut the blastula of a frog in half, however, we would find several features not visible before. There are cells of many different sizes. The **dorsal** or upper cells, for example, are smaller than the **ventral** cells. Also there is a mass of still smaller cells (Fig. 20.5) at the dorsal side.

Richard F. Trump

Richard F. Trump

Figure 20.3 Examples of asexual reproduction. The geranium stem (above), which has been rooted in damp sand, and the strawberry plant (at left) are examples of asexual reproduction in the plant world. The strawberry plant develops new individuals by means of "runners." The parent plant (1) forms a lateral stem, or runner, which roots when it comes in contact with the ground, at which points new plants develop (2 and 3). These can be detached and grown as separate individuals.

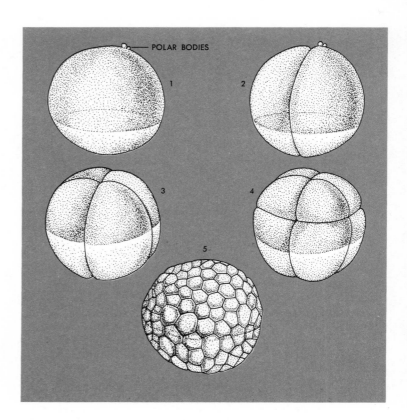

POLAR BODIES

Figure 20.4 These early cleavage stages in the development of the frog show the one-, two-, four-, and eight-cell stages, and the external appearance of the early blastula.

Figure 20.5 A section through the blastula of a frog shows the central cavity, the larger ventral cells, which are filled with yolky materials for nutrition, and the smaller dorsal cells.

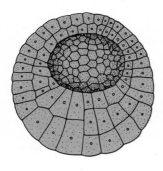

The region between the large and small cells will be the point of change, or **gastrulation,** which eventually will lead to the sphere changing into a tadpole, and then into a frog.

Gastrulation is a period of cell movement. The embryo becomes completely reorganized. By marking the blastula so that selected regions can be recognized later, it is possible to see where certain cells take up their final residence, and what they eventually give rise to (Fig. 20.6). In a general way, all the small cells of the ventral pole go to the interior of the embryo, while the larger cells of the dorsal pole spread over the outside of the embryo. This is accomplished by a movement of cells.

An opening forms in the ventral region, and the cells turn in at this point, moving internally. The opening, which began as a slit, eventually becomes a full circle, with the cells that were once external being rolled inwardly. The opening surrounds the mass of the **yolk,** which is finally enveloped, leaving a tiny slit. This becomes the anus of the developing frog. In the meantime, the yolk is being absorbed by the growing organism.

Gastrulation, therefore, has given the embryo shape and organization by a rearrangement of cells. Cell division continues at the same time. The reorganization continues, and the gastrula becomes transformed into a **neurula** (Fig. 20.7). This is characterized by the formation of the **neural folds** on the top of the embryo. Continued growth and unfolding, followed by a fusion of the two sides of the fold, produce the **neural tube.** This structure later develops into the brain and spinal cord.

At the end of about four days, the single fertilized frog egg has become a recognizable tadpole. As Fig. 20.8 shows, the main internal organs are beginning to form, while externally the tadpole has gills with blood running through them, a tail, an olfactory (smelling) organ, and identifiable regions for eyes and mouth. In this condition it is released from the egg membranes and becomes a free-swimming tadpole. The obvious dramatic changes are over. The remainder of the growth period is that of filling in the details. To achieve the same stage of development as that described for the frog, a human embryo requires about 1 month.

The filling in of details is, itself, equally interesting, but is of a more localized nature as one or another organ is formed. One facet of this period of development is that some parts grow faster or slower than others; furthermore, the relative rates of growth of different parts and regions differ from time to time. At 2 months of fetal age, the human head is equal in length to the rest of the body. Arms grow faster at an early period than do legs. The body, on the other hand, grows at a steadier rate

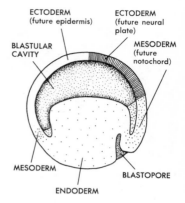

Figure 20.6 This section through an embryo at the beginning of *gastrulation* shows the cells beginning to move inward by a process of *invagination*. The regions of the gastrula are labeled to indicate that particular groups of cells will eventually contribute to the formation of particular parts of the body as development continues. Compare this with Fig. 20.10.

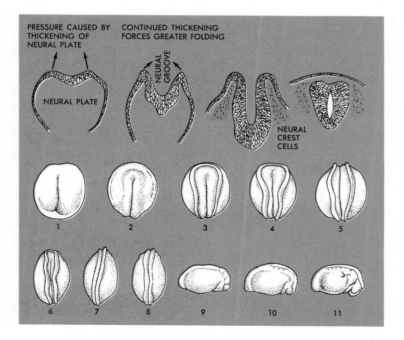

Figure 20.7 The formation of the neural tube and the outward appearance of the neurula of the frog are shown here. Above, the folding of the ectoderm and eventual pinching off of the neural tube; below, successive stages of neurula formation as viewed externally.

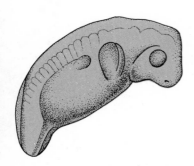

until maturity is reached. In this way the organism reaches its adult form (Fig. 20.9).

DIFFERENTIATION   As the embryo grows, cells are increasing in number, invaginating, and regrouping. Each change in the embryo takes place at a certain time, gradually altering a formless mass of cells into an organism of distinctive characteristics. In the frog, some of these changes can be followed with the naked eye. But other, less obvious events are also occurring and giving the embryo character and individuality. Internal as well as external organs are being formed. Within each, individual cells are beginning to assume the shapes and structures related to their adult function. This is the process of **differentiation.** We can think of this process occurring at the level of the organism itself, at the level of an organ, or at the cellular level. No matter what the level, differentiation places a stamp of uniqueness on the whole organism or its parts.

Basically, differentiation is genetically controlled, but environment also plays a role. For example, the cells destined to become part of the liver, heart, or kidney are determined by origin, position, and local environment, even though each cell has the same set of genes as all other cells (Fig. 20.10). The effect of specific environments can be seen acting on the development and differentiation of a fertilized *Fucus* egg (Fig. 20.11). After the first division, one of the two cells will form the thallus, the other the rhizoid. The rhizoidal end of the undivided egg is indicated early by the formation of a small protuberance. But as the illustration indicates, the position of the protuberance is determined by light, temperature, and pH. Once the protuberance develops, the developmental pattern of the embryo is set.

Figure 20.9 Relative sizes of different portions of the human body are shown in this diagram as development proceeds from the early fetal stage to adulthood.

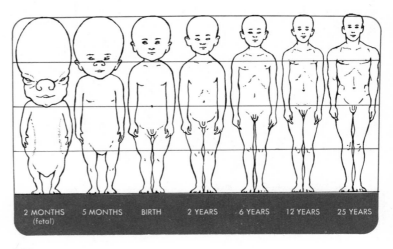

| 2 MONTHS (fetal) | 5 MONTHS | BIRTH | 2 YEARS | 6 YEARS | 12 YEARS | 25 YEARS |

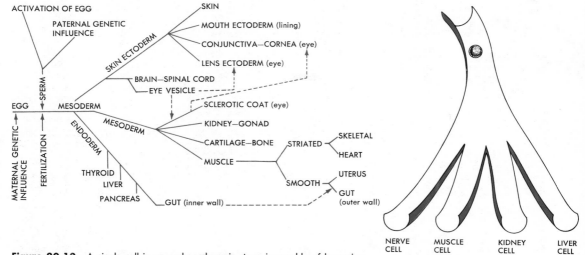

Figure 20.10 A single cell in an early embryonic stage is capable of becoming a nerve, muscle, liver, or kidney cell. What it becomes depends on its place of origin in the embryo and the local environment among other things. The chart shows the time course of development, during which a single cell (the egg) becomes an organized mass of different kinds of cells as the result of cell division and differentiation.

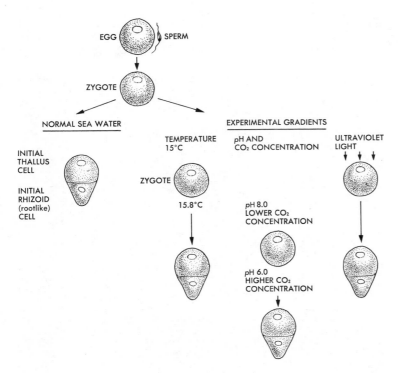

Figure 20.11 The *Fucus* (brown alga) egg has its polarity or direction of development determined by a number of environmental variables. The rhizoidal cell forms in the warmer region, where the pH is lower and the CO_2 concentration higher, and on the side opposite from a source of ultraviolet light.

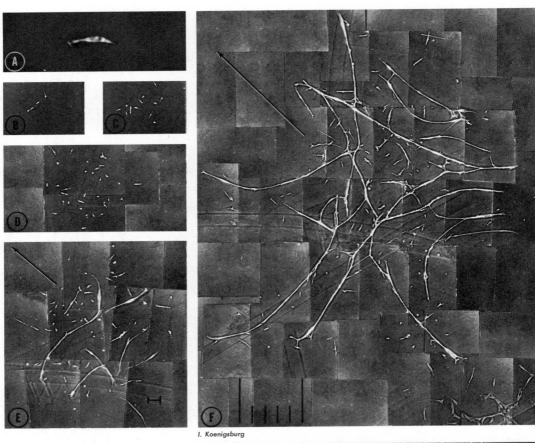

I. Koenigsburg

Figure 20.12 Muscle cells
(myoblasts) can be grown in
tissue culture. A through G show
the same culture at various time
intervals. A shows the single cell
before division, which gives rise
to all other cells in the culture.
B, C, and D show an increase in
the number of cells through
division. E, F, and G show the
same culture at later stages,
after the cells have aggregated
into bundles and differentiated
into striated muscle cells
(see Fig. 20.13).

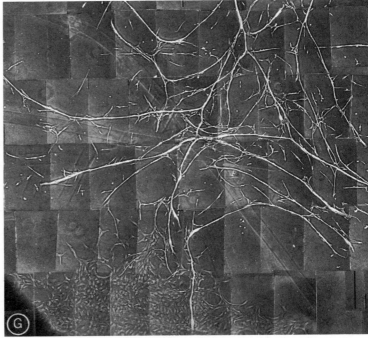

At the cellular level, differentiation causes cells to lose their generalized form. Muscle cells begin to elongate and acquire contractile fibers (Figs. 20.12 and 20.13). Their tendency to do so is established chemically before the eye can detect it visually, and is beautifully demonstrated by tissue culture techniques. Heart cells change internally and begin a rhythmic beating that will continue throughout the life of the organism. A simple cell thus becomes a specialized cell. In doing so, the differentiated cell loses its ability to do other things as its specialized ability increases. For example, the more specialized the cell, the less likely it is to divide. In any individual cell, therefore, growth and differentiation appear to be mutually exclusive states of activity. Growth involves the ceaseless formation of new cells, while differentiation makes specialists out of various cells of this general mass. Figure 20.14 shows the result at a cellular level. Each of the cells of the gut has a different function, and each has the necessary internal structures needed to perform that function.

We are now faced with the problem of trying to explain *why* differentiation occurs. The problem in its simplest form is this: the exactness of DNA replication and of chromosomal segregation in mitosis makes it reasonably certain that all somatic cells in an organism will have the same genotype. Then why do they come to have different phenotypes? Let us take three cells in the gut concerned with digestion: **mucous** cells, **parietal** cells, and **chief** cells of the **gastric mucosa** (stomach lining). Figure 20.14 shows that they are quite different in internal structure, even though they have a common origin from a group of deeper cells. Their functions are also different. They produce, respectively, mucus, hydrochloric acid, and pepsin as their major product. Their enzymes must, consequently, be different.

As we have already learned, enzymes are proteins, and genes are responsible for the production of proteins. Therefore, if cells

Figure 20.14 The diagram below shows four kinds of cells found in the lining (gastric mucosa) of the stomach of a bat. Two of the cells form mucus (polysaccharide). The parietal cell forms hydrochloric acid, and the chief cell forms enzymes (protein). All these substances participate in the digestive processes of the stomach.

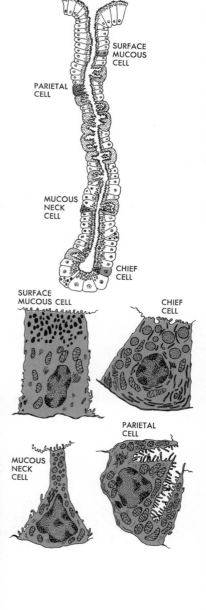

Figure 20.13 When shown at higher magnifications, muscle cells from the culture in Fig. 20.12 reveal the cross-striations that develop and enable the cells to contract.

E. Leitz, Inc.   I. Koenigsburg

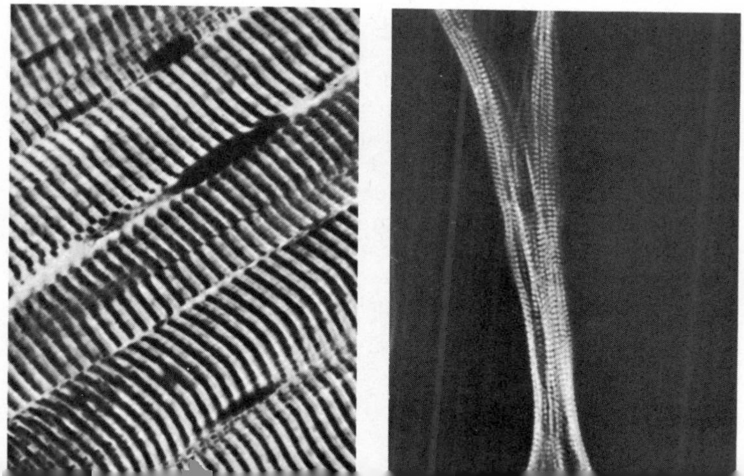

of the same genotype have different kinds of proteins (enzymes are only one kind of protein that can differ), cellular differentiation must be due to *differential gene action*. In other words, not all genes are active all the time; those that are active determine the phenotype of the cell. We do not know what determines when a gene will be active or inactive, but we do know that genes respond differently to different environmental conditions.

In the bacterium, *Escherichia coli,* for example, there is a gene that controls the formation of an enzyme called **tryptophan synthetase.** The function of this enzyme is to catalyze a reaction that combines two chemicals, **indole** and **serine,** into **tryptophan,** an essential amino acid. If the colon bacterium is grown in a medium containing tryptophan, no enzyme is formed. If tryptophan is removed from the medium, the enzyme is immediately synthesized. From these observations we can conclude that the gene controlling enzyme formation is inactivated in the presence of tryptophan, and is active in its absence.

Another gene in *E. coli* controls the formation of a different enzyme, **beta-galactosidase.** Its function is to break down a sugar, **galactose,** so that it can be used as a carbon source within the cell. When galactose is present, the enzyme is being formed; when galactose is absent, so, too, is the enzyme.

Here, then, are two genes whose activity depends on the presence or absence of a particular chemical. But their activities are opposed to each other. In one instance, tryptophan inhibits enzyme formation; in the other, galactose promotes enzyme formation. The two examples, on the other hand, show that environmental differences can act on a uniform genotype to evoke different phenotypic responses.

Are the gene inhibitions of a temporary nature as the ones discussed appear to be? Or can they be of a more permanent type? At least a partial answer can be given by certain experiments performed in the frog. It is possible surgically to remove a nucleus from one cell and transplant it to another cell. If the cells are of the same sort, say from a blastula, nothing unusual happens. However, let us now transplant nuclei from progressively more differentiated cells and put them into an egg that has had its own nucleus removed. Will these nuclei permit the egg to continue its development? If nuclei from blastula cells are so transplanted to an egg, the egg proceeds with development and everything is normal. If nuclei from cells of the gastrula are similarly transplanted, the egg will develop through the blastula stage, but stop development abruptly at the onset of gastrulation. Nuclei from more differentiated cells, such as from adult liver, will cause a similar change at gastrulation. It would appear, then, that nuclei as well as cytoplasm can become differentiated.

This is very beautifully demonstrated in the larvae of *Drosophila.* As Fig. 20.15 shows, the chromosomes in the cells of the salivary gland are very different in character from those of ordi-

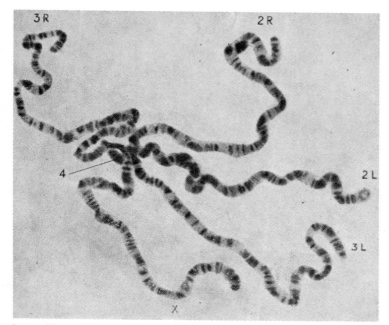

Figure 20.15 The chromosomes of *Drosophila melanogaster,* as they appear in the cells of the salivary gland. Each chromosome can be recognized by a distinctive banded structure. The chromosomes are united to each other at their centromeric regions. Chromosomes 2 and 3 have median centromeres, and their right and left arms are identified. The *X*-chromosome and the tiny chromosome 4 do not have two arms.

nary somatic cells, but they are also found in other portions of the larval body—in the gut and rectum, for example. In each type of cell, and at different times during larval development, the chromosomes have a characteristic appearance. Puffs (Fig. 20.16) appear or disappear. It is now believed that the appearance of a puff means that the gene at that location is active, and is producing RNA. Genes, therefore, can be active or inactive at different times, and the environment as well as the stage of

Figure 20.16 A portion of a single giant chromosome from a midge, *Rhyncosciara angelae,* appears differently at different stages of larval development. The arrows indicate comparable bands in each of the figures. When a particular region puffs out and becomes diffuse in appearance, it is believed that the gene or genes in that region are active in making messenger RNA.

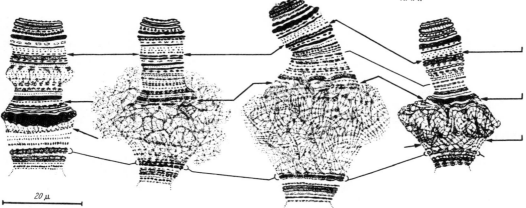

20μ

311

development determines how they behave. As a very rough guess, it is likely that in a given cell at a given time only one-tenth of the genes are active. The remainder will be active in other cells and at other stages of development or activity.

INTEGRATION The problems of growth and differentiation at the cellular level are among the most intriguing facets of modern biology; development in its total aspect, however, is still more challenging, but also more perplexing. The whole course of development is characterized by a unity and a harmony of structure and behavior and maintenance. The egg, in its own way, is a complete and total organism. So, too, are seeds, embryos, larvae, and pupae. They are as complete as the mature individual to which they will give rise in the sense that they are fully functional entities developing as a whole and not simply as a collection of parts.

For lack of a better word, we can refer to this harmony of existence as **integration,** even though we do not understand how it is maintained. So far as we know, life is a series of chemical reactions taking place within an organized structure. Integration, therefore, must begin at the molecular and cellular level, and then be expressed at the level of tissue, organ, and organism. Structure and behavior at all levels must be compatible with the functioning of the whole organism. Among the integrative mechanisms are those relating to our sense organs, nervous system, blood supply, and hormonal condition. For example, an inadequate supply of growth hormone produced by the pituitary gland at the base of the brain leads to dwarfism; an oversupply leads to gigantism. The body, at maturity as well as during development, is maintained by a series of checks and balances.

Perhaps we can best point out the significance of integration by asking some as yet unanswered questions. What determines the size of one organ as compared with the size of other organs? Why do your nose, fingers, and toes reach a certain length and stop growing? What determines "mature" size? Why does aging begin at a certain age? Some of these aspects of development may be answered at a cellular level, but others may involve higher levels of integration. If some day we can comprehend these aspects of regulative control, we may be well on our way to an understanding of the biology of cancer and aging.

summary

An organism inherits from its parent or parents a set of cellular structures capable of performing or directing certain chemicals reactions. **Development** is the process whereby a fertilized egg or a group of cells is transformed into a recognizable organism. Both the genes and the environment participate in determining the direction of development.

Development consists of three major phases: (1) **growth,** (2) **differentiation,** and (3) **integration.** Growth is an increase in mass. It includes cell enlargement or cell division, or, more commonly, it includes both proceeding together. Differentiation is a process that transforms an unspecialized cell into a cell having a special structure and function. The more complex the organism, the greater is its need for specialized cells and organs, and the greater is the degree of differentiation. Integration, or regulation, is a term that describes all the control mechanisms which keep the organism functioning as a unit.

for thought and discussion

1 Suppose that fertilization, instead of uniting single cells and forming a zygote, involved a mass of cells each of which had to be fertilized by a single sperm. What difficulties would be encountered? What would be the genetic consequences?

2 As you found in an earlier chapter, immortality was lost when sexual reproduction became a way of life for most organisms. What does this statement mean to you? Think deeply. What are the philosophical implications?

3 A chicken is said to be an egg's way of making more eggs. Does this statement make biological sense? Does it make sense if it is applied to you as an individual?

4 Be certain that you understand the difference between "growth" and "differentiation." What would happen if a zygote of a frog or a daisy underwent one process but not the other? Can you think of any reasons why the cells of an organism should not divide continuously until the adult number of cells was attained, and then go through the processes of differentiation all at once?

5 Cancer may develop in only one organ of the body, but the cancerous cells may spread and invade other organs. Why don't normal cells do this?

6 You fall and skin your knee. The skin grows back and covers the wound, but only just covers it. What biological processes are involved in the repair job?

7 We have implied that chemical differentiation must precede visible morphological differentiation. Why?

8 Think back to the chapter on chromosome structure and ask yourself, "Can the proteins of the chromosome play a role in determining whether genes are active or not in a differentiated cell?" Can you think of a way to test this question?

9 There is an insect hormone, "ecdysone," that controls molting. Remembering that *Drosophila* salivary gland chromosomes undergo puffing when particular genes are active, can you devise an experimental plan to test for the effects of the hormone on genes?

10 The developing limb bud of a frog contains recognizable cells of different kinds: those that will eventually form skeleton, muscle, inner skin (dermis), and outer skin (epidermis). If the limb bud is removed and the cells are separated from each other (this can be done with the enzyme trypsin) and randomly dispersed in a cell culture, they will reaggregate and form something like a limb bud, with each particular cell in its right place—skeletal cells inside the mass, then muscle, dermis, and, finally, epidermis on the outside. Cell position is an example of a regulatory process. From the way the cells behave, regulation is cellular and is not lost by dissociating the cells. Can you give any reasons for this regulation? Use this problem to illustrate how you would go about understanding this kind of regulation in a scientific manner. What are your facts? What hypotheses can you develop? How can you test them?

selected readings

CORNER, G. W. *Ourselves Unborn.* New Haven: Yale University Press, 1944. A classic book dealing with the development of the human individual.

MARKERT, C. L., and H. URSPRUNG. *Developmental Genetics.* Englewood Cliffs, N.J.: Prentice-Hall, Inc., 1971. An excellent treatment of the genetic control of developmental processes.

MEDAWAR, P. B. *The Uniqueness of the Individual.* New York: Basic Books, Inc., 1957. A discussion of the interaction of heredity and development in the determination of individual uniqueness.

MOORE, J. A. *Heredity and Development,* 2nd ed. New York: Oxford University Press, 1972. A readable account of the interrelations of hereditary and developmental processes.

TRINKAUS, J. *Cells into Organs: The Forces that Shape the Embryo.* Englewood Cliffs, N.J.: Prentice-Hall, Inc., 1969. Excellent treatment of cell movement and cell adhesion in relation to the formation of organs.

WADDINGTON, C. H. *Principles of Development and Differentiation.* New York: The Macmillan Company, 1966. A modern view of the problems of growth and development.

the evolution
of inherited
patterns
chapter 21

So far we have discussed a number of basic concepts in biology: the cell theory, cellular structure and behavior; the chemistry of biological reactions and their control; modes of inheritance and the physical basis of heredity; and the realization of inherited patterns of development. Let us now try to examine these concepts from the points of view of **unity** among organisms, **diversity** among organisms, and **continuity** (Fig. 21.1).

As we stressed earlier in this book, the cell is the basic unit of organization. The individual cells of a tree, a cow, and a person are more alike than unlike. Only the viruses represent a major departure if we consider them as organisms. Nearly all cells fall within definite size limits, and as a rule they contain a single nucleus in a mass of cytoplasm bounded by a plasma membrane. In most cells the nucleus also is bounded by a membrane. Only bacteria and blue-green algae depart from this general arrangement.

On the structural level, most cells also contain the same organelles; ribosomes, chromosomes, mitochondria, Golgi apparatus, and endoplasmic reticulum are common to all higher plant and animal cells. While animal cells lack plastids and a cell wall, most plant cells lack a centriole, but the similarities of plant and animal cells outweigh their differences. We also find unity at the chemical level. Metabolism is essentially the same in all cells, and

Figure 21.1 An example of continuity through time is shown by a comparison of a fossil lobster claw (left), which is millions of years old, and the claws of the common lobster, *Homarus vulgaris*. The claws are not greatly different from each other.

the key molecules of nucleic acids, proteins, fats, and carbohydrates are much the same wherever they are found. The enzyme cytochrome C performs the same function in all organisms from bacteria to man; this can hardly be pure happenstance. Finally, cell division in both somatic and reproductive cells is also similar throughout the plant and animal world.

Unity also exists at higher levels of organization. Despite their differences in color, shape, and size, flowers are built on the same basic plan (Fig. 21.2); unity also characterizes the structure of their stems, roots, and leaves. The arm of a man, the foreleg of a dog, and the wing of a bird are again the same, despite their differences. We emphasize this by referring to them as **homologous** structures. The science of comparative anatomy, founded by the great French naturalist, Georges L. Cuvier (1769–1832), deals with this kind of information and reveals essentially how structures are related. Furthermore, we can see by examining the developmental stages of animals that structures arise from similar sites within the embryo, and in the same sequence. It is for this reason that the embryos of related animals show a marked similarity (Fig. 21.3).

DIVERSITY If we now shift our view from unity to diversity, we are confronted by a bewildering array of differences. Diversity exists at all levels. Table 21.1 on page 319 shows that the nucleotide composition of DNA from various organisms varies widely. On another scale, this can be demonstrated within a particular molecule, hemoglobin, which is made up of two identical halves, each half consisting of an alpha (α) and a beta (β) peptide chain.

316

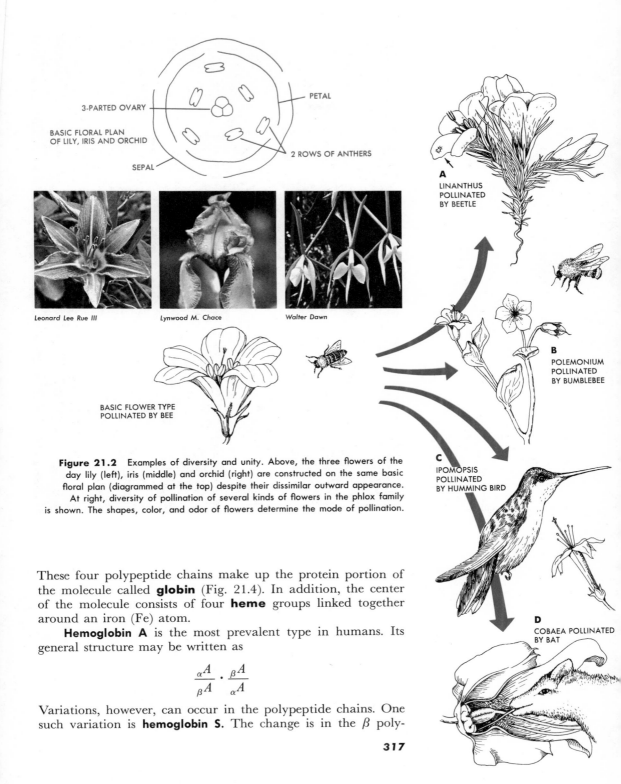

3-PARTED OVARY

BASIC FLORAL PLAN
OF LILY, IRIS AND ORCHID

SEPAL

PETAL

2 ROWS OF ANTHERS

Leonard Lee Rue III

Lynwood M. Chace

Walter Dawn

BASIC FLOWER TYPE
POLLINATED BY BEE

Figure 21.2 Examples of diversity and unity. Above, the three flowers of the day lily (left), iris (middle) and orchid (right) are constructed on the same basic floral plan (diagrammed at the top) despite their dissimilar outward appearance. At right, diversity of pollination of several kinds of flowers in the phlox family is shown. The shapes, color, and odor of flowers determine the mode of pollination.

A
LINANTHUS
POLLINATED
BY BEETLE

B
POLEMONIUM
POLLINATED
BY BUMBLEBEE

C
IPOMOPSIS
POLLINATED
BY HUMMING BIRD

D
COBAEA POLLINATED
BY BAT

These four polypeptide chains make up the protein portion of the molecule called **globin** (Fig. 21.4). In addition, the center of the molecule consists of four **heme** groups linked together around an iron (Fe) atom.

Hemoglobin A is the most prevalent type in humans. Its general structure may be written as

$$\frac{_{\alpha}A}{_{\beta}A} \cdot \frac{_{\beta}A}{_{\alpha}A}$$

Variations, however, can occur in the polypeptide chains. One such variation is **hemoglobin S**. The change is in the β poly-

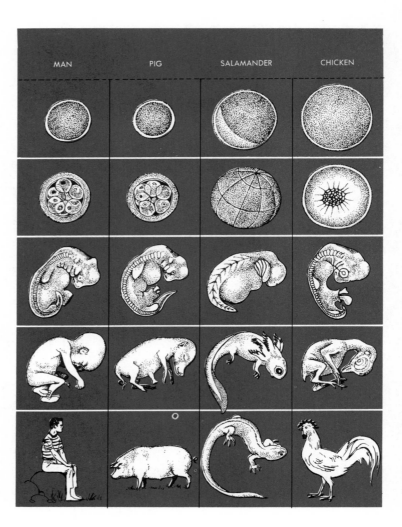

Figure 21.3 There are striking similarities, as well as dissimilarities, in the development of the four vertebrates shown here. The diversities become more apparent as the embryo progresses toward adult form.

peptide, and the homozygous and heterozygous types, respectively, are

$$\frac{_\alpha A}{_\beta S} \cdot \frac{_\beta S}{_\alpha A} \quad \text{and} \quad \frac{_\alpha A}{_\beta A} \cdot \frac{_\beta S}{_\alpha A}$$

Individuals possessing these hemoglobins are anemic (**sickle-cell anemia,** so called because the red blood cells are sickle-shaped), and the homozygous individuals die early. Only one amino acid has been changed to alter hemoglobin *A* to hemoglobin *S*. There are about 560 amino acids in the hemoglobin molecule, yet the change of one amino acid can mean the difference between a normal and an anemic condition.

318

TABLE 21.1. **Base Ratios of Several Well-Known Organisms***

	Adenine	Thymine	Guanine	Cytosine	Ratio of $\frac{A + T}{C + G}$
Man	29.2	29.4	21.0	20.4	1.53
Sheep	28.0	28.6	22.3	21.1	1.38
Calf	28.0	27.8	20.9	21.4	1.36
Salmon	29.7	29.1	20.8	20.4	1.43
Yeast (fungus)	31.3	32.9	18.7	18.1	1.19
Staphylococcus	31.0	33.9	17.5	17.6	1.85
Pseudomonas	16.2	16.4	33.7	33.7	0.48
Colon bacterium	25.6	25.5	25.0	24.9	1.00
Vaccinia virus	29.5	29.9	20.6	20.0	1.46
Pneumococcus	29.8	31.6	20.5	18.0	1.88
Clostridium	36.9	36.3	14.0	12.8	2.70
Wheat	27.3	27.1	22.7	22.8	1.19

* Values are arbitrary but accurate as ratios.

Figure 21.4 Diagrammatic representation of the hemoglobin molecule of man, which consists of four polypeptide chains linked together in the center. The Greek letters alpha (α) and beta (β) indicate the positioning of the *alpha* and *beta* polypeptide chains. The heme and iron (Fe) portion of the molecule is represented by the central dot.

On the level of organisms, diversity is more obvious. We can readily tell plants from animals. We can also distinguish various kinds of mammals and various breeds of dogs. Diversity enables us to distinguish not only among races of human beings, but also among individual human beings. In fact, of the 3 billion humans inhabiting the planet, no two are *exactly* alike (Fig. 21.5). Even identical twins have their minor dissimilarities of structure, behavior, and ability. If we now realize that there are about 2 million known different species of plants and animals, the magnitude of the diversity comes home to us.

Organic life is, therefore, characterized by both unity and diversity. Which is the more prominent depends on our point of view. At times these two aspects of life seem to oppose each other, but many things said in this book tell us that unity and diversity have a common basis. Heredity and environment, for example, are the sources of both unity and diversity in identical twins. Although unity implies common ancestry and, in particular, sets of similar genes, superficial unity may arise because of a particular environment. Diversity, on the other hand, implies different genes, different ancestry, and diverse environments.

Figure 21.5 Of the many people in this crowd, no two are exactly alike.

Ewing Galloway

CONTINUITY Life does not begin anew with each generation. Life, in fact, does not cease between generations, but rather, is passed on without interruption. When reproduction is asexual, a living fragment of one individual develops into a complete organism. This can be readily demonstrated in the flatworm, *Planaria* (Fig. 20.2). Each

319

piece from a single animal has the ability to form a complete organism. In sexual reproduction, the egg and sperm are the living cellular bridges that span the gap between generations.

The basis of continuity resides in the ability of a cell to replicate itself. As the German physician Rudolf Virchow stated in 1858, every cell comes from a pre-existing cell. The living cells now existing are only the temporary ends of long chains of cells that extend backward in time to that period in the history of life when the first cell, or cells, arose. On a molecular level, the basis of continuity resides in the ability of the cellular organelles to replicate themselves exactly. We have already discussed how DNA can do this, but centrioles, mitochondria, and plastids also have this ability to replicate themselves. Perhaps the most remarkable thing about the whole process of replication is that life at various levels of organization—molecules, organelles, cells, organisms, species, and communities of species—is characterized by this power of replication. Life is immortal even if individuals are not.

THE THEORY OF Although life is continuous, it is
EVOLUTION also continuously changing. It is characterized by unity, but also by diversity. The theory of evolution was advanced to account for these three aspects of life. By embracing unity, diversity, and continuity, the theory states that all living organisms have arisen from common ancestors by a gradual process of change that leads to diversification. The theory also denies the validity of the doctrine of **special creation,** which advanced the belief that organisms were immutable, that is, they were created as they now appear and that no change has taken place.

The doctrine of special creation dominated the thinking of men during ancient and medieval times and up to the 19th century. This was, in part, a religious belief, but it was also due to misconceptions that man had about the age of the Earth. Before the 15th and 16th centuries, man had a limited idea of the enormity of the universe. The Ptolemaic theory of the universe placed the Earth at the center of a limited universe. Man thought of himself as the creature for whom the Earth and all plants and animals were created. It was natural, then, to place man's home, the Earth, at the center of the universe. Copernicus and Galileo, who lived in the 1500s and 1600s, destroyed this idea. We know now that our Solar System is out near the edge of a galaxy (the Milky Way), which is composed of 100 billion stars (Fig. 21.6), and that there are a billion or more galaxies in addition to our own. Space, in fact, is so vast we have difficulty comprehending its immensity.

It was the geologists who demonstrated that the Earth did not spring into being in the year 4004 B.C., as Archbishop Ussher

Figure 21.6 An artist's conception of the Milky Way, our home galaxy, which consists of about 100 billion stars. The Sun is located near the edge of the galaxy.

of Ireland once calculated the Earth's age to be. Today it is calculated not in thousands of years, but in billions; it is about 4.5 billion years old. The oldest known fossils are about 3 billion years old. Life, therefore, has had a long time to change, a long time during which the great chain has never been disrupted from the beginning to the present.

What evidence do we have that life has evolved? It is so overwhelming that evolution cannot be rationally disbelieved. Evidence can be found in the physiology, biochemistry, and anatomy of organisms; but let us choose examples that are familiar, say domesticated plants and animals. The dog was probably the first animal domesticated by man. As Fig. 21.7 shows, the various breeds of dogs are quite different from each other, in size, body conformation, facial shape and expression, hair, the way they bark, instinctive behavior, and so on. Yet all breeds arose from a wolf-like animal that man domesticated early in his rise to civilization. By selective breeding man has been able to isolate certain variations and produce pure breeds. Mongrels, showing a mixture of various breeds, result from a random, not a selective, pattern of breeding.

The development of the many breeds of dogs illustrates the

Figure 21.7 The English setter (top), Scottish terrier (middle), and boxer (bottom) represent three breeds that have been developed by man through artificial selection.

course of evolution. It is, however, an artificial evolution in the sense that man directed it for his own purposes; man instead of the environment becomes the selective agent. Then, again, man is part of the environment, and his purposefulness is as natural to him as a trunk to an elephant. The basic principles are the same, however, whether evolution is artificial or natural. For one thing, there must be variation among the breeding population in order for selection to take place. If all organisms were alike, selection could have no effect. There must also be a pattern of breeding; that is, the parents must be selected by the breeder rather than by random picking. And once a particular variation is achieved and breeds true in succeeding generations, the parents of one breed are kept reproductively isolated from other breeds. On a larger scale and with a longer time period, the same events occur naturally and give rise to new forms of plants and animals.

What man has done through selective breeding of dogs he has also done with horses, cattle, swine, and poultry. One need only open a seed catalog to see the variations man has selected to produce the great diversity in our vegetable and cereal crops, and in our garden flowers.

What proof do we have that organisms have changed as one period of time succeeded another? Although early evidence came from fossil remains in rock formations (Fig. 21.8), we can visualize this change more clearly by observing alterations of a more subtle nature as they take place in the laboratory. The bacterium, *Staphylococcus aureus,* is a **pathogen** (disease-inducing organism) that affects humans. It can be grown in a test tube. If we add

Figure 21.8 This winged ant was fossilized in amber. Amber itself is fossilized pitch of coniferous trees.

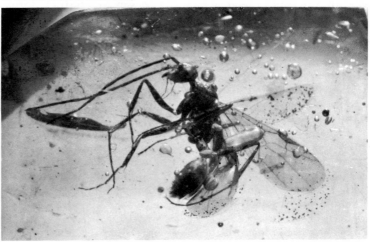

a minute amount of streptomycin (an antibiotic) to the medium in which this organism is growing, we find that the great majority of cells are killed. About 1 in 1 million cells survives, however, and the survivors continue to grow. They are resistant to the antibiotic. If we continue to add streptomycin to the media, thus gradually increasing the total amount, we find that each additional amount of streptomycin kills the great majority of remaining cells, but that each time some survive.

If we now test the cells that survive high concentrations of streptomycin, we may encounter one or both of two situations. One kind of cell will be highly resistant to streptomycin; another kind may actually require streptomycin as a nutrient. We can separate these two kinds of cells by placing them in a medium lacking streptomycin. Those that are streptomycin-dependent die or will not grow. But both types of cells—the resistant and the dependent—are different from those in the original culture. By breeding techniques it can be shown that some of their genes are different. They have **evolved.**

The situation in the laboratory is an "artificial" one controlled by the investigator. The fact that the same situation has occurred without the purposeful intention of man, however, is indicated by the fact that some of our hospitals are infected by a staphylococcus that is antibiotic-resistant. Since antibiotics came into general use in the mid- and late 1940s, the changed nature of some staphylococci is a relatively recent phenomenon.

When we wish to reconstruct the changing nature of organisms over long spans of time, we must turn to the fossil record. Our source of information is from sedimentary rocks. The surface of the Earth is continually being eroded by wind and water, and the fragments are carried away by rivers as sand and silt (Fig. 21.9). When deposited as sediment in quiet waters, the sand and silt are often compressed and hardened into sedimentary rock. An animal or plant that dies and becomes embedded in the sediment may be preserved well enough to be recognizable millions of years later when the rocks are eroded and exposed (Figs. 21.9 and 21.10). The Grand Canyon is such an exposed mass of sedimentary rock.

In general, the bottom layers of sedimentary rock should contain earlier forms of life than the top layers do. This is what has been found. Scientists can thereby follow a succession of life. By determining the age of the rocks, it is then possible to calculate the time required to bring about such changes and to visualize the lines of succession of life.

Determination of lines of succession requires one major assumption—that is, later organisms must show a relation to the earlier ones that gave rise to them. We can see this beautifully demonstrated in the evolutionary history of the horse. The Cenozoic era has been called the Age of Mammals, and it was during the Eocene period, approximately 60 million years ago, that

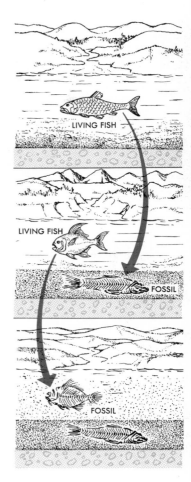

Figure 21.9 In this schematic representation, a fish dies (top) and is covered over by sediments (middle) and fossilized. Thousands of years later the living fish in the middle panel dies and its remains are covered by a second layer of sediments. Both layers are compressed and hardened into rock. A fossil may be simply the impression of an organism in rock, or it may be a portion of the organism (generally a hard part, such as bone) that is preserved.

Figure 21.10 The cliffs of the Grand Canyon were formed by the erosion of rock by the river. The oldest rocks and most ancient fossils are found at the lower levels of the canyon.

American Airlines

eohippus (*Hyracotherium*), the "dawn-horse," made its appearance. Fossil remains show that *eohippus* was not very much larger than a modern cat, 10 inches high at the shoulder and weighing 8 or 9 pounds. Its teeth reveal that it was a browsing animal; that is, it fed on the leaves and twigs of trees. It walked on three toes, even though the forefoot had four toes. It had disappeared in Europe by mid-Eocene, and in North America by late-Eocene. Before doing so, it left descendants, and these in turn left other descendants, with the modern horse the living member of a long line of ancestors. During the 60 million years, a change in tooth structure took place, showing that the horse became a grazing animal; that is, it fed primarily on grass. Also during this period the number of toes was being reduced, resulting in a single toe, or hoof. Other lines of horses were also evolved. Some were even smaller than *eohippus,* others larger; tooth structure varied widely and not all became one-toed. Only the line that led to the present-day horse persisted. All other lines became extinct. Only the fossil record remains.

THE COURSE OF EVOLUTION The theory of evolution states that organisms arose through a process of gradual change from more primitive ancestors. We cannot now and never will be able to reconstruct the entire evolutionary history of all living things. The fossil record is not complete. Some organisms were not suitable for fossilization, and often the conditions for fossilization were not right. There are, as a consequence, great gaps in our knowledge of past life on this planet. However, by making use of what fossils have been found, and by assuming that similarities of structure provide evidence of relationships, the major sequences of life are known.

Table 21.2 is a geological timetable, providing information about changes in environment and the rise and fall of plant and animal groups. The oldest rocks of which we have knowledge do not contain fossil remains. These were formed in the Archeozoic era. Life apparently began in the Proteozoic. We make this assumption not because of fossil evidence, of which there is very little, but because the Cambrian rocks of the Paleozoic era contain abundant fossil remains. In fact, by the end of the Cambrian period most modern groups except the chordates were already established.

The first organisms were unicellular plants and animals: the **protists.** These had their origin in the seas. There were few, if any, freshwater forms, and the land masses were barren. Multicellular algae and fungi among the plants, and the coelenterates (jellyfishes), sponges, mollusks, arthropods, and echinoderms among the animals soon followed. All these are **invertebrates,** lacking a vertebral column. The first chordates (fishes), which lead eventually to man, appeared in the middle of the Ordovician period, and there is some evidence that they arose in freshwater. The fishes increased greatly in the Silurian and Devonian periods. At about the same time, the spiders and insects took their place on the evolutionary stage. Invasion of the land by plants occurred in the Silurian period, and by the Devonian, the gymnosperms (ancestors of the present-day pines and spruces) had become established. Amphibians, lungfish, and sharks also arose during the Devonian period.

During the Carboniferous and Permian periods great forests of gymnosperms and seed ferns covered the land. These, together with the lycopods and horsetails, led to the formation of the great coal beds of the world. The fallen plants piled up in great layers, were compressed into strata, and were carbonized into coal as the volatile substances evaporated. By the end of the Permian, the lycopods and horsetails declined, but they persist today as miniature remnants of a once great flora (Fig. 21.11).

The reptiles appeared in the upper Carboniferous, or Pennsylvanian, period when insects and amphibians abounded. The rise of the Appalachian Mountains signaled the end of the Paleozoic and the beginning of the Mezozoic era. The Mezozoic is the Age of Reptiles, the time of the rise and fall of the dinosaurs. It is also the time when the first birds appeared, the earliest ones bearing teeth and being decidedly reptilian in character (Fig. 21.12). Mammals, too, came into existence, the earliest of them being egg-laying, like their reptilian ancestors.

The Mesozoic also witnessed the rise of the flowering plants.

Figure 21.11 The present-day lycopods and horsetails are remnants of a once vast tree-like flora that contributed heavily to the formation of the coal beds of the world.

R. H. Noailles

TABLE 21.2. Geological Time Scale

Eras (years of duration)	Major Divisions	Periods (years from present)	Epochs	Dominant Organisms	Events of Biological Significance	Geological and Climactic Phenomena
	Quaternary	2 million	Recent	Age of Man and Herbs	Rise of civilized man	
			Pleistocene		Extinction of great mammals and many trees	Periodic glaciation
Cenozoic (60 million)	Tertiary	Late Tertiary	Pliocene	Age of Flowering Plants, Mammals, and Birds	Rise of herbs; restriction of forests; appearance of man	Climactic cooling; temperate zones appear; rise of Cascades, Andes
			Miocene		Culmination of mammals; retreat of polar floras; restriction of forests	Cool and semiarid climate; rise of Himalayas, Alps
		Early Tertiary (60 million)	Oligocene		Worldwide tropical forests; first anthropoid apes; primitive mammals disappear	Climate warm and humid; rise of Pyrenees
			Eocene		Modernization of flowering plants; tropical forests extensive; modern mammals and birds appear	Climate fluctuating
Mesozoic (125 million)	Late Mesozoic	Cretaceous (125 million)		Age of Higher Gymnosperms and Reptiles	Rise and rapid development of flowering plants; gymnosperms dominant but beginning to disappear; rise of primitive mammals	Rise of Rockies and Andes; great continental seas in North America; Europe climate fluctuating
					Extinction of great reptiles.	Climate very warm
	Early Mesozoic	Jurassic (157 million)			First known flowering plants, gymnosperms prominent but primitive ones disappear; dinosaurs and higher insects numerous; primitive birds and flying reptiles	Great continental seas; rise of Sierras; climate warm
		Triassic (185 million)			Gymnosperms increase; first mammals; rise of dinosaurs	Climate warm and semiarid

TABLE 21.2. (Continued)

Era	Period	Age of	Life forms	Geological events
Paleozoic (368 million) — Late Paleozoic	Permian (223 million)		First modern conifers; rise of land vertebrates	Periodic glaciation; rise of Appalachians, Urals
	Pennsylvanian (271 million)	Age of Lycopods, Seed Ferns, and Amphibians	Primitive gymnosperms dominant; extensive coal formation in swamps	
	Mississippian (309 million)		Lycopods, horsetails, and seed ferns dominant; some coal formation; rise of primitive reptiles and insects	Shallow seas in N. America
Middle Paleozoic	Devonian (354 million)	Age of Early Land Plants and Fishes	Rise of early land plants; rise of amphibians; fishes dominant	Shallow seas in N. America
	Silurian (381 million)	Age of Algae and Higher Invertebrates	First known land plants; algae dominant; first air-breathing animals (lungfish and scorpions)	
Early Paleozoic	Ordovician (448 million)		Marine algae dominant; corals, star fishes, bivalves; first vertebrates (fishes)	Shallow seas in N. America
	Cambrian (553 million)		Algae dominant; many invertebrates	Shallow seas in N. America
Proterozoic (900 million)	(1,500 million)	Age of Primitive Marine Invertebrates	Bacteria, algae, worms, crustaceans prominent	Formation of Grand Canyon, Laurentians; sedimentary rocks
Archeozoic (550+ million)	(2,000 million)	Age of Unicellular Forms	No fossils; organisms probably unicellular; origin of first life	Rock mostly igneous
	(10 billion) ?			Beginning of present universe and the Solar System

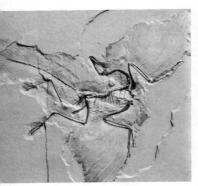

Figure 21.12 A fossil of *Archaeopteryx*, the oldest known bird. It lived during the Jurassic period, some 157 million years ago.

Figure 21.13 The duck-billed platypus has many features of the mammal group, yet it lays eggs that are hatched outside the body. Consequently it lacks the mammary glands characteristic of mammals.

The modern cone-bearing gymnosperms were present, the seed ferns were disappearing, and first the dicotyledonous and then the monocotyledonous plants emerged. The latter two groups were to become dominant members of our flora.

The mammals came into prominence in the Cenozoic. Of the egg-laying type, only the duckbill platypus and the echidnas remain (Fig. 21.13). The other mammals were placental, giving rise to living young. The rise of the anthropoid apes and then of man (toward the end of the Cenozoic) began the present period. Meanwhile, the great deciduous forests developed, later to give way to herbs and grasses. The world of today was then ushered in.

The fate of most species of living things over the great sweep of time since life began has been extinction, without descendants. This is as much a part of evolution as is persistence. The present flora and fauna, therefore, represent those species that managed somehow to keep the thread of life going. This thread of life, from its earliest beginnings to the present time, is illustrated by Fig. 21.14. The main avenues are reasonably clear, although not every connection is known for certain. For example, the ancestors of both the chordates and the flowering plants remain in doubt. Both groups appear suddenly in the fossil record, with transitional forms absent, not yet discovered, or possibly misinterpreted.

summary

Organisms differ from each other, but they also show a common unity of metabolism, cell structure, and developmental patterns. Both **unity** and **diversity** result from inheritance; and since life

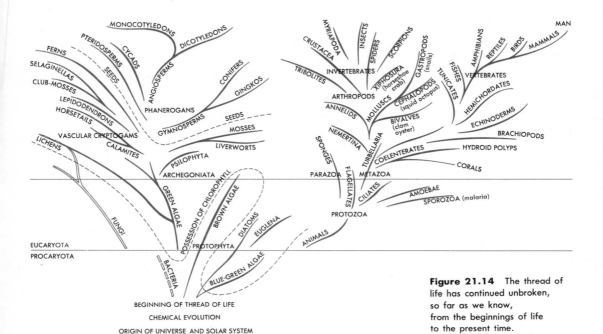

MONOCOTYLEDONS
DICOTYLEDONS
FERNS
PTERIDOSPERMS
CYCADS
SELAGINELLAS
SEEDS
CLUB-MOSSES
ANGIOSPERMS
CONIFERS
LEPIDODENDRONS
GINGKOS
HORSETAILS
PHANEROGANS
VASCULAR CRYPTOGAMS
SEEDS
LICHENS
GYMNOSPERMS
MOSSES
CALAMITES
LIVERWORTS
PSILOPHYTA
FUNGI
ARCHEGONIATA
GREEN ALGAE
POSSESSION OF CHLOROPHYLL
BROWN ALGAE
DIATOMS
EUGLENA
EUCARYOTA
PROTOPHYTA
ANIMALS
PROCARYOTA
BACTERIA
BLUE-GREEN ALGAE

MYRIAPODA
INSECTS
CRUSTACEA
SPIDERS
SCORPIONS
MAN
TRIBOLITES
INVERTEBRATES
XIPHOSURA (horseshoe crab)
GASTROPODS (snails)
AMPHIBIANS
REPTILES
BIRDS
MAMMALS
ARTHROPODS
CEPHALOPODS (squid octopus)
FISHES
VERTEBRATES
ANNELIDS
MOLLUSCS
TUNICATES
HEMICHORDATES
NEMERTINA
BIVALVES (clam oyster)
ECHINODERMS
SPONGES
TURBELLARIA
COELENTERATES
BRACHIOPODS
PARAZOA
METAZOA
HYDROID POLYPS
FLAGELLATES
CORALS
CILIATES
AMOEBAE
SPOROZOA (malaria)
PROTOZOA

BEGINNING OF THREAD OF LIFE

CHEMICAL EVOLUTION

ORIGIN OF UNIVERSE AND SOLAR SYSTEM

Figure 21.14 The thread of life has continued unbroken, so far as we know, from the beginnings of life to the present time.

from one generation to another is passed on by way of cells, there is also **continuity.** All three aspects of life are understandable if we assume that life has evolved through millions of years, giving rise to the present forms of plants and animals.

Evolution is change. For the process to be effective, variation and selection are necessary. Man has imposed evolution artificially among domestic plants and animals by acting as the selecting agent. Nature does it more slowly; we now realize that life has existed on Earth for more than 2 billion years. Present species of plants and animals are the present ends of long chains of continuous life that extend backward in time to the moment when life first began on our planet. The most striking evidence of this is found in sedimentary rocks, where the remains of former life appear as fossils.

Past ages of the Earth show that there is a progression from one form of life to another. Older rocks have more primitive organisms than younger rocks. Through comparative anatomy, a history of life can be reconstructed, but gaps in the fossil record prevent us from reconstructing a *complete* history of all organisms.

1 How do you suppose Archibishop Ussher arrived at the date 4004 B.C. as the beginning of all things?

2 Consider two points of view: (1) that the Earth is the center of the

for thought and discussion

329

universe and that all plants and animals are on Earth for man's use (a view prevalent up to the 19th century); and (2) that we exist near the edge of a galaxy and revolve around a star in a universe of millions of other galaxies; further, that all life is related to us through evolution. Does acceptance of one or the other of these points of view affect your thoughts about yourself: who you are, where you came from, what is to become of mankind?

3   Why can selection have no effect if there is no variation?

4   The Earth is about 5 billion years old. How is this age determined? Look up information on the carbon-14 dating process. Why is the process reliable only to about 30,000 years in the past? What is carbon-14? How does it differ from the more common carbon-12?

5   Why are there no fossils in igneous rocks? If fossils are not found in a certain layer of sedimentary rocks (where they normally occur), can we conclude that there was no life when the layer of sedimentary rock was formed? Explain.

6   How can the teeth of a fossil animal reveal anything about its dietary habits?

7   Clover is a plant that shows variation in the height of individual plants: some are tall, some intermediate, some creeping. Imagine two fields of clover, one closely grazed for several years by sheep, the other ungrazed. At the end of this time would you expect the genotypes of the surviving plants to be the same or different in the two fields? Why? How would you test your hypothesis?

8   The American horses of today originated in the Old World, but abundant horse fossils are found in North America. Why do you suppose that only one kind of horse survived to the present time? What could have caused their disappearance in America but not in the Old World?

9   Do the facts that cells are the basic unit of organization through most of the plant and animal kingdoms, and that cells of all kinds have quite similar modes of metabolism, have any evolutionary significance?

10   Consider your classmates and yourself. What features do you all share in common? What features show a great deal of diversity? What does this kind of comparison tell us about development, ancestry, and evolution?

11   Would you expect to find as much variation among a group of sexually breeding organisms as among a group arising through asexual reproduction? Explain.

12   If extinction is a more likely evolutionary fate than persistence, can man escape extinction? What aspects of man's structure and/or behavior could promote and hasten extinction? What aspects could favor persistence?

13   Make a list of common plant or animal species that display considerable diversity. Make a list in which no diversity is present.

14 "The purpose of every unit—organism, cell, or chemical compound—is to be itself for a short period and then to become something else. . . ." G. Ehrensvärd. What does this statement mean for the units mentioned? What does it mean when you apply it to yourself?

selected readings

DE BEER, SIR GAVIN. *Atlas of Evolution.* Nashville, Tenn.: Thomas Nelson Inc., 1964. An expensive volume, but one of the finest and most beautifully illustrated books on evolution. It covers all of the plant and animal kingdom, including man.

SIMPSON, G. G. *Horses.* Garden City, N.Y.: Doubleday & Company, Inc., (Natural History Press), 1951. An excellent presentation of one of the most completely known fossil histories of an organism.

WHITE, J. F. (ed.). *Study of the Earth.* Englewood Cliffs, N.J.: Prentice-Hall, Inc., 1962. Includes discussions of fossils and how one can determine the age of rocks.

causes and
results
of evolution
chapter 22

One of the most common observations we could make, but one so obvious that we rarely give it thought, is that organisms in their native habitat seem to belong there. The organism and the environment fit each other: a cactus in the desert, a polar bear in the frozen north, an ameba in a wayside pool, an earthworm in rich garden soil. Indeed, our everyday language reflects our thinking about this: we say that someone is "in his element" when the situation is such as to give the individual an opportunity for full expression of his talents. Or, we say that someone is like "a fish out of water" when the individual and the situation are mismatched.

While some organisms exist only in a restricted kind of environment, others seem relatively indifferent to their surroundings. A fish cannot exist out of water. Yet water is not the only limiting factor. Temperature, salinity, available food, and amount of oxygen must also be considered. Man, on the other hand, lives in a wide variety of environments: in the dry and treeless desert, the hot and humid tropics, and in the frozen Arctic. Man's ability to invade and live in difficult environments depends on his ability to protect himself from harmful aspects of the environment with housing and clothing, for example, or to change the environment, as he does in high-flying aircraft, spaceships, in submarines, and in deep mines (Fig. 22.1). Air

Figure 22.1 The American astronaut E. H. White, II, "walks" in space at a rate of about 17,000 miles an hour. A line tethers him to the spaceship so that he will not wander off and be lost. His ability to survive in this airless, frigid environment depends upon his space suit, within which a controlled environment is maintained.

conditioning and irrigation are but two means that enable him to make an environment more suitable to his needs as an organism.

The simple fact is that continued existence of a species requires that it be adapted to an environment. Failure to adapt can lead only to extinction of the species. Individuals, of course, can make temporary adjustments to changes in the environment, and can live for a time in an unsuitable environment. But if the species is to persist, its members must grow, maintain themselves, and reproduce generation after generation. We have already discussed this problem in another context. We showed that when certain plants are moved to lower or higher altitudes they fail to reproduce and sometimes die. Again, we emphasize a most important fact: the species and the environment must be in harmony with each other if the species is to persist.

An environment itself is seldom constant. There are daily and seasonal changes in temperature, precipitation, and other forms of weather, and many destructive forces such as hurricanes, floods, and prolonged dry spells. The daily rhythm of light and darkness is one of the most important aspects of our environment on the Earth. Why do chrysanthemums bloom in late Summer and Fall rather than in the Spring? Why do birds produce their young in the Spring? The answer, in part, is found in the changing length of day and night, and the response of organisms to a changing regime of light intensity. We know that light or darkness changes the behavior of organisms profoundly. But temperature and the amounts of food, water, and oxygen are also important.

In response to seasonal aspects of the environment, seeds and spores carry plants over periods of cold or dryness; some animals hibernate, others migrate, while still others may change their coat color as winter approaches (Fig. 22.2). We also know that an environment may be altered "permanently." Although the breadfruit tree now grows only in the tropics, fossils of this plant are found beyond the Arctic Circle. At one time in the past, the climate must have been warm in this region. Mountains form and, as they do, the environment around them is altered. If organisms cannot adapt to the new conditions, they must migrate to more suitable climates, change so that they are better adapted, or die.

Environmental changes are not only physical in nature. Some changes are biological, such as groups of organisms competing for living space. Fossil-bearing rocks record the failure of many organisms to survive the competition for "elbow room." The dinosaurs appear to be a notable example. We still do not understand fully the reason for their disappearance, but the blunt fact is that extinction is a more likely fate of a species than is continued survival over long periods of time. In the continuing fight for living space, man is an important biological factor. As

Figure 22.2 The rock ptarmigan is shown here in its brown summer plumage (left), and as it is changing from its white winter to its brown summer plumage (right).

he hunts, farms, and in other ways alters the land to suit his way of life, he destroys much around him: the dodo bird, the auk, and the ivory-billed woodpecker have disappeared within the last century, and conservation is required to protect other animals and plants.

Over the span of many hundreds, thousands, and millions of years, many species have died out, but some have continued to survive. These have given rise to the species living today. The fact that we can write these words, and you can read them, means that we are just the present-day members of an unbroken chain of organisms extending far back in time. *Homo sapiens* has existed for 1 million years or more, but for how much longer he will continue to inhabit the planet, no one can say. Environments change, and so do organisms. Those that have survived have been adaptable, and our understanding of evolution rests upon our understanding of the *causes of adaptation.*

CHARLES DARWIN'S In the previous chapter we stated
THEORY OF that life was continuous, but also
NATURAL SELECTION that it was constantly changing.
We also said that within the continuity of life we find both unity and diversity. The transmission of genes determines unity in that a cat always produces kittens,

not some other kind of animal. But since genes can undergo change (mutate) and also recombine through meiosis, diversity is continuously introduced into the life of a species. It is this diversity which, through a gradual process of change, leads to the evolution of species on a grand scale, and to the adaptation of the species in a more restricted sense. If we phrase this thought somewhat differently, we can state that all adaptive changes that are inherited are evolutionary changes. The reverse, however, is not true; all evolutionary changes are not adaptive. It is important to make this distinction.

Theories attempting to account for the evolution of organisms go back to the Greeks, but our current point of view was developed principally by Charles Darwin, the great English naturalist. His book, *The Origin of Species,* which appeared in 1859, is probably the single most influential volume of biology ever written. He termed his ideas on evolution the **theory of natural selection.** We can understand this best by following Darwin's line of reasoning.

The first point, or assumption, made by Darwin was that the number of individuals in a species tends to increase in geometric ratio, generation after generation. This means that each generation has more offspring than there are parents. Thus, if a single bacterium divides and produces two bacteria, that generation of two will produce four bacteria; the four will give rise to eight, the eight to 16, the 16 to 32, and so on. A single maple tree produces thousands of seeds, each of which is capable of producing more maple trees. A single female codfish annually sheds several million eggs, each of which can hatch into a young codfish.

The fact remains, however, as Darwin clearly knew (assumption two), that the number of individuals in each generation remains fairly constant. Great fluctuations in numbers do not generally occur. Some individuals are eaten by predators, and there is competition among those that survive, competition for the basic essentials of life: water, light, food, and "elbow room." Some organisms thrive amid competition, others do not. The individuals that survive and become the parents of the next generation are, therefore, only a small number of the original population.

Darwin made a third assumption: among the individuals belonging to a single species there is diversity, or **variation.** This is an observation you can make for yourself by comparing the cats and dogs of your neighborhood, or the plants in a garden or flower bed. It is more difficult to make the same observation among wild plants and animals, but the variations are equally numerous.

On the basis of these assumptions, Darwin developed his theory of natural selection. He argued that, if some variations were more favorable for survival than others, these gave certain individuals a better chance of passing these variations on to the next generation. That is, these individuals are more likely to be the parents of the next generation. Individuals with less favorable variations stand a lesser chance. Gradually, over a period of time, the favorable variations would tend to accumulate, and the species would undergo a gradual evolution. Species with short generation times could evolve faster than those with longer times, since over a given period of time the number of generations would be greater.

Only those variations that are heritable are of any significance in evolution. Many variations are caused by environment, but these play no role in the evolution of species. Furthermore, favorable variations do not guarantee survival to a reproductive age. Survival or death of an individual is often a matter of chance, and many favorable variations are lost in the course of evolution. Also, a favorable variation in one environment may be unfavorable in another, or be of no use at all. The ability to swim is of no value to an individual who lives in a desert. We must, therefore, think of variations in terms of the organism in a given environment, and define as "favorable" those changes which in the long run increase the individual's chance of producing more and better adapted offspring. If the individual fails to participate in production of the next generation, any unique variation it has will be lost when the individual dies.

Let us take a closer look at what Darwin meant by the term *natural selection.* We have just stated that all the individuals of one generation do not contribute their genes to the next generation. In terms of human populations, some individuals are sterile, some choose not to produce offspring, and some die before they reach reproductive age. The parents of one generation are, therefore, a selected part of that generation, and their hereditary materials are a selected part of the hereditary material of that generation (Fig. 22.3). As Bruce Wallace and Adrian Srb of Cornell University have stated:

Figure 22.3 In three generations of individuals, the number of individuals per generation remains the same. (The nonreproducers are shown in black.) If there is a heritable difference between the reproducers and the nonreproducers, a gradual change will take place in the population.

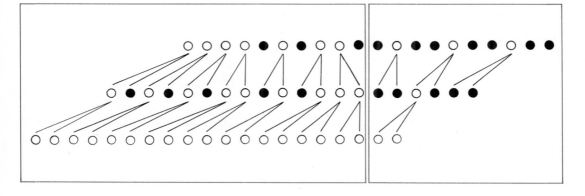

The disparity between parents as one group of individuals and the rest of the population as another *is* natural selection. . . . Thus, when we say that natural selection results in the adaptation of organisms to their environment and evolutionary changes in populations, we are simply saying that the continual contrast, generation after generation, between reproducing individuals as one group and the remainder of the population as another results in adaptation and evolution.

Natural selection, therefore, is a random process in the sense that it has no predetermined goals. Variations are produced in low numbers and environmental changes occur gradually. But if both changes occur, then evolutionary change is unavoidable; it will take place, as, indeed, we believe it has been doing since life first arose on our planet.

We need now to inquire in more detail into the source of variation, its fate in succeeding generations, and its effect on the adaptability of populations.

THE SOURCE OF VARIATION The heritable material of an individual is its DNA (RNA in some viruses). It is found primarily in the nuclei of cells, where it forms the chromosomes. Short segments of DNA form the genes. As yet we do not know, except in a few instances, how many nucleotide pairs make a gene, but we must assume that genes differ among themselves in number of nucleotide pairs if only because proteins differ in the number of amino acids they contain. We do know, however, that a gene is made up of a particular sequence of nucleotide pairs, and that this sequence is the genetic alphabet. It spells out the message through RNA, which, in turn, takes part in the formation of a particular protein. The uniqueness of every individual, plant or animal, is determined by the proteins it contains and also, of course, by its genes.

As indicated in an earlier chapter, DNA is a large molecule so constructed that it can replicate itself in a most exact fashion, one cell generation after another. This is the basis of biological continuity and biological unity. DNA is also responsible for biological diversity. If this is so, diversity, or variation, has its origin in the fact that changes can occur in the sequence of nucleotide pairs within a gene. Gene A, for example, then becomes a, or A_1, A_2, A_3, and so on, depending on where within the gene a change occurred in the sequence of nucleotide pairs. A gene, therefore, can change into many forms. Each one is a mutation, and the several mutants arising from a particular gene are called **multiple alleles.** Each mutation alters the action of a gene by changing the gene product, or protein. This, in turn, changes the organism in some manner, allowing us to detect that a mutation occurred. From these facts, you will recognize that the existence of a gene can be detected and studied only when at least two alleles of the same gene can be compared.

How often do genes mutate? Genes are very stable structures, but we now recognize that each gene has its own rate of mutation. This can be seen in Table 22.1, which shows the rates of mutation of a number of genes in corn (*Zea mays*), *E. Coli,* and man. The seed color gene, *A*, mutates to the *a* allele at a rate of 492 per million (or 10^6) cells, or gametes. *Wx*, on the other hand, is exceedingly stable, and no mutations to *wx* were found in this particular experiment. The rate of change of the other genes tested was intermediate.

Bacteria or viruses provide a convenient way for us to study mutations, since millions of cells or virus particles can be grown in a matter of hours or days. Some mutation rates are as low as 1 in 1 billion (10^9) cells. In human beings, mutation rates are difficult to determine. We can recognize them only by knowing how frequently particular mutations make their appearance in a population. The mutation that gives rise to hemophilia, a disease in which the blood fails to clot readily, occurs once in about every 50,000 persons. This is a fairly high rate of mutation.

If gene *A* mutates to *a*, and this change is an alteration in the nucleotide sequence, then we should expect that *a* can mutate to *A*. This is known to occur, and can be expressed as $A \rightleftharpoons a$, with the change of $A \rightarrow a$ being a forward mutation rate, and $a \rightarrow A$ being a backward mutation rate. The two rates need not be the same, and, in fact, generally are not.

TABLE 22.1. **Spontaneous Mutation Rates of Specific Genes in Several Organisms**

Organism	Gene	Rate
Maize	R	492 per 10^6 gametes
	I	106 per 10^6 gametes
	S	1 per 10^6 gametes
	Wx	0 per 10^6 gametes
E. coli	Leucine-1	0.07 per 10^9 cells
	Leucine-2	1.42 per 10^9 cells
	Arginine-2	0.37 per 10^9 cells
	Tryptophan-6	5.61 per 10^9 cells
Man	Achondroplasia (dwarfism)	41 per 10^6 gametes
	Hemophilia	32 per 10^6 gametes
	Albinism	28 per 10^6 gametes
	Total color blindness	28 per 10^6 gametes
	Infantile amaurotic idiocy	11 per 10^6 gametes

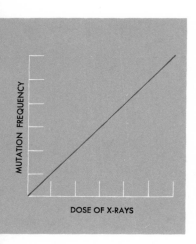

Figure 22.4 If a given dose of x-rays induces *X* mutations, twice the dose will produce *2X* mutations. This relationship is indicated by the straight line.

Figure 22.5 The dose of ultraviolet (UV) light remains the same for all wavelengths. The solid line represents the degree of absorption of UV of different wavelengths by the DNA of the cell; the broken line represents the frequency of induced mutations. The similarity of the two curves indicates that the UV must be absorbed in order to be effective as a mutation-inducing agent.

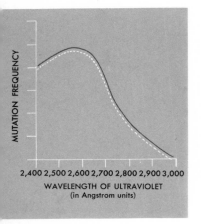

We do not know how many genes there are in any particular organism. The number in human beings has been estimated to be between 10,000 and 50,000. But if each gene mutates at a given rate, it means that every cell has a good chance of containing a mutant gene, newly arisen. If these arise in somatic cells, it may be detected, but it is not passed on to future generations. If arising in eggs or sperm, however, they can be passed on. A reasonable guess would be that one out of every 100 gametes has a new mutation in it. Since these are being passed on to future generations, the individuals of each generation have a wealth of genetic diversity.

Each individual, therefore, is likely to be different from others in the population. This fact reinforces three points already made:

1. Gene mutations provide the initial diversity that makes evolution possible. If there were no diversity, clearly there would be no evolution. Furthermore, since genes recombine in meiosis, the number of possible combinations of genes increases the amount of diversity to an enormous degree. For example, if a species has 1,000 heterozygous genes in its population (a conservative estimate), the number of possible combinations is $2^{1,000}$, a number so large that it exceeds the number of atoms in the universe.

2. Since individuals differ genetically, they will also differ in their degree of adaptability to a given environment. Some will be more, others less, adaptable.

3. Since only a selected number of individuals in a population gives rise to the next generation, the chances are good that the genetic nature of the population as a whole will change with each generation. If the change goes continuously in a given direction, the population will evolve in that direction.

We shall return to these points later in the chapter.

The mutations just discussed arise naturally in populations. We speak of them as **spontaneous mutations** to distinguish them from those artificially induced. X-rays, radioactive materials released by a nuclear bomb, ultraviolet light, and a wide variety of chemicals can induce mutations. The greater the amount of x-rays received by the cells, the greater is the number of mutations induced (Fig. 22.4). We can think of x-rays as small rifle bullets that pass through the genes and alter them during passage. Ultraviolet radiation, on the other hand, behaves differently. It is absorbed (captured) by DNA, and the absorbed energy of ultraviolet makes the DNA unstable, thus bringing about change. Some regions of ultraviolet are more effective than others in inducing mutations. The region around 2,600 Å is particularly effective. The relation of wavelength to mutation induction is shown in Fig. 22.5.

340

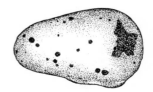

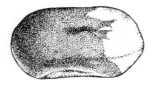

X-rays are very energetic rays, and readily pass through cellular substances. They can induce mutations in any organism. Ultraviolet, however, is quickly absorbed by cellular substances, and it is a good mutation-inducing agent for viruses, bacteria, and single layers of cells.

Many chemicals cause mutations. Some are merely destructive substances, damaging any part of the cell including DNA. Others are more selective in their action. One of these (BUDR, 5-bromouridylic deoxyriboside) is closely related to the nucleotides in DNA and can replace thymine specifically. When it does so, the DNA becomes unstable and tends to mutate.

Interestingly enough, the mutations induced by these agents are the same as those that arise spontaneously. No agent has been found that gives rise to a particular mutation, so we are unable as yet to mutate one particular gene and leave all other genes unaffected. Certain genes, however, can cause other genes to mutate in a specific way. One such mutator gene is found in corn, and is called **Dotted (Dt)**. When *Dt* is in a cell in the presence of gene *a*, which governs color, gene *a* will mutate to *A* at a given frequency. As Fig. 22.6 shows, the mutations show up as colored spots on an uncolored seed. Table 22.2 shows how the number of *Dt* genes in a cell influences the mutation rate $a \rightarrow A$.

Other changes can occur in cells and produce genetic changes that are inherited like mutations, but that are really changes in chromosome structure rather than changes in nucleotide sequences. Some of these are losses of genes, called **deletions.** In *Drosophila* a group of *Notch* mutants, which produce a nick in the wings, and *Minute* mutants, which reduce the overall size of the fly, are in most instances due to losses (Fig. 22.7). The *Notch* effect appears when a particular piece of chromatin on the *X*-chromosome is missing, but the *Minute* mutants are scattered over all the chromosomes, and appear to be due to losses of chromatin forming tRNA.

Figure 22.6 In these two kernels of corn, gene *a* has mutated to A (dark areas). In the top kernel the mutations occurred late in the development of the seed and the spots of A are, consequently, small. In the bottom kernel one or more mutations occurred early, giving a large patch of dark color.

Figure 22.7 Top: the normal wing of *Drosophila melanogaster* (right) is compared with the reduced wing size found in the *Minute* phenotype (left). Bottom: several types of wing alterations result from *Notch* "mutations."

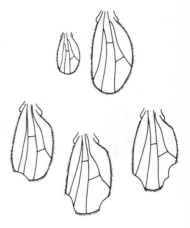

TABLE 22.2. Mutations of a $\rightarrow$ A Occurring in Maize Seeds (Triploid Tissue) When the Number of Dt and a Genes Is Varied

Genetic Composition of Plant	Number of Mutations per Seed
aaa/Dt dt dt	7.2
aaa/Dt Dt dt	22.2
aaa/Dt Dt Dt	121.9
aa^{p}a^p/Dt dt dt	3.20 (calculated)
aaa^p/Dt dt dt	5.64
aaa/Dt dt dt	8.16

After M. M. Rhoades.

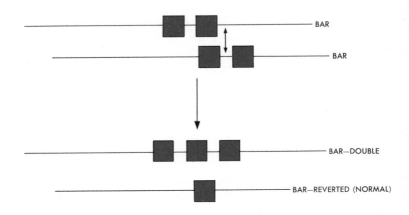

Figure 22.8 The *Bar* "mutation" is a duplication of a segment of chromatin. It causes a change in the shape of the eye and reduces the number of facets per eye (see Fig. 22.9). Occasionally, crossing-over occurs (as diagrammed here), giving a normal chromosome without duplication, and a chromosome in which the Bar region (represented by the dark blocks) is triplicated. This results in the *Bar-double* phenotype in which the eye is even more severely altered.

BAR

BAR

BAR—DOUBLE

BAR—REVERTED (NORMAL)

The *Bar* gene, so-called because it affects eye shape in *Drosophila*, is the result of a gain in genetic material. The piece **duplicated** is in the *X*-chromosome, and by manipulation it is possible to increase the number of duplications (Fig. 22.8). Each added piece reduces still more the size of the eye (Fig. 22.9).

The gain or loss of whole chromosomes also produces a mutation-like effect. **Down's syndrome** (Mongolian idiocy) in human beings appears when the small 21 chromosome is present three times in each cell, rather than in the normal diploid condition. Loss of an *X*-chromosome, to give an *XO* instead of an *XX* condition, leads to the defective physical development of female individuals **(Turner's syndrome)**. An added *X* in males, to give an *XXY* instead of an *XY* situation, also leads to a comparable defective development **(Klinefelter's syndrome).**

Gains or losses of chromosomes, like deletions and duplications, produce their effects by upsetting genetic balance. Think

Figure 22.9 The eye of *Drosophila melanogaster* is shown in the normal form (left) and when the *Bar* region is duplicated or triplicated. The diploid chromosome situation for each one is represented above. The homozygous *Bar* and heterozygous *Double Bar*/normal have the same number of *Bar* regions, but the arrangements of 2/2 and 3/1 have a different effect on eye structure and facet number.

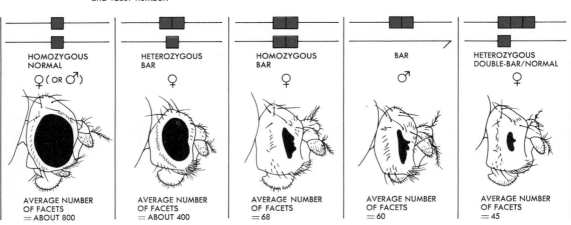

HOMOZYGOUS NORMAL
♀ (OR ♂)
AVERAGE NUMBER OF FACETS = ABOUT 800

HETEROZYGOUS BAR
♀
AVERAGE NUMBER OF FACETS = ABOUT 400

HOMOZYGOUS BAR
♀
AVERAGE NUMBER OF FACETS = 68

BAR
♂
AVERAGE NUMBER OF FACETS = 60

HETEROZYGOUS DOUBLE-BAR/NORMAL
♀
AVERAGE NUMBER OF FACETS = 45

of a normal set of genes, single in the haploid state and double in the diploid state, as that array of genes needed for normal expression of development. Gains or losses of genes will shift expression away from normality. Their role in evolution is still being investigated. Duplications at least provide a possible source of added genetic material, which, over a period of time and through mutation, might come to serve some useful genetic purpose.

When whole sets of chromosomes are added, to give triploids (3*n*), tetraploids (4*n*), and so on, the changed appearance of the organism is slight (generally of an increased size). Such organisms are called **polyploids,** in contrast to normal haploids and diploids. Since one third of the flowering plants are polyploid, polyploidy has played a significant role in their evolution. This phenomenon is rare among animals, however.

THE RATE OF VARIATION Let us assume that gene *A* and its allele *a* exist in a population. Our knowledge of Mendelian inheritance tells us that all the individuals in the population will be *AA, Aa,* or *aa.* If *A* or *a* do not mutate to other allelic forms of the gene, no other genotypes for this gene are possible. If these individuals breed together in random fashion, and no other events change the proportion of *A* to *a*, then the frequencies of *A* and *a* will remain the same from one generation to the next.

This is known as the **Hardy–Weinberg principle,** and we can demonstrate its validity in the following way. The frequency of *A* in a population is equal to the frequency of *AA* individuals plus one half the *Aa* individuals. Let us designate this by the letter *p.* Similarly, the frequency of *a* is equal to the *aa* individuals plus one half the *Aa* individuals. This frequency is designated by the letter *q.* Since all the *AA, Aa,* and *aa* individuals in a population equal 100 per cent, or 1.00, then $p + q = 1.00$. In a breeding population, *A* and *a* sperm and *A* and *a* eggs will be produced, and random fertilization will occur as indicated below:

Eggs	Sperm	Resulting Individuals	Frequency
A	*A*	*AA*	$p \times p$ or p^2
A	*a*	*Aa*	$p \times q$
a	*A*	*Aa*	$p \times q$
a	*a*	*aa*	$q \times q$ or q^2

$\left. p \times q \atop p \times q \right\} 2pq$

The equation $p^2 + 2pq + q^2$ is another expression of the use of the binomial. It represents the proportion of *AA*, *Aa*, and *aa* individuals in a population, and for all succeeding generations. Let us assume that there are nine times more *A* genes than *a* alleles; p^2 is then equal to 0.81 (0.9 × 0.9); $2pq = 0.18$ (0.9 × 0.1 × 2); and $q^2 = 0.01$ (0.1 × 0.1). If these individuals breed together randomly, they will give rise to the next generation, in which the proportion of genes remains the same.

The Hardy–Weinberg principle is a theoretical one, but it shows that variation—once arisen—will persist in a population if it is found in the breeding individuals. It assumes, however, random breeding, equal survival of all genotypes to reproductive age, a large population, no additional mutations, and absence of migration of individuals in and out of the population. Few populations exist in this ideal state. Mutations do occur, although many are lost because the individuals possessing them are not among the selected parents of the next generation. But even if lost, they can occur again and again (Fig. 22.10). Over long periods of time, measured by geologic time, a mutation rate as low as one in 10^6 can ensure continued variability upon which natural selection can act. Those mutations that are clearly harmful—that is, harmful in the sense that the individual possessing

Bristol Laboratories

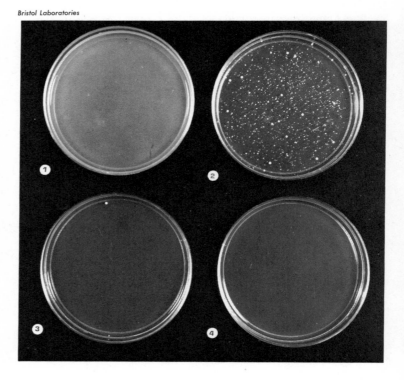

Figure 22.10 Cultures of the bacterium *Staphylococcus aureus.* Plate 1: without penicillin; the colonies are so numerous they merge to form a solid layer over the surface of the dish. Plate 2: medium contains 0.016 mg per ml of penicillin, which has killed most of the bacteria; most of these colonies are probably not strongly penicillin-resistant due to mutation, but some may be. Plate 3: contains 0.032 mg per ml and all bacteria have been killed except two colonies (top and bottom of dish) which are probably penicillin-resistant. Plate 4: contains 0.063 mg per ml and all bacteria have been killed. Resistance to penicillin can result from the mutation of several genes, causing different degrees of resistance.

them has a lowered reproductive capacity—tend to be eliminated. Beneficial ones tend to be retained, and gradually increase in the population.

The retention of variability in a population is often aided by the fact that heterozygous Aa individuals are favored over the homozygous AA or aa. Why this is so is not always clear, but in one well-known instance it is clearly demonstrated. Human hemoglobin A, the common type, is found in all populations, and is controlled by gene A. In Africa, among certain tribes, a mutant type, A_s, causes the appearance of a changed hemoglobin. It is recognizable because the red blood cells collapse into a sickle-shaped configuration. As we saw earlier, hemoglobin S can cause anemia. It has no effect in AA_s individuals because their blood is a mixture of A and S hemoglobins; but it brings on severe and lethal anemia in A_sA_s individuals. Malaria is also a prevalent disease in Africa, and an AA genotype stands a good chance of contracting the disease and dying early in life. The A_sA_s will die of anemia. Only the AA_s heterozygotes have good prospects of being the selected parents of the next generation. They do not suffer from anemia, and the malarial parasite cannot exist in AA_s blood. Yet both genes must be present for the heterozygote to be formed. Should malaria be wiped out as a disease, the environment would be drastically altered, and the AA homozygote would then be at a greater advantage.

This kind of variability is called **polymorphism** (many forms). The type described above is of a kind restricted to one gene, but many instances are known where polymorphism involves many genes. The advantages are obvious. Where the environment over a wide area is highly uniform, polymorphism is likely to be absent, for one kind of genotype would be selected. But this is rare. Most environments are a mixture of innumerable microenvironments, differing from each other in temperature, light, moisture, and amount and kind of food. Each one favors the persistence of a given genotype, and a wide-ranging population of organisms is almost certain to consist of as many sub-populations as there are microenvironments. These populations overlap and interbreed, but they tend to possess a large amount of variability. Only when small populations become isolated, and a certain type of variation becomes fixed throughout the population, do we find the development of new species.

RESULTS
OF VARIABILITY
AS ADAPTATIONS
The great diversity of life around us, ranging in size and form from viruses and bacteria to redwoods, whales, and man, is the result of heritable variations that have survived. Just being alive is a measure of evolutionary success. Being alive and able to reproduce carries such success to another generation further in time.

You are here today because when life first arose on Earth some combination of variations continuously met the challenge of the environment successfully.

During the past, possibly, 3 billion years, the combination of variations underwent continuous change, sometimes slowly, sometimes quite rapidly as measured by geologic time. Different forms of life, exhibiting different modes of existence, came into being. Some have persisted, others have fallen by the wayside. The fossil record in the rocks tells us a partial story of these evolutionary "failures." But these are failures only in the sense that they are not alive today. In their time, before their bodily remains were fossilized, they were a success; they were alive. Yet it is a mistake to think of all fossils as evolutionary failures or as temporary successes. In our discussion of the evolution of the horse, we saw that some lines of evolution came to a dead end. However, the line that gave rise to the modern horse was an evolutionary success even though the modern horse differs from its earlier ancestors. The chain of life was never broken at the same time that the horse was undergoing change.

Consider the huge dinosaurs. So far as we know, they disappeared from the Earth without giving rise to any present-day species. As such they were temporary evolutionary successes. But the reptiles as a whole, the class to which the dinosaurs belonged, were more successful. Some forms gave rise to the birds, others to the mammals, and still others to the reptiles of today. Evolutionary or adaptive success is, therefore, to be measured and judged only in terms of reproductive success. The failure of parents to produce offspring represents the end of the line for the genetic material found in those individuals.

It is difficult, if not impossible, to assess the results of genetic variability in terms of adaptation, except in very simple cases. Adaptation is concerned with the whole business of living, from the egg through birth, reproductive age, and death. All the structures of an individual and all the physiological processes are part of and contribute to adaptability, and these are controlled by many genes. It is true, of course, that human beings can lose an arm or eye, or lack the ability to produce insulin (a diabetic), and still live and reproduce, but we are less adaptable when these organs or processes are missing.

We have already discussed a simple case of adaptation, that of *Staphylococcus aureus,* to an environment suddenly changed by the introduction of antibiotics. Let us examine this somewhat further. This time we shall use the common colon bacillus, *Escherichia coli,* a nonpathogenic form that you can experiment with if you choose. It will respond to antibiotics, much as does *Staphylococcus,* by displaying resistance.

Bacteria such as *E. coli* multiply by simple division. One cell becomes two, and the two daughter cells will, in general, be genetically identical. If we now spread a large number of bacteria onto an agar medium containing streptomycin, most of the cells will die. These are nonresistant cells. An occasional colony will form, however, each colony being the result of repeated divisions of a single cell when the plating was done. Cells from this colony can be repeatedly transferred to a streptomycin-containing medium, and they will continue to grow. These cells are resistant. Genetic tests show that resistance is determined by one or, possibly, several genes.

A question then arises. Was the genetic variation in the form of streptomycin resistance induced by the streptomycin? Or did the mutation occur in the absence of streptomycin, and only reveal itself when the antibiotic was added to the culture medium? The answer is that streptomycin did not induce the mutation; the mutation arose spontaneously and in the absence of streptomycin. This can be demonstrated in the following way.

A culture of *E. coli* is spread on a medium lacking streptomycin. Colonies will occur, as with the staphylococcus in Fig. 22.10. If we then gently press a sterile velvet disc onto the culture plate, this disc can be removed and pressed onto another petri dish containing a medium containing streptomycin. The colonies of bacteria will be transferred by the nap of the velvet in the exact pattern as found in the original dish. Most of the colonies—those formed by nonresistant cells—will die. Occasional colonies will arise, however, and from their position on the plate they can be identified with their sister colonies on the non-streptomycin-containing medium. These sister colonies can again be tested. They will be found to be streptomycin-resistant even though they had never previously been exposed to the antibiotic.

The resistant cells, of course, could not have anticipated that sometime they would be exposed to streptomycin. This is only one of hundreds of substances that can kill *E. coli*. In the absence of streptomycin, the resistant trait is a useless one so far as we know. Yet it appeared. Adaptation, therefore, is a makeshift event, having no purpose until an unforeseen change in the environment allows the variation to be expressed. The frequency of such mutations is low, one in 10^8 or 10^9 cells possessing the variation. However, the fact that such random variations occur indicates that most organisms have a greater storehouse of variability than will ever find adaptive expression.

The case of adaptation just described is a simple one, easily understood in genetic terms. A somewhat more complicated instance is that known as **industrial melanism** among moths. As a phenomenon, it is not restricted to any particular species of moth—it occurs in about 70 different species, most of which do not interbreed—and it can be described as the acquisition of protective coloration. Let us consider the case of *Biston betularia*.

As Fig. 22.11 shows, the usual moth is a light-colored specimen. When not in flight, the moth rests on lichen-encrusted trees, a position in which it blends into the background. It is not readily

Figure 22.11 These two moths, each having a different color form, inhabit an industrial area of England. The light colored form is the ancestral one; the dark form, resulting from randomly occurring mutations, increased in numbers in the soot-covered areas. Birds, which preyed on the moths, found it more difficult to detect the dark colored form.

R. H. Noailles

visible to predators. About the middle of the 19th century, a few melanic (dark) forms made their appearance in the vicinity of the industrial city of Manchester, England. At this time, the soot from factories and homes was being deposited throughout the neighborhood, killing off the lichens and blackening the tree trunks. Against this background, the melanic form blended easily, while the normal lighter form was relatively conspicuous. The dark form spread rapidly, and was also noticed in the vicinity of other industrialized areas. It was given the variety name of *carbonaria.*

It seems clear that the melanic color is an adaptation to a changing environment. To prove this, H. B. D. Kettlewell, an English biologist, showed that the moths are eaten by birds. He released a number of light and melanic individuals in sooty and nonsooty areas, noticed their resting places, and then after a period of time counted their numbers. In the sooty areas, the light form was eaten more frequently; in nonsooty areas, the melanic form was more readily detected by birds. Ease of concealment, and thus survival, is, therefore, a function of both body coloration and the character of the physical environment.

Coloration is generally governed by a number of genes. In genetic tests that have been conducted, melanism is governed by dominant genes, and this accounts for the rapid spread of melanism. Why would the spread be less rapid if melanism were controlled by recessive genes?

Other examples of adaptation are so complicated that we stand little chance of getting at the genetic basis of them (Fig. 22.12). In Australia, for example, the male bower bird goes through an elaborate nuptial ceremony. He builds an intricate bower or nest, collects bright stones or other objects for display, sings in a special way, and performs a nuptial dance, all this to make himself acceptable to a female of the species. Is this necessary for reproduction, which is the ultimate criterion of adaptive success? Is the mating ritual of the male a means of recognition, or is it tied closely to a physiological response related to reproduction? Answers are not easy, but we can guess that if the mating ritual of the male repelled rather than attracted the female, the mating ritual would quickly disappear as harmful adaptive behavior.

In general, organisms possess a wealth of genetic variation. This arises by mutation and without foresight or purpose. If the variation is in harmony with the environment, in such a way as to promote reproductive success, it will stand a chance of being perpetuated. This is the source of the diversity of life. If the variation is harmful, it will be eliminated. If it arises at the wrong time, it will probably be lost. Furthermore, an adaptive variation in one environment may be useless or even harmful in another. Evolution, therefore, appears to be governed by chance, the chance that the right kind of variation appears in the right kind of environment and in the right kind of organism.

O. S. Pettingill, Jr. from National Audubon Society

O. S. Pettingill, Jr. from National Audubon Society

Figure 22.12 This sequence of photographs shows the courtship display of the Laysan albatross on Midway Island in the Pacific. It can be assumed that behavior of this sort, which often reaches a highly complex form of ritual, is a heritable trait and a necessary prelude to successful reproduction.

THE ORIGIN OF LIFE   If the diversity of life now present on the Earth evolved from past forms of life, and if the unity of life, expressed through homologous structures and similar pathways of metabolism, indicates a common ancestry, then we must assume that life in all its aspects has a developmental history. If this is so, then there was a beginning some time in the past.

All of us know that the world around us can be divided into two more or less distinct systems: the living and nonliving. If there was a beginning, did these systems arise at the same time, or did one precede the other? Older theories advanced the idea that life was specially created from the nonliving. The creation myths of ancient peoples in Babylonia, Greece, Egypt, India, and the early Indian civilizations of the Americas all held that life arose spontaneously from the nonliving by supernatural or divine intervention.

Such ideas are also part of our great Judaic-Christian traditions. However, the work of Pasteur showed that life does not arise spontaneously *under present conditions,* and Darwin's theory of evolution tells us only how life evolved, not how it originated. Until recently, scientists could not deal meaningfully with the question of the origin of life. They had no way of getting beyond the point that all organic molecules—that is, all those containing carbon atoms—were the products of living cells. If this were so, they could not conceive of any way by which nonliving matter could become living matter. But in 1923, A. I. Oparin, a Russian biologist, suggested how this might be possible. His book, *The Origin of Life,* was the beginning of new approaches to this perplexing problem.

THE PLANET EARTH   The part of the universe we can observe contains millions of galaxies. Our own galaxy, the Milky Way, contains about 100 billion stars. The Sun, around which the Earth revolves, is a star of medium size located near the edge of the Milky Way. The Solar System consists of the Sun at the center, plus the nine planets revolving around it and held in position by gravitational attraction. The Solar System is about 9 billion miles in diameter, and the order of the planets, moving from the Sun outward, is Mercury, Venus, Earth, Mars, Jupiter, Saturn, Uranus, Neptune, and Pluto. There are 31 smaller satellites revolving around the planets, for example, our Moon.

The age and origin of the universe are unknown. One idea is that it has no beginning and no end. This is a difficult idea to handle intellectually; it is also believed that about 10 billion years ago a dense core of primordial matter exploded. The effects of the explosion are still evident in the universe, for the galaxies are moving outward and away from each other at enormous speeds. The expanding material eventually thinned out, cooled, and then condensed into the present galaxies, stars, and planets. Condensation of such enormous masses of material produces heat, and our Sun and the planet Earth became molten as condensation proceeded. The Sun, with a diameter of about 850,000 miles, generated enough heat to remain in a gaseous state, with temperatures that range from 6,000°C at the surface to about 25,000,000°C at its center. The most prevalent gas is hydrogen, and at the temperature of the Sun's core hydrogen fuses and forms helium, the reaction being the same as that occurring in the explosion of a hydrogen bomb. When the hydrogen is exhausted through transformation into helium, our Sun will become a dead star, no longer a source of light and heat for the Earth. Life on Earth will then cease.

The Earth is too small to generate enough heat to sustain nuclear reactions, but it achieved a molten state at first and then gradually cooled, forming a crust at its surface. As it cooled, the crust cracked, wrinkled, and formed an irregular surface. This process is still going on today and accounts, in part, for the earthquakes that occur periodically. Water was formed during the cooling process, but it could not, of course, exist as a liquid until the Earth's surface cooled enough to allow it to condense and form the oceans. But once this happened, the stage was set for the beginnings of life.

IN THE BEGINNING . . .   If we compare the living with the nonliving, three things stand out:

1. Life is built around the carbon atom.
2. Water is a major component of all living systems.
3. Life exists within a rather narrow range of temperatures.

Chapters 6 through 13 deal primarily with the chemistry of the carbon atom in various configurations. Other chapters have dealt with cellular structures, each one of which has carbon as a central atomic ingredient. Carbon is, therefore, unique among the elements in being very much a part of the living systems known to us. This uniqueness lies in the ability of the carbon atom to form stable molecular configurations. It can be bonded equally well with hydrogen, oxygen, nitrogen, and other elements. In a variety of combinations it forms molecules as small as carbon monoxide and as large as proteins and nucleic acids, which have molecular weights of many millions. The variety of combinations, considering both size and structure, into which carbon can enter is almost infinite, but of equal importance is the fact that these molecules are exceedingly versatile in function and in their ability to capture, retain, and transfer energy. If life exists on other planets in other solar systems, biologists believe that such life must be based on the carbon atom.

Of comparable significance is the water molecule. Life as we know it could not have originated in its absence, and the patterns of life we see today, and that have existed in the past, have been determined by the presence and relative abundance of water. The importance of water lies in its two major properties: it is a molecule of great stability, and it is the "universal solvent" of biological systems.

Water exists as a gas, a liquid, and a solid, but it is as a liquid that water plays a biological role. The human body is about 70 per cent water; a jellyfish is about 98 per cent water. Water is the solvent in which metabolic reactions go on; chemical substances ranging in size from ions to huge molecules exist as water solutions or suspensions. Water is the major vehicle for the transport of substances in and out of the body, from one part of the body to another, and in and out of cells. But water is not so stable that it is inert. We have already seen that if participates in numerous reactions, such as the hydrolysis of carbohydrates, fats, and proteins, and in the process of photosynthesis.

Temperature is also a limiting factor for the existence of life. If the environment is too hot, molecules cannot achieve great size and complexity. Increased heat may speed up the formation of biologically important molecules, but it also increases the rate of breakdown. Decomposition of molecules cannot outrun the rate of synthesis if structures are to be built up. The environment cannot be too cool either. Although the synthesis of molecules might go on, although slowly, breakdown would also be slow. As a result, nature would arrive at a point of stagnation. There would be a great complexity of molecules, but there would be little change. Life is ceaseless change. There is constant activity at the molecular level. The breakdown of molecules competes with synthesis; simplicity of molecular structure competes with complexity; and all the while energy is being introduced, transferred, and converted by means of molecular systems.

Above all else, life is associated with carbon atoms, water, and energy. How were these arranged, resulting in a substance said to be *living?* Since it happened in the past, why isn't it happening now? Charles Darwin considered this problem in 1871, and he wrote:

> It is often said that all the conditions for the first production of a living organism are now present, which could ever have been present. But if (and oh! what a big if) we could conceive in some warm little pond, with all sorts of ammonia and phosphoric salts, light, heat, electricity, etc. present, that a protein compound was chemically formed ready to undergo still more complex changes, at the present day such matter would be instantly devoured or absorbed, which would not have been the case before living creatures were formed.

The point of view expressed by Darwin is not far different from that which we would offer. When the Earth was first formed conditions must have been so harsh that no living thing could have survived, had there been living things around. But the Earth was going through its own evolution, and conditions changed. What were these conditions? Darwin's "warm little pond" was probably one of them. Since oxygen and carbon dioxide are largely by-products of life, the early atmosphere probably did not contain them to any great degree. But hydrogen, ammonia (NH_3), and methane (swamp gas, CH_4) probably were plentiful. Water was also abundant, and was warmed by the heat of the gradually cooling Earth. Energy, too, was available in the form of lightning and in the form of ultraviolet light from the Sun.

If these were the conditions found at the surface of our planet during its prelife state, then a model experiment can be carried out to test the idea that organic molecules can form spontaneously in such conditions. This was done by Stanley Miller and Harold Urey at the University of Chicago in 1953. They exposed a solution of hydrogen, water, ammonia, and methane to electrical discharges and to ultraviolet light for a period of 24 hours, then analyzed the solution. A number of low molecular weight carbon compounds were formed, many of them acids, but most interestingly and surprisingly, a number of amino acids were also formed. These are the building blocks of proteins. By the addition of other materials believed to have been present on the Earth during its prelife state—particularly hydrogen cyanide (HCN)—purines and pyrimidines, the building blocks of nucleic acids, appeared in the solution. Only the carbohydrates and fats seemed to be lacking.

Darwin's "warm little pond" could, therefore, become a rich broth of chemical compounds, increasing in concentration as time passed. All these compounds contained carbon in various combinations with itself and with other elements. Those that were stable persisted; those that were unstable or reactive broke down or were reconverted to other molecular forms. Natural selection took place at the chemical level, and we must assume, therefore, that a long period of chemical evolution preceded biological evolution. The broth, furthermore, was sterile, since no life as yet existed.

The broth thickened with time, and then somewhere, somehow, a self-replicating system arose. The geological record indicates that this probably occurred about 3 billion years ago. Our knowledge at this point is fragmentary, but primitive self-replicating systems with sources of energy have been produced in the laboratory. Once formed, such a system could draw from the broth the nutrients it needed for "survival," and eventually encase itself in a semipermeable membrane, thus permitting various molecules to enter and leave the system on a selective basis.

The system—we may, indeed, call it a primitive cell—was heterotrophic; its source of energy was from preformed compounds outside the cell. However, such a cell and its descendants would soon exhaust its immediate environment of nutrient materials, and we can only assume that natural selection would lead to the evolution and survival of more complicated cellular machinery capable of making some of its own nutritional substances. Enzymes would consequently be required (but amino acids required for their formation were available). In addition, we must assume that a shift in the source of light energy took place. As a result of the use of water and carbon dioxide, molecular oxygen (O_2) was formed and released from the surface of the Earth as a gas. Forming a layer around the Earth, the oxygen would screen out much of the ultraviolet light. The evolution of compounds that could capture the energy of visible light would lead eventually to the process of photosynthesis and, consequently, to the major oxygen source for the living world to come. Autotrophy was then initiated. The fossil record supports such a succession of events. Biological evolution was now well on its way.

You might now say, and with good reason, that this story of the origin of life is highly speculative and far from proved. You would, of course, be quite correct. We have no knowledge of many of the intermediate steps; the jump from the Miller–Urey broth, or even from a primitive self-replicating system, to an integrated living cell is an enormous one. But two additional items support the argument. In the first place, there is an increasing number of experimental facts being accumulated, all of which point in this direction. And, second, time has been available for these processes to occur. A period of from 2 to 3 billion years passed from the time that the Earth first formed until life made its appearance. Most biologists believe that if conditions favoring the emergence of life are present, then life will inevitably arise in time as a natural consequence of changes in matter and energy. So among the many galaxies, stars, and planets of the universe, there is a high probability that life exists elsewhere, and that we are not alone in the immensity of space and time.

summary

Adaptation can be viewed as a state of being or as a process. As a state of being, adaptation is the sum total of all the characteristics—including the element of chance—that permit an organism to live and reproduce. As a process, adaptation, or **natural selection,** is the manner by which evolutionary success is achieved. Since environmental changes occur, adaptation involves a continual adjustment of organisms to an environment. The bases for adaptation are the heritable variations in a population and the fact that not all individuals in a population contribute their genes to the next generation.

Under such conditions, change is inevitable and evolution

proceeds. Therefore, all heritable adaptive changes contribute to evolution, but all evolutionary changes are not necessarily adaptive. The fossil record indicates that many species are now extinct, even though at one time these species were adaptively successful. Adaptation, consequently, is only a temporary success, and adaptation in one set of circumstances does not guarantee equal success under a different set of circumstances.

Darwin's theory of evolution through natural selection most satisfactorily explains the successive changes that have occurred as one generation of organisms succeeds another. More recent studies have demonstrated that variations result from changes in the base sequence of DNA, and from gains or losses of chromosomes or parts of chromosomes.

Life is believed to have originated from a nonliving state about 3 billion years ago. We are now beginning to have some understanding of how this might have happened, but we have far from a full knowledge of all the events.

for thought and discussion

1 Make a list of species particularly well suited to the environment in which they exist. Try to analyze one species and show how it is adapted. What factors of the environment must be considered? Do you know of any organisms that would not be adapted if the care of man were relaxed or stopped? What features necessary for adaptation are missing in these organisms?

2 How can the rising up of a mountain chain such as the Appalachians or the Rockies affect the environment?

3 "A variable environment strongly promotes rapid evolution and may, in fact, be essential for speeding up evolutionary change." Explain this statement.

4 Evolution is essentially irreversible. Why?

5 The dolphin is a mammal, but in many of its features and behavior it is very similar to a fish. Is there any reason for this degree of parallelism?

6 The following animals all fly: butterfly, sparrow, bat, and flying fish. Are their flying organs homologous or analogous?

7 Man has been called a "lethal factor" in the environment for other organisms, and possibly for himself. What does this statement mean? What can be done about it? Could any other organism be so labeled? Explain.

8 It has been estimated that it may take 1 million years to form a new species of animal. What does this tell us about the causes of evolution?

9 There is much talk today about the human population explosion. What caused it in the first place? What will be the consequences if such a trend goes unchecked? Does it illustrate any facet of Darwinism?

10 We have stated that polyploidy is rare in sexually reproducing animals. Why should this be so? Why can plants tolerate polyploidy more successfully than animals?

11 Most mutations that arise are deleterious. Why should this be so?

12 What factors in the environment would keep a population constant in numbers?

13 Two similar species of birds occupy the same environment. When do they compete with each other, and when don't they compete?

14 Why are there only two species of elephants and one species of man in existence, but thousands of species of *Drosophila*?

15 Why should a deletion or a duplication be likely to produce a dominant effect?

16 Why is reproductive isolation, by a geographical barrier, for instance, necessary for the formation of a new species?

17 A *rassenkreis* (race circle) is a continuous circle (often over a wide area of the range of an organism) of changing races. Assume that the races range from A to Z, with the ranges of A and Z overlapping. Assume also that A can interbreed with B, B with C, C with D, and so on, but that A cannot interbreed with Z. Can you explain what has happened?

18 In what ways can chance influence the evolutionary picture?

19 Suppose in a population breeding according to the Hardy–Weinberg principle that A for some reason now mutates to a at a slow rate, but a does not mutate back to A. What will be the consequences? What will they be if a, for some reason, breeds more rapidly than A?

selected readings

EHRENSVARD, G. *Life: Origin and Development.* Chicago: University of Chicago Press, 1962. An interesting and clear discussion of how life might have originated on Earth.

GREENE, J. C. *The Death of Adam.* New York: New American Library, Inc., (Mentor Books), 1959. A history of the ideas of evolution from earliest time to Darwin.

HUXLEY, J. *Evolution in Action.* New York: Harper & Row, Publishers, 1953. A series of fine essays on evolution as a process.

SIMPSON, G. G. *The Meaning of Evolution.* New York: New American Library, Inc., (Mentor Books), 1949. One of the finest discussions of the processes of evolution.

STEBBINS, G. L. *The Processes of Organic Evolution.* Englewood Cliffs, N.J.: Prentice-Hall, Inc., 1966. An excellent and up-to-date account of why evolution takes place.

WALLACE, B., and A. M. SRB. *Adaptation.* Englewood Cliffs, N.J.: Prentice-Hall, Inc., 1964. A clear, well-written account of the meaning of adaptation and its relation of genetics and evolution.

the origins
of man
chapter 23

Man is an animal. If you were to dissect him, you would find
the usual organs—heart, liver, lungs, stomach, and so on. They
would differ very little, except in size, from similar organs in cats,
horses, mice, and monkeys (Fig. 23.1). At the cell level of structure
it is almost impossible (except for a specialist) to distinguish
among the liver cells of these animals, including man. If you were
to study the physiology of these organs—their respiration and
enzyme activity, for example—the same general chemical structure and behavior characterizes all of them. However, if you were
to compare the nuclei of the cells of these various animals, you
would find that each has a different number of chromosomes.
Closer examination, by genetic techniques, would reveal that the
genes in these chromosomes are also different, although many
might well be similar. It is a long evolutionary road from fish
to man, but these two groups share about 10 per cent of the same
genes. It is the differences among the genes that make us human
beings instead of some other animal.

Where does man fit into the animal kingdom? He has a
vertebral column, which extends from his skull to the lower part
of the back, which is made up of individual vertebrae, and
through which runs the spinal cord. Man, therefore, belongs to
the phylum Chordata. He is also a mammal, possessing mammary glands, and is put in the class Mammalia. It is true that

Figure 23.1 The skeleton of a man and that of a rearing horse reveal the broad similarities of bone number and arrangement. At the same time, modifications of the pelvis, tail, head, and appendages characteristic of each species can be seen.

Figure 23.2 The general evolutionary relationships among vertebrate animals, including man, are shown here.

he is a peculiar, almost hairless mammal, but so too is the whale, which has even less hair. Also, man walks upright on two legs, but kangaroos and some monkeys and apes are occasionally **bipedal** as well. But monkeys, apes, and men have many features in common. They are grouped in the order Primates, along with tree shrews, tarsiers, and lemurs. The common features of monkeys, apes, and man put them into a suborder Anthropoidea, and within this grouping the family of man, Hominidae, is found. The relationship of man and several of his animal relatives is shown in Fig. 23.2.

Although man clearly is an animal, it is equally obvious that he is a unique animal. When we try to analyze this uniqueness, we can best sum it up by saying that man has developed a **culture,** something that no other living thing has done. The term "culture" is, of course, a general one. Man is a maker and user of tools, a planner, an inventor of symbols—spoken and written languages. Man also has a system of values. And he can do things with a purpose, not only for himself, but for future generations as well. All these add up to a cultural inheritance, a unique phenomenon that is different from the biological inheritance common to all living things.

THE BEGINNINGS OF MAN Man is an adaptive creature, as much a product of evolution as any other living organism. Adam is a symbolic first man, but man's actual origin goes far back in geological history. We do not yet know all the steps of his evolution, and the search for fossil man continues, but the main pathways are generally known and accepted.

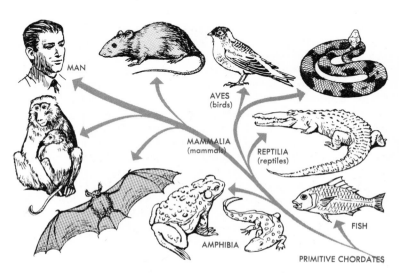

Figure 23.3 shows some of man's primate relatives. None, however, is a direct ancestor, and it is not correct to state that present-day apes were man's ancestors. Rather both man and the modern apes had a common ancestor that was ape-like.

The first human fossil to be discovered was the skull fragments of Neanderthal man, found in 1856 near Düsseldorf, Germany (Fig. 23.4). Rudolf Virchow, the German physician, did not believe it to be a fossil man, but a skull of a modern man showing pathological deformities. Other discoveries, however, left no doubt about the antiquity of *Homo neanderthalensis*, and he is now known to have occupied parts of Europe some 70,000 to 40,000 years ago. He had a primitive culture. He buried his dead with food and weapons for an afterlife. Ultimately, he was probably overrun by other races of men who came in from eastern Asian regions. But he did not, apparently, give rise to modern man. The reconstruction of *Homo neanderthalensis* shows him as not too different from some of our more brutish appearing friends (Fig. 23.5), yet he belonged to a species different from modern man, *Homo sapiens*.

Fossil remains show that Neanderthal man was a relatively late arrival on the evolutionary stage. A long period of history preceded him. The geology of the Pleistocene period, the Ice Age, can be used to date various fossil remains. The Ice Age was characterized by various periods when the ice advanced **(glacial)** and when it retreated **(interglacial)**. The earliest of these fossils, dated from the late Pliocene into the first interglacial period (2,000,000 to 700,000 years ago), have been called *Australopithecines* (*Australis* meaning "south"; *pithecus* meaning "ape"). Many fossils of these pre-men have been found. As a group, they were upright in stature, bipedal in their walk, about 5 feet tall, low browed, and had large teeth and powerful jaws. Their cranial capacity was about 450 to 550 cm³, just a bit larger than that of a chimpanzee (350 to 450 cm³), but much smaller than that of modern man (1,200 to 1,500 cm³). Were they really men, or apes, or something between? It is hard to say. The australopithecines are quite ape-like, but L. S. B. Leakey, the African anthropologist, believes that some of them, found with australopithecine fossils but possibly differing from them, fashioned crude stone tools for hunting. One group of these early men has been named *Zinjanthropus,* after the area of Zinj, in Tanganyika, where fossil remains of this group have been found. Age determinations suggest that the fossils are about 1,750,000 years old. Another fossil find recently made of an apparently more human-like being than the australopithecines are believed to be more than 2,600,000 years old, making them the oldest known "human" fossils. This may seem a very long time ago, but in geological terms it is very recent. Present records suggest that these individuals died out about 400,000 years ago (mid-Pleistocene).

Other fossilized human, or near-human, remains have been

The American Museum of Natural History

Figure 23.3 Two members of the primate group to which man belongs: the chimpanzee (above), and an Old World Rhesus monkey with her offspring.

Photo by Lynwood M. Chace

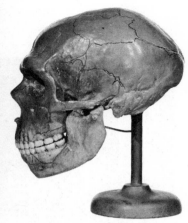

Figure 23.4 The fossilized skull of an extinct species of man, *Homo neanderthalensis.*

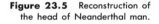

Figure 23.5 Reconstruction of the head of Neanderthal man.

found in Java (*Pithecanthropus,* meaning "ape-man"), in China (*Sinanthropus,* meaning "Chinese man"), and in other parts of Africa and in Europe. They have been grouped as *Homo erectus.* Their relation to the australopithecines is still uncertain, but it is generally assumed that they were ancestral to modern man. *H. erectus* lived about 600,000 to 300,000 years ago. One thing we do know for certain is that, with the passage of time, these fossil remains show an increase in cranial capacity. *Sinanthropus* had a cranial capacity of 750 to 1,000 cm^3, while Cro-Magnon man, living in southern France and Spain about 50,000 to 75,000 years ago, had a brain case as large as today's adult. In addition, Cro-Magnon man must also have had a well-developed culture, if we are to judge from the remarkably beautiful paintings that he drew on the walls of his caves (Fig. 23.6). Certainly he must have been human in all respects.

When did modern man make his appearance? This is as hard to answer as the question of his direct ancestors, but 50,000 to 100,000 years ago is a reasonable guess. This was the time of the Wisconsin Ice Age. By then man was a skilled tool maker, had artistic ability, and was a domesticator of animals. Unfortunately, we are as much in the dark about his place of origin as we are about his time of origin. About all we can say is that he lived in many parts of Europe, Asia, Africa, and inhabited the large islands of the Pacific Ocean, and that the change from australopithecine to *H. erectus* to *H. sapiens* was a gradual one taking place over a long period of time. Out of these have come today's races of men.

THE RACES OF MAN One useful definition of a species is that it consists of a group of populations actually or potentially capable of interbreeding, but not generally capable of cross-breeding with other species. Only one species of man, *Homo sapiens,* has survived, although there were probably other species who lived in the past. Human beings, however, fall into several races, superficially distinct from each other in skeletal and surface features but overlapping in many other characteristics, and quite capable of mixing and producing fertile offspring. The different races arose through the isolation of groups by geographic or social barriers, and because of these barriers different groups of genes were isolated. The origin and perpetuation of the races of man can be explained that simply.

The number of races of man depends upon how closely one defines the term "race." It is a scientifically difficult task, and one that is made more difficult by emotions. Within each race—such as the white, or **Caucasian,** race—there is a great deal of variation. When we take all racial variations into account, it becomes clear that man is highly variable, and that the races

Figure 23.6 Reconstruction of Cro-Magnon artists decorating the walls of a cave at Font de Gaume, in southern France. From what we can tell, cave painting of animals was performed as an act of magic intended to bring about successful hunting.

overlap each other in many ways. This must mean that there is a tremendous store of genetic variation in mankind which is unevenly distributed among the peoples of the earth. Intermarriage between races means that distinctions are even more difficult to make.

The major, but not the only, races of man are three: **Caucasoid** or white, **Negroid** or black, and **Mongoloid** or yellow. Their distribution, despite a vast amount of migration, is basically geographical: the Caucasoids inhabit Europe and migrated to the Americas; the Negroids are African; the Mongoloids, Asian. Let us consider some of the characteristics of each, and the existence of **subraces.**

The Caucasoid whites get their name from the Caucasus region of eastern Europe. They are characterized generally as follows: they have large, heavy bones, well-developed muscles, hair that is wavy or straight, but never woolly or kinky, much body hair, heavy beards in males, a straight narrow nose, blue to brown eye color, and skin that ranges from white to light brown. Within the race there are many subraces (Fig. 23.7); Mediterranean, Nordic, Alpine, Lapp, Irano-Afghan, Indian (Asian, not North American), and some Africans are among them. As Table 23.1 shows, each has certain features or groups of features that distinguish them as "types."

The Negroid race is equally variable both in stature and facial features. Very probably the Negroid race is a mixture of types. The skin varies from dark brown to black, hair is black and woolly or kinky, the nose is broad and flat, the ears are small,

Figure 23.7 Three types of Caucasians: a Swiss mountain man (top); an Indian (Asian) ivory carver (middle); and a Portuguese country girl.

lips thick, upper jaw protruding, and beard and body hair are sparse. Although concentrated in central and southern Africa, the Negroid race is also found in Malaysia, the Philippines, and many of the Oceanic islands. Where they encounter members of the Caucasian race, intermixture has taken place to a great extent, leading to many different blends of racial characteristics.

TABLE 23.1. Characteristics of Some Caucasoid Subraces

Source	Characteristics
Mediter-ranean	Main subrace of Caucasoids. Light build, medium stature, rather long head, and dark complexion. Originated near Palestine and migrated westward along the shores of the Mediterranean into Greece, Italy, southern France, Spain (north shore), and into Egypt and Arabia (south shore). The Basques differ in having a narrow face and narrow, prominent nose. The Irano-Afghan differ in having a high head, long face with high-bridged nose, and taller stature.
Alpines	Round head, thick-set build, sallow complexion, and broad nose. Represented by some of the French, Bavarians, Swiss, and northern Italians.
Nordic	Tall, blond, long head, long face, narrow nose, deep chin. Represented by inhabitants of Sweden and eastern Norway. The Anglo-Saxon type of North Germany and England differs in being heavier boned, more rugged, rounder head, broader nose, and more prominent cheekbones. Invasions of Europe by Angles, Saxons, Jutes, Danes, Goths, and Vandals spread the Nordic features widely.
Lapps	Round head, short stature, forehead steep with no brow ridges, face short, flat, and broad, small jaws, small teeth, projecting cheekbones, hair generally dark and straight, legs short but arms long, hands and feet small. Live in forested highlands of Sweden, northern coasts of Norway, and the tundra area of northern Finland.
Indians (Asian)	Difficult to characterize because of many variants, but darker skinned than the Mediterraneans, of medium stature, small face and chin, dark somewhat wavy hair. The Ceylonese are somewhat darker, while the Sikhs are taller, more heavily built, and have strong beards. Includes most of the Indian population.

In stature, the Negroid race includes the extremes. The pygmy, or Negrillo, of equatorial Africa, the bushman of southern Africa, and the Hottentot of western South Africa are 5 feet or under in height. The Negritos of Asia and the Oceanic Islands are also short. But the Negroes of the upper reaches of the Nile River—of the Dinka tribe—average more than 6 feet in height, and are commonly over 7 feet.

The Mongoloid race is characterized by a yellowish skin, straight black hair, flat face and nose, high cheekbones, little development of the brow ridges, and the **epicanthic fold** of the upper eyelid that gives the eyes a slit-like appearance. The body tends to be short and thick, and there is a sparseness of beard and body hair. The Mongoloid race includes the Chinese, Japanese, Koreans, Siamese, Burmese, Tartars, Tibetans, the Dyaks of Borneo, Eskimos, some Filipinos, and American Indians.

Wherever the Mongoloid has met the white or black race, mixing has occurred. The American Indian, who probably crossed the Bering Strait from Asia 10,000 to 20,000 years ago, is not typically Mongoloid. Possibly he is a product of interbreeding between the Mongoloid and the Ainus of Japan, an ancient white stock that may have been the ancestor of the Mongoloid race. We must recognize that man has always been a mobile animal, moving constantly over the face of the earth. It is not surprising, then, that races having little variation, such as the Australian aborigines, occur only where isolation has existed for long periods of time. Mixing of races has been the rule; hence the many subraces (Fig. 23.9).

CLIMATE AND RACE   We have said that man is the product of evolution. He should, therefore, reflect the selective pressures exerted on him by the environment. Through natural selection, some heritable characters are favored while others are selected against, particularly if the environmental conditions are severe. The white race, generally occupying the temperate zone, does not show extreme specialization. Where extremes of heat and cold are found, the races occupying these areas follow three general rules of adaptation to climate. These rules have been derived from studies of certain animal populations, but they seem to be applicable to man as well.

Gloger's Rule. This states that pigmentation is greatest in warm and humid areas. Certainly, the black tribes of Africa reflect this adaptation, and no dark races are found in the cold zones.

Bergman's Rule. This states that animals are smaller in warmer than in cooler zones. This may seem to be contradicted by the Tungus of eastern Siberia, but if one interprets the rule in terms of body build rather than stature, agreement is found.

Ewing Galloway

Ewing Galloway

Figure 23.8   Two members of the Negro race: a young Zulu from Rhodesia (top); and a Maori tribal chief, with tattooed face and wearing a cloak of fine feathers, from the islands of the southwest Pacific.

Figure 23.9 This group of young Hawaiian women shows the results of racial mixing. The group derives from Hawaiian, Japanese, Chinese, English, Scotch, and Irish ancestry.

The pygmies of Africa, for example, are a slender race with long extremities while those peoples around the Arctic Circle are thick-set, short, well padded with fat, and have flat faces. It is thought that the Tungus arose during the intense cold of the last ice age, about 25,000 years ago. In terms of surface to volume, they present the least surface area, and therefore radiate very little heat. The pygmies, however, have a large surface area relative to the volume of their bodies (Fig. 23.10).

Allen's Rule. This states that animals living in cold areas have shorter extremities than animals living in warm areas. This is true for man as well. The Eskimos, for example, have short toes, legs, fingers, and noses. White races of the desert, such as the Arabian Berbers, have long slender limbs and, often, prominent noses.

We should not, of course, apply these rules too closely. The body of man is relatively unspecialized. He has interbred widely, is quite mobile, and his mastery in making clothing, housing, tools, and fire gives him a control over the environment possessed by no other animal.

THE GENETICS OF RACE The genetics of man is no different from that of other mammals.
Human DNA is like that of other
organisms, differing only in the number and kinds of genes. The
rules of transmission of heritable traits apply to man as well. The
methods of human genetics, however, may seem somewhat different. Although biologically possible, it is not socially possible to
breed human beings selectively as we do fruit flies or mice. But
studying the pedigrees of human families is no different from a
study of other organisms, even though one generally has to work
with fewer numbers of individuals, and the generation times are
quite long (Fig. 23.11).

Nevertheless, much is known of man's inheritance. On a
racial basis, however, details are available only for blood types.
The major groupings of individuals are A, B, AB, and O. These
correspond to six genotypes: *AA, BB, AB, AO, BO,* and *OO. AA*
and *AO* produce the same phenotype, as do *BB* and *BO.* When
plotted on a racial basis, it is found that these traits are not
randomly distributed. For example, type B is not found in the

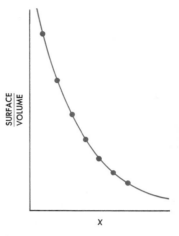

Figure 23.10 This graph
shows how the surface-to-volume
ratio decreases as a body of
constant shape (any shape,
including the human body)
increases in size. (X = size,
and increases from left to right.)

Figure 23.11 Pedigree of European royalty and the transmission of the gene for
hemophilia. Every female carrier and every hemophilic male can be traced back to
Queen Victoria of England. The mutation probably originated in her. The gene
responsible for the hemophilic trait is on the X-chromosome. The transmission can
be understood if it is remembered that the human female is *XX* and the male *XY*
(the Y-chromosome carries no known genes on it).

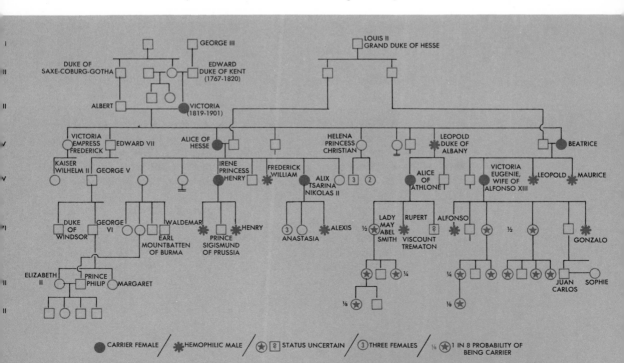

American Indian, the Spanish Basque population, or among the aboriginies of Australia. These races are only remotely related, so the absence of this blood type—and the responsible gene—can be attributed to the fact that these races arose through migration by small groups lacking B. Continued isolation preserved the genetic difference. This seems a more reasonable explanation than that type B was eliminated by rigorous selection.

Other genes, as might be expected, show similar disrupted distributions. Two more might be mentioned, both similar in their effect on populations. A disease known as **thalassemia,** genetically determined, is confined to peoples of Mediterranean (generally Italian) origin. The homozygous (*tt*) thalassemics die early of severe anemia. One would imagine that this would tend to eliminate the responsible gene from a population, and that its frequency would be determined only by its mutation rate, $T \rightarrow t$. However, *t* had a selective advantage: the heterozygotes (*Tt*) did not contract malaria, but the homozygous normal (*TT*) did. An equilibrium would tend to become established so that the rate of deaths from anemia (in the *tt*'s) would be balanced by the rate of death from malaria (in the *TT*'s). The heterozygote, *Tt,* having 50 per cent abnormal hemoglobin and 50 per cent normal hemoglobin, is a favored genetic combination. In the absence of malaria, the *t* gene would be selected against at a rapid rate. The situation is similar to that earlier described for hemoglobin *S*. This trait, however, is confined almost entirely to the Negroids, with only rare occurrences among some of the Caucasoids of India and the eastern Mediterranean countries.

THE EVOLUTION OF MODERN MAN
Modern man had his origin in the Paleolithic, or Old Stone Age. This was a food-gathering stage. His way of life was not appreciably different from that of his primate ancestors except that man had developed tools to assist in killing game and to defend himself (Fig. 23.12). The population was probably made up of small groups, and the distinctiveness of their tools suggests that isolation was the rule. The flow or exchange of genes, through interbreeding, was probably slight, even though man was nomadic in his habits, providing ideal conditions for rapid evolutionary change and distinct racial types. As hunters and food gatherers, probably only a single small group occupied any given area.

Such groups still exist, although all are likely to lose their uniqueness. Before civilized man changed his way of life, the Eskimos, the Fuegians of South America, the Australian aborigines, and the Bushman of Africa had not progressed much beyond the Paleolithic (Old Stone) stage. But ease of travel has made it difficult for isolation to persist, and these peoples will one day blend with the general population around them.

Probably by the end of the Pleistocene (10,000 to 25,000

Figure 23.12 Implements of later Stone Age man: an ax hammer (1); a flint ax with one edge polished (2); a flint saw with one edge notched (3); a flint dagger (4); a flint knife (5); an arrowhead (6).

years ago), the major races of men were established, and a gradual change from food gathering to food producing took place. Possibly this change came about first among the inhabitants along lake shores or coastal areas, for the fisherman is a more stable inhabitant than the hunter. The time was about 7000 B.C. The principal area probably was in the Middle East, although some would argue for southeastern Asia. From these places the herdsmen and farmers spread into Europe, North Africa, India, and China, mixing with native groups as the migrations proceeded.

Food producing requires the domestication of plants as well as animals. With the exception of the cultivation of rubber and quinine trees, every domesticated plant and animal had been grown by man before recorded history. Vast improvements in quality and productiveness have been accomplished, but new domesticated species have been rare. While the dog or the pig was probably the first domesticated animal, the basis of the earliest food-producing communities was possibly the tuber-producing perennial plants with a high starch content. Later came cultures of seed producers such as the cereals—rice, wheat, and corn. Both kinds of plants are easily grown in large quantities, can be stored easily, and are easily transported. This being so, it would appear that agriculture, as a practice, arose independently in several regions. Certainly it would be reasonable to suppose that the corn-based economy of the American Indian—

367

first the Mayan and Incan civilizations of South America, and later that of North America—was independent of the agricultural development of the Middle East. The cereals give stability to communities, the stored grains lasting over periods of poor harvests.

The domestication of animals appears not to have been a necessary step toward the establishment of communities. The South American Indians, with their highly developed culture, had virtually no animals as either a source of food or as beasts of burden. As a food gatherer, early man was most likely a carnivore, adding to his meat diet such roots and berries as he could find. As he became more settled he shifted to a vegetarian diet, although his taste for meat remained and was satisfied initially by game and fish.

With the establishment of communities capable of sustaining themselves through agriculture, man turned his increased leisure time to other things. Out of this has come the culture that sets man apart from his fellow animals. Central to the development of culture was the development of the art of communication, particularly that of the written language. It is this art which forms the basis of our cultural inheritance, the means whereby history and tradition are preserved, and rapid instruction becomes possible. The stored information on the magnetic tape of a computer is simply the latest step in this long process of communicative art (Fig. 23.13).

Figure 23.13 Ever since man invented language he has been a storer of information. At right: a baked clay hexagonal prism covered with cuneiform (wedge) writing, describes the siege of Jerusalem by Sennacherib, king of the Assyrians, in 686 B.C. Left: many miles of magnetic tape are used to store information in a minimum of space.

Westinghouse Photo

The Granger Collection

In the future the evolution of our cultural inheritance will play a greater role in the development of civilization than will the evolution of our biological inheritance. One of the most profound aspects of this cultural inheritance has been the development of science. Out of man's curiosity of the world about him and his desire and need to control the environment, he has reached a point where he can control his own evolution if he so chooses. He can do this wisely or carelessly, but he cannot escape the fact that he cannot long exist in an unfavorable environment. It is for this reason that solutions are needed for the population explosion, the degree of purity of the air we breathe and the water we drink, the controlled use of our natural resources, the proper distribution of food throughout the world, and the preservation of our wilderness areas.

summary

Man is a vertebrate mammal belonging to the Primate group, which includes the tree shrews, lemurs, tarsiers, monkeys, and apes. He is a product of evolution like any other animal, and his evolutionary history can be traced back through fossils more than 1 million years. As man gradually evolved, he assumed an upright stature and a bipedal walk; he learned to use fire and to make tools. He was first a hunter, then a farmer, and later a builder of cities and nations. In the process he domesticated plants and animals for his use, and gained an increasing measure of control over his environment.

Among his unique features are his hands, which have opposable thumbs, his much-enlarged brain, and his use of a complex language for communication. These features have given rise to human cultures; man today is evolving more rapidly in a cultural than in a biological way.

Man is one species consisting of many races. The races arose (probably) from small isolated groups in which certain variations were fixed by natural selection. The mobility of man and continued interbreeding between races tends to break down the distinction between races.

for thought and discussion

1   What physical and behavioral traits does man share with his primate relatives?

2   What role does modern medicine play in affecting the genetic variability of mankind?

3   Do you consider human values such as *honesty, generosity,* and *compassion* to have an evolutionary significance? Explain. Does one find these attributes expressed in other animals?

4   Some of you may have read the novel, *The Lord of the Flies.* Can you interpret this book in biological terms?

5 Works of art are expressions of human behavior and feeling. Are they cultural or biological expressions? How do you distinguish between these aspects?

6 Our Constitution states that Americans are born "free" and "equal." Comment on this from a biological point of view. From a cultural point of view.

7 Many animals exhibit territorialism, or the tendency to defend a territory against invasion by other members of the same species. What biological significance is there in this behavior? Is man a territorial animal? If your answer is yes, is this an acquired cultural attribute or the retention of a biological attribute?

8 The enlargement of man's brain to its present size was a late development in human evolution. It has been suggested by some authorities that increasing brain size followed the development of upright stature, grasping hand, and opposable thumb. Does this seem to be a logical sequence? Is it more probable that these features of man developed together? Explain your answers.

9 The dolphin is an intelligent animal with a large brain. Do you think that it is capable of evolving a culture? Give your reasons.

10 Change and chance, rather than purpose and plan, are characteristic of biological evolution. What does this statement mean to you? Can you defend it if it is applied to man? Can the same be said of cultural evolution?

11 What role has science played in the development of our culture? Consider this question particularly in terms of language, education, and philosophy. Can science have any effect on our biological evolution? Explain.

12 The literary works of man are reflections of his cultural experience. They have their origin in man's ability to think abstractly and to symbolize his thoughts in the form of words. Below are selected literary excerpts, all expressing a point of view of nature as seen through the eyes of the writer. Evaluate these in the light of your knowledge of biology in particular, and of science in general. Do you agree with the statements? Are they valid scientifically? If they are valid, what scientific thought is being expressed? How would these thoughts be expressed by a scientist? If they are not valid, to what extent do you feel that "poetic license" is justified?

(a) *Man is incomprehensible without nature, and nature is incomprehensible apart from man. . . .* Hamilton Wright Mabie.

(b) *And a mouse is miracle enough to stagger sextillions of infidels. . . .* Walt Whitman.

(c) *One secret of success in observing nature is the capacity to take a hint; a hair may show where a lion is hid. One must put this and that together, and value bits and shreds. Much alloy exists with the truth. The gold of nature does not look like gold at first sight. It must be smelted and refined in the mind of the observer. And one must crush mountains of quartz and wash hills of sand to get to it. To know the indications is the main matter. . . . John Burroughs.*

(d) *Mountains are earth's undecaying monuments. . . . Nathaniel Hawthorne.*

(e) *Man can have but one interest in nature, namely to see himself reflected or interpreted there. . . . John Burroughs.*

(f) *Nature, like a loving mother, is ever trying to keep land and sea, mountain and valley, each in its place, to hush the angry winds and waves, balance the extremes of heat and cold, of rain and drought, that peace, harmony and beauty may reign supreme. . . . Elizabeth Cady Stanton.*

(g) *Nature is the most thrifty thing in the world; she never wastes anything; she undergoes change, but there's no annihilation . . . the essence remains. . . . Thomas Binney*

selected readings

BAKER, H. G. *Plants and Civilization.* Belmont, Calif.: Wadsworth Publishing Company, Inc., 1965. A fine volume on the dependence and relations of plants to man.

BATES, MARSTON. *Man in Nature* (2nd ed.). Englewood Cliffs, N.J.: Prentice-Hall, Inc., 1964. Deals with man as an animal, his ancestors, the varieties of man, problems of domestication, health and disease, and the environment.

BERRILL, N. J. *Man's Emerging Mind.* New York: Fawcett World Library (Premier Books), 1965. The story of man's progress through time.

BRODRICK, A. H. *Man and His Ancestry.* New York: Fawcett World Library (Premier Books), 1964. The story of man and his primate relatives.

DOBZHANSKY, TH. *The Biological Basis of Human Freedom.* New York: Columbia University Press, 1956. A philosophical discussion of the implications of modern biology in understandable nonscientific terms.

DOBZHANSKY, TH. *Heredity and the Nature of Man.* New York: New American Library, Inc. (Signet Science Library), 1964. An excellent book dealing with the nature of beauty, the variety of human differences, the meaning of race, the genetic problems of mankind, and the future of man.

DUNN, L. C., and TH. DOBZHANSKY. *Heredity, Race and Society.* New York: New American Library, Inc. (Mentor Books), 1946. A sound description of the basis of human differences.

EISELEY, LOREN. *The Firmament of Time.* New York: Atheneum Publishers, 1960. The history of science as it aids in our understanding of man.

HARRISON, R. J. *Man, the Peculiar Animal.* Baltimore, Md.: Penguin Books Inc., 1938. Man's structures and functions are contrasted with those of other animals.

SWANSON, C. P. *The Natural History of Man.* Englewood Cliffs, N.J.: Prentice-Hall, Inc., 1973. A consideration of man's changing view of himself as science gives him a changing concept of reality and himself.

index